Advances in Mathematical Chemistry and Applications

AF849321

Advances in Mathematical Chemistry and Applications

Volume 2

Edited By

Subhash C. Basak

International Society of Mathematical Chemistry
1802 Stanford Avenue, Duluth
MN 55811 and UMD-NRRI
5013 Miller Trunk Highway
Duluth MN 55811
USA

Guillermo Restrepo

Laboratorio de Química Teórica
Universidad de Pamplona
km 1 vía Bucaramanga
Pamplona, Norte de Santander
Colombia

&

José L. Villaveces

Universidad de los Andes
Carrera 1 No 18A-12
Bogotá, D. C.
Colombia

ELSEVIER

AMSTERDAM • BOSTON • HEIDELBERG • LONDON • NEW YORK • OXFORD
PARIS • SAN DIEGO • SAN FRANCISCO • SINGAPORE • SYDNEY • TOKYO

Elsevier
Radarweg 29, PO Box 211, 1000 AE Amsterdam, Netherlands
The Boulevard, Langford Lane, Kidlington, Oxford OX5 1GB, UK
225 Wyman Street, Waltham, MA 02451, USA

Copyright © 2015 Bentham Science Publishers Ltd. Published by Elsevier Inc. All rights reserved.

No part of this publication may be reproduced or transmitted in any form or by any means, electronic or mechanical, including photocopying, recording, or any information storage and retrieval system, without permission in writing from the publisher. Details on how to seek permission, further information about the Publisher's permissions policies and our arrangements with organizations such as the Copyright Clearance Center and the Copyright Licensing Agency, can be found at our website: www.elsevier.com/permissions.

This book and the individual contributions contained in it are protected under copyright by the Publisher (other than as may be noted herein).

Notices
Knowledge and best practice in this field are constantly changing. As new research and experience broaden our understanding, changes in research methods, professional practices, or medical treatment may become necessary.

Practitioners and researchers must always rely on their own experience and knowledge in evaluating and using any information, methods, compounds, or experiments described herein. In using such information or methods they should be mindful of their own safety and the safety of others, including parties for whom they have a professional responsibility.

To the fullest extent of the law, neither the Publisher nor the authors, contributors, or editors, assume any liability for any injury and/or damage to persons or property as a matter of products liability, negligence or otherwise, or from any use or operation of any methods, products, instructions, or ideas contained in the material herein.

ISBN: 978-1-68108-053-6

British Library Cataloguing in Publication Data
A catalogue record for this book is available from the British Library

Library of Congress Cataloging-in-Publication Data
A catalog record for this book is available from the Library of Congress

For Information on all Elsevier publications visit our website at http://store.elsevier.com/

ELSEVIER
Book Aid International
Working together to grow libraries in developing countries
www.elsevier.com • www.bookaid.org

Bentham e Books

Cover Art

The cover represents an Erlenmeyer flask made from symbols of chemistry and mathematics of almost 2000 years. The symbols are organized in a chronological order starting with the Platonic solids at the bottom. Two equations at the top, the Schrödinger equation and the Wiener index equation, represent a balance between continuous and discrete mathematics used in current mathematical chemistry. The mosquito at the mouth of the flask and the chiral mosquito repellent represent practical applications of mathematical chemistry. The cover was designed by Guillermo Restrepo and Subhash C. Basak.

CONTENTS

CHAPTERS

FOREWORD

In the middle 1960's, there arose a significant number of approaches to the quantitation of molecular attributes influencing their behavior and function. These approaches took the form of some physical properties and various approaches to quantitate molecular structure. Since molecular structure is an attribute most directly correlateable with function, this approach prospered and led to a number of useful methods at that time. These included molecular orbital theory, graph theory, information theory and some spatial quantitative methods. As we begin the second half of this century of study, we have on hand a plethora of methods to achieve these goals. The extent of this burgeoning has given rise to a descriptive categorical term, *Mathematical Chemistry*.

This volume appears on the scene to celebrate the emergence of *Mathematical Chemistry* in its second half century. It brings to the molecular scientist several new approaches plus the introduction of variations on existing ideas. The editors of this text have wisely and thoroughly gathered together a broad representation of contributions. About a half of the book contains contributions of scientists active in the life sciences. These chapters focus primarily on methods of creating models of drug function relative to molecular structure. This realm of applications of mathematical chemistry has contributed significantly over the years to the process of drug design. We would still be in the empirical realm of trial and error to find a new drug, without the role of structure-activity models using many good methods.

The other half of the chapters describes current work, new applications, and new approaches to quantitation of the essential aspects of molecular structure. The imaginative ideas flowing through these chapters are quite impressive. It creates a text that should be on the shelf of every scientist working quantitatively at the molecular level. There is a stimulation arising from the book that will excite the creativity in many readers. Such a book is essential in every field of science where periodic overview and exposure to new ideas is essential.

Congratulations to the authors who have contributed to the book and congratulations to the editors who have selected the mix of articles and authors.

Lemont B. Kier
Emeritus Professor
Dept. of Medicinal Chemistry
Virginia Commonwealth University
USA

PREFACE

I would like to emphasize strongly my belief that the era of computing chemists, when hundreds if not thousands of chemists will go to the computing machine instead of the laboratory for increasingly many facets of chemical information, is already at hand. There is only one obstacle, namely that someone must pay for the computing time.

Robert Sanderson Mulliken

In the first quarter of the nineteenth century the experimental proof for the interdependence of the composition and properties of chemical compounds resulted in the theory that they are mutually related, so that like composition governs like properties, and conversely.

Wilhelm Ostwald

We are glad to introduce our readers to volume 2 of the eBook Advances in Mathematical Chemistry and Applications. This volume, like volume 1, has thirteen chapters written by well-known experts from around the world. As shown in the two figures of the preface of volume 1, chapters of volume 1 and 2 of this eBook deal with several mathematical approaches to chemistry, *e.g.* topology, graph theory, information theory, to name but a few. Readers interested in a particular mathematical area, say topology, are directed to those figures to know which other chapters report novel topological approaches in chemistry.

In the first chapter entitled "Topological Efficiency Approach to Fullerene Stability - Case Study with C_{50}", by Ante Graovac, Ali Reza Ashrafi, and Ottorino Ori, the authors use topological approaches for classifying fullerene isomers, a group of structures important both from the mathematical and chemical points of view. In this approach two shape descriptors: topological compactness and topological sphericality turn out to be of key importance for the ordering.

Andrés Bernal, Eugenio Llanos, Wilmer Leal, and Guillermo Restrepo, authors of the second chapter, entitled "Similarity in Chemical Reaction Networks: Categories, Concepts and Closures", claim that chemical identity is the result of chemical reactivity and analyze chemical entities by studying reaction networks. In such formalism the identity of a substance is given by those related to it, which gives place to an interpretation in the framework of Category Theory. The similarity between substances is studied through topological spaces, with closures as key concepts for similarity.

The third chapter, "Discrimination of Small Molecules Using Topological Molecular Descriptors", by Chandan Raychaudhury and Debnath Pal, reviews the use of numerical topological graph invariants to differentiate molecular structures of small size, a subject of utter importance in chemo and bioinformatics.

"The Periodicity of Molecules" is the title of the fourth chapter, by Fanao Kong, Weiqiang Wu, Na Ji, and C. L. Calson, where the authors review their work on periodic tables for di, tri, and tetra-atomic molecules. These tables are further used to estimate properties of molecules, such as bond length, binding energy, ionization potential, among others.

The fifth chapter, "The GRANCH Techniques for Analysis of DNA, RNA and Protein Sequences", by Ashesh Nandy, reviews two classes of methods, *viz*., graphical methods and graph invariant techniques, for the characterization of sequences of DNA, RNA, and proteins, which constitute the GRANCH (graphical representation and numerical characterization) methods. The author shows how these characterizations can be used in the study of viral proteins as an example of application.

Jarad B. Niemi and Gerald J. Niemi are the authors of the sixth chapter, entitled "Linear Regression, Model Averaging, and Bayesian Techniques for Predicting Chemical Activities from Structure". The chapter reviews several statistical approaches to deal with situations where there are more chemical structure measurements or available molecular descriptors/ properties than the number of chemicals with measured property values, which are commonly found in many current Quantitative Structure-Activity Relationship (QSAR) and Quantitative Structure-Property Relationship (QSPR) studies.

"Marine Algal Toxicity Models with *Dunaliella Tertiolecta*: *In Vivo* and *In Silico*", by Melek T. Saçan, Marjana Novic, M. Doğa Ertürk, and Nikola Minovski, is the seventh chapter of this volume. In their chapter, authors model toxicity of phenolic compounds to *Dunaliella Tertiolecta* using counter-propagation artificial neural networks and point out to the importance of D. tertiolecta, a marine alga, for toxicity determination of chemical substances.

In the eighth chapter of this volume, "Anti-Tubercular Drug Designing Using Structural Descriptors", Manish C. Bagchi and Payel Ghosh discuss the use of molecular descriptors, including graph invariants, and robust statistical methods like ridge regression in the formulation of models for the discovery of drugs active against tuberculosis, a contemporary problem to human health worldwide.

"Integrating Bioinformatics and Systems Biology for Exploring Novel Lipid Pathways in Infectious Diseases" is the title of the ninth chapter, written by Sonali Shinde, Vineetha Mandlik, and Shailza Singh. The authors develop mathematical models for delving into the complexity of biological interactions related to Schistosomiasis and Trypanosomiasis, two neglected tropical diseases affecting a large segment of world population. Their specific focus is on the lipid metabolism of the two parasites causing the diseases.

Santiago Medina, Susana Casas, Mariana Restrepo, Alejandro Alvarez, Adriana J. Bernal, and Andrés F. González are the authors of the tenth chapter, entitled "Applications of Molecular Docking and Molecular Dynamics on the Inhibition of Quorum Sensing Systems". Given that bacteria communicate through chemical signals allowing the regulation of a variety of physiological functions and group behavior (quorum sensing), a key concept in medicinal chemistry is to inhibit such a communication through novel molecules. In this chapter the authors describe some computational approaches used to find such inhibitory molecules for several bacteria.

The eleventh chapter, entitled "Designing Models for Metalloenzymes", by James F. Weston, discusses how quantum chemistry plays an important role in designing small organometallic catalysts. Weston particularly analyzes quantum mechanical approximations for the process of catalysis, showing the advantages and disadvantages of those approximations.

"The Multi-Factor Coupled Protein Folding: Insights from Molecular Dynamics Simulations" is the title of the twelfth chapter, where Xiaomin Wu, Gang Yang, and Lijun Zhou review factors that are important for protein folding, a process of paramount importance in understanding the physiological function of proteins. The factors analyzed are space limitation, where entropy constraints play a central role; solvent effects, where solvation models are discussed; physicochemical properties and their relationship with non-covalent interactions; and structural modifications, such as mutation and protonation.

Christoph Flamm, Bärbel M. R. Stadler, and Peter F. Stadler are the authors of the thirteenth chapter, entitled "Generalized Topologies: Hypergraphs, Chemical Reactions, and Biological Evolution", where they discuss how the topological concepts of similarity, neighborhood, connectedness and continuity pervade chemical and biological situations such as chemical reaction networks and evolutionary processes. By finding, however, that those concepts are not laid down to the mathematical formalism of topology, they undertake the task of doing it by avoiding the topological concept of open set and instead taking advantage of the mathematical richness of closure operators.

We sincerely hope the second volume of the book, analogous to the first volume, will be highly appreciated by researchers and scholars worldwide.

Subhash C. Basak
International Society of Mathematical Chemistry
University of Minnesota
USA

Guillermo Restrepo
Universidad de Pamplona
Colombia

José L. Villaveces
Universidad de los Andes
Colombia

List of Contributors

Adriana J. Bernal	Mycology and Plant Disease Laboratory (LAMFU), Department of Biological Sciences, Universidad de los Andes, Bogotá, Colombia
Alejandro Alvarez	Grupo de Diseño de Productos y Procesos (GDPP), Department of Chemical Engineering, Universidad de los Andes, Bogotá, Colombia; Mycology and Plant Disease Laboratory (LAMFU), Department of Biological Sciences, Universidad de los Andes, Bogotá, Colombia
Ali R. Ashrafi	Department of Pure Mathematics, Faculty of Mathematical Sciences, University of Kashan, Kashan 87317 – 51167, I. R., Iran
Andrés Bernal	Grupo de Química Teórica, Universidad Nacional de Colombia, Bogotá, Colombia
Andrés F. González	Grupo de Diseño de Productos y Procesos (GDPP), Department of Chemical Engineering, Universidad de los Andes, Bogotá, Colombia
Ante Graovac	Faculty of Science, University of Split, Nikole Tesle 12, HR-21000 Split, Croatia
Ashesh Nandy	Centre for Interdisciplinary Research and Education, 404B Jodhpur Park, Kolkata 700068, India
Bärbel M.R. Stadler	Max Planck Institute for Mathematics in the Sciences, Inselstraße 22, D-04103 Leipzig, Germany
C. L. Calson	The Institute of Chemistry, Chinese Academy of Sciences, Beijing, 100190, China
Chandan Raychaudhury	Bioinformatics Centre, Indian Institute of Science, Bangalore 560012, India
Christoph Flamm	Institute for Theoretical Chemistry, University of Vienna, Währingerstraße 17, A-1090 Vienna, Austria
Debnath Pal	Bioinformatics Centre, Indian Institute of Science, Bangalore 560012, India
Eugenio Llanos	Corporación Colombiana del Saber Científico SCIO, Bogotá, Colombia
Fanao Kong	The Institute of Chemistry, Chinese Academy of Sciences, Beijing, 100190, China
Gang Yang	Resources and Environmental Department, Southwest University, Chongqing 400715, P. R. China; Engineering Research Center of Forest Bio-preparation, Ministry of Education, Northeast Forestry University, Harbin 150040, P. R. China

Gerald J. Niemi Natural Resources Research Institute and Department of Biology, University of Minnesota, 5013 Miller Trunk Highway, Duluth, MN 55811, USA

Guillermo Restrepo Laboratorio de Química Teórica, Universidad de Pamplona, Pamplona, Colombia

James F. Weston Departamento de Química, Universidad de los Andes, Carrera 1[a], N° 18-A-12, Bloque Q, 8°Piso, Bogotá, D.C., Colombia, S. A.

Jarad B. Niemi Department of Statistics, Iowa State University, Ames, IA 50011, USA

Lijun Zhou Resources and Environmental Department, Southwest University, Chongqing 400715, P. R. China

M. Doğa Ertürk Boğaziçi University, Institute of Environmental Sciences, 34342, Hisar Campus, Bebek, Istanbul, Turkey

Manish C. Bagchi School of Bioscience & Engineering, Jadavpur University, Kolkata 700032, India

Mariana Restrepo Mycology and Plant Disease Laboratory (LAMFU), Department of Biological Sciences, Universidad de los Andes, Bogotá, Colombia

Marjana Novic National Institute of Chemistry, Hajdrihova 19, 1000 Ljubljana, Slovenia

Melek T. Saçan Boğaziçi University, Institute of Environmental Sciences, 34342, Hisar Campus, Bebek, Istanbul, Turkey

Na Ji The Institute of Chemistry, Chinese Academy of Sciences, Beijing, 100190, China

Nikola Minovski National Institute of Chemistry, Hajdrihova 19, 1000 Ljubljana, Slovenia

Ottorino Ori Actinium Chemical Research, Via Casilina 1626/A, 00133 Rome, Italy

Payel Ghosh Department of Biotechnology, University of Pune, Pune-411007, India

Peter F. Stadler Institute for Theoretical Chemistry, University of Vienna, Währingerstraße 17, A-1090 Vienna, Austria; Max Planck Institute for Mathematics in the Sciences, Inselstraße 22, D-04103 Leipzig, Germany; Bioinformatics Group, Department of Computer Science, and Interdisciplinary Center of Bioinformatics, University of Leipzig, Härtelstraße 16-18, D-04107 Leipzig, Germany; RNomics Group, Fraunhofer IZI, Perlickstraße 1, D-04103 Leipzig, Germany; Center for noncoding RNA in Technology and Health, University of Copenhagen, Grønnegårdsvej 3, DK-1870 Frederiksberg, Denmark; The Santa Fe Institute, 1399 Hyde Park Rd., Santa Fe, New Mexico, USA

Santiago Medina	Grupo de Diseño de Productos y Procesos (GDPP), Department of Chemical Engineering, Universidad de los Andes, Bogotá, Colombia
Shailza Singh	Computational and Systems Biology Laboratory, Lab 9, New Building, NCCS, Pune -411007, India
Sonali Shinde	Computational and Systems Biology Laboratory, Lab 9, New Building, NCCS, Pune -411007, India
Susana Casas	Grupo de Diseño de Productos y Procesos (GDPP), Department of Chemical Engineering, Universidad de los Andes, Bogotá, Colombia
Vineetha Mandlik	Computational and Systems Biology Laboratory, Lab 9, New Building, NCCS, Pune -411007, India
Weiqiang Wu	The Institute of Chemistry, Chinese Academy of Sciences, Beijing, 100190, China
Wilmer Leal	Fundación Instituto de Inmunología de Colombia FIDIC, Bogotá, Colombia; Universidad del Rosario, Bogotá, Colombia; Laboratorio de Química Teórica, Universidad de Pamplona, Pamplona, Colombia
Xiaomin Wu	Key Laboratory of Plant Resources and Biology, College of Life Sciences, Huaibei Normal University, Huaibei 235000, P.R. China; Engineering Research Center of Forest Bio-preparation, Ministry of Education, Northeast Forestry University, Harbin 150040, P. R. China

ACKNOWLEDGEMENTS

The editors of the eBook "Advances in Mathematical Chemistry and Applications" and Bentham Science Publishers would like to gratefully acknowledge the dedicated work of the following distinguished scientists who reviewed the chapters of this book and gave valuable suggestions as well as constructive criticisms to the authors and editors that substantially enhanced the quality of this volume.

REVIEWERS OF EBOOK CHAPTERS

Apurba Bhattacharjee	Department of Medicinal Chemistry, Division of Experimental Therapeutics, Walter Reed Army Institute of Research, 503 Robert Grant Avenue, Silver Spring, MD 20910-7500, USA
Dorota Bielińska-Wąż	Department of Radiological Informatics and Statistics, Medical University of Gdańsk, Tuwima 15, 80-210 Gdańsk, Poland
Danail Bonchev	Center for the Study of Biological Complexity, Virginia Commonwealth University, Richmond, VA 23284-2030, USA
Rainer Brüggemann	Leibniz-Institute of Freshwater Ecology and Inland Fisheries, Berlin, Germany
Eduardo Castro	INIFTA, Departamento de Química, Facultad de Ciencias Exactas, Universidad Nacional de La Plata, La Plata, Argentina
Pratim K. Chattaraj	Department of Chemistry and Center for Theoretical Studies, Indian Institute of Technology, Kharagpur 721302, India
Snigdhansu Chatterjee	School of Statistics, University of Minnesota, Minneapolis, MN 55455, USA
Ivan Gutman	Faculty of Science, University of Kragujevac, 34000 Kragujevac, Serbia
Wieslaw Nowak	Institute of Physics, Nicholas Copernicus University, ul.Grudziadzka 5, 87-100 Torun, Poland
Guillermo Restrepo	Laboratorio de Química Teórica, Universidad de Pamplona, Pamplona, Colombia
Vellarkad N. Viswanadhan	Department of Computational Chemistry, Jubilant Biosys Limited, Bangalore 560022, India
Gang Yang	Resources and Environmental Department, Southwest University, Chongqing 400715, P. R. China; Engineering Research Center of Forest Bio-preparation, Ministry of Education, Northeast Forestry University, Harbin 150040, P. R. China

CHAPTER 1

Topological Efficiency Approach to Fullerene Stability - Case Study with C_{50}

Ante Graovac[1], Ali Reza Ashrafi[2] and Ottorino Ori[3,*]

[1]*Faculty of Science, University of Split, Nikole Tesle 12, HR-21000 Split, Croatia;* [2]*Department of Pure Mathematics, Faculty of Mathematical Sciences, University of Kashan, Kashan 87317 – 51167, I. R. Iran and* [3]*Actinium Chemical Research, Via Casilina 1626/A, 00133 Rome, Italy*

Abstract: An innovative theoretical mechanism leading to the classification of the stability of fullerene isomers is presented. This approach is based on the action of suitable topological potentials impacting on molecular stability, namely topological compactness and topological sphericality indexes, providing a fast and general ranking algorithm. Present results point out that electronic properties of sp^2 carbon systems are deeply rooted in the topology of their atomic network.

Keywords: C50 fullerene, chemical distance, chemical graph, DFT, energy ranking algorithm, fullerene stability, graph invariants, long-range connectivity effects, minimal vertices, schlegel graphs, sp2 carbon systems, topological compactness, topological efficiency index, topological modeling, topological potential, topological sphericality, wiener index, wiener-weights, π-electron energy.

INTRODUCTION

The relevant role of topology in modeling chemical systems and their properties has been intensively demonstrated, graph theory representing nowadays one of the most established and effective tools in theoretical chemistry. Since the first collection of topology-based computational methods applied to chemistry, edited in 1976 by Balaban [1], an enormous amount of new topological *graph invariants* (topological indices) has been introduced and applied for Quantitative

***Corresponding author Ottorino Ori:** Actinium Chemical Research, *Via* Casilina 1626/A, 00133 Rome, Italy; Tel/Fax: +39-06-2055084; E-mail: ottorino.ori@gmail.com

Subhash C. Basak, Guillermo Restrepo & José L. Villaveces (Eds)
Copyright © 2015 Bentham Science Publishers Ltd. Published by Elsevier Inc. All rights reserved.
10.1016/B978-1-68108-053-6.50001-6

Structure-Activity Relationship (QSAR) and Quantitative Structure-Property Relationship (QSPR) studies focusing on structure-dependent chemical behavior of molecules. Predicting chemical properties of an untested compound from its molecular structure is in fact one of the main tasks fulfilled by QSAR methods, providing precious information about chemical hazards for example, while reducing cost, time and animal testing. Periodically, the atlas edited by Todeschini and Consonni [2] updates the international scientific community on the latest advances in this sector on both the practical and theoretical side. QSAR algorithms have been recently expanded by Putz [3] with innovative concepts, like the *minimum spectral path principle* and some new correlation factors to measure chemical–biological interactions, constituting the so-called SPECTRAL-SAR method. Successful SPECTRAL-SAR applications concern, for example, ecotoxicity hazard assessment of various chemicals.

It is worth noticing that topological methods are not confined to molecular *finite* systems being, from the computational point of view, naturally extendable to *infinite* systems like polymers [4] or, more recently, to bidimensional structures [5-7]. The elegant procedure developed in the seminal article [4] allows in fact the calculation of the specific π-electron energy and energy gap for a *homologous series* of polymers depicted as *infinite* chemical graphs. That method is based on the properties of the distance matrix D of the polymer, and it shows a good correlation between the energy of the system and a peculiar topological index, the Wiener index W, that corresponds to the half-sum of D entries (these concepts will be illustrated in the next chapter), revealing in such a way the strong inter-dependence between *long-range connectivity properties* of a chemical system and its *chemical properties*. These fundamental results opened the door to property-predicting *topological methods* suitable for infinite chemical structures. The present authors extended the range of applications of topological invariants by studying bidimensional infinite structures like graphenic planes [5, 6] with a particular focus in comparing pristine hexagonal lattices and nanocones [7], both these systems being made by the same number of hexagonal rings with the exception of a single pentagon in the center of the graphenic mesh. This apparently "small" difference in connectivity plays instead an important role in modifying the (topological) stability of these two competing *infinite* systems, favoring the nanocone growing structure. Further applications of graph theory on the properties of Stone-Wales topological defects in graphenic fragments lead to the prediction of non-trivial phenomena, including the specific wave-like linear propagation of Stone-Wales defects, which have been analyzed in [8] just from a *topological perspective* by assigning to the Wiener index W the role of *topological potential* Ξ of the chemical system ($\Xi=W$), according to *topological modeling* (TM) approach recently reviewed in [9]. TM achieves interesting results on the relative structural stability of relevant classes of sp^2 carbon systems like fullerenes and fulleranes. On fullerenes in particular, it has been observed [10] that the physically stable buckminsterfullerene, the ''buckyball'' C_{60} I_h with icosahedral symmetry has the minimum W value among 1,812 isomers (3,532 when enantiomeric pairs are considered as two distinct cages), evidencing in this way the role of topological invariant W in selecting stable isomers of a given fullerene.

This result points out the relevance of the concept of *topological compactness* according to the following topological minimum principle on W: *stable sp^2 structures tend to minimize their W index*. This principle represents a simple topological justification to the isolated pentagon rule (IPR).

Detailed studies on C_{28} fullerene confirmed this principle [11]; however the topological index W is not capable of selecting the stable isomer of a generic C_n cage. Correlations between Wiener index and molecular energy have been intensively investigated by other authors for higher fullerenes (see article [12] and related bibliography), showing that, although W performs quite well in the C_{80} case, it fails however in sieving stable IPR isomers of C_{84} and C_{86} fullerenes.

Next section will show in which way the *topological potential* Ξ is improved to allow an efficient selection of stable chemical structures. Before entering into computational details, the main computational assumption underlying TM approximation has to be remembered at this point: *for a given chemical system the topological potential* Ξ *expresses the long-range connectivity effects, arising from the atomic network topology, on the electronic properties, allowing the fast selection of potentially stable configurations to be compared with ab-initio (DFT) calculations.*

Fullerenes are a perfect case evidencing the need for simple tools to identify the most stable isomers in a given set of molecules. With higher fullerenes the number of IPR isomers increases in fact very rapidly. For example, with $n = 100$ the well-known spiral algorithm [13] produces 450 IPR isomers (862 if enantiomers are distinguished), a large set of isomeric molecules quite difficult to be handled by quantum-chemical methods. Chemical theories not yet suggested a practical *stability rule* in order to assess the stabilities of fullerenes of different dimensions and symmetries. *Devising such a rule is not an easy task*, as the mechanism of formation of fullerenes most probably depends on a complicated kinetics, involving various steps not only in gas but also in the condensed phase. The strong demand for fast and powerful computational methods arising from current investigations finds an effective answer (summarized in this article) in the elegant tools provided by the topological modeling approach. The good match encountered so far between TM predictions and *ab-initio* methods provides evidence regarding the effectiveness of topological indices in describing the physico-chemical and structural features of finite and infinite carbon systems. In the next section a concise introduction to topological modeling is given.

DISTANCE-BASED INVARIANTS – THE WIENER INDEX

Graph-theoretical approaches to chemical investigations are based on the representation of a *molecule* of n carbon atoms by a graph of n nodes, the *chemical graph*, obtained by depleting hydrogen atoms until just a topological *network* made by n carbon atoms will remain, being the basic structural information stored in the adjacency matrix of the graph itself. The *hollow* molecule of a C_n fullerene is therefore conveniently depicted by a *planar cubic* graph having n vertices, all 3-connected, composing $12+h$ carbon rings, including a fixed number of 12 pentagons and a variable number of h hexagons with $h \geq 2$ or null; forcing the number of carbon atoms to be $n = 20 + 2h$ starting from the dodecahedron with $h = 0$ and $n = 20$, the smallest fullerene with D_{3d} symmetry and no hexagons.

Larger cages $n = 24, 26, 28, \ldots$ are built by adding two carbon atoms for each newly inserted hexagonal face, with the sole forbidden case with $h = 1$, $n = 22$.

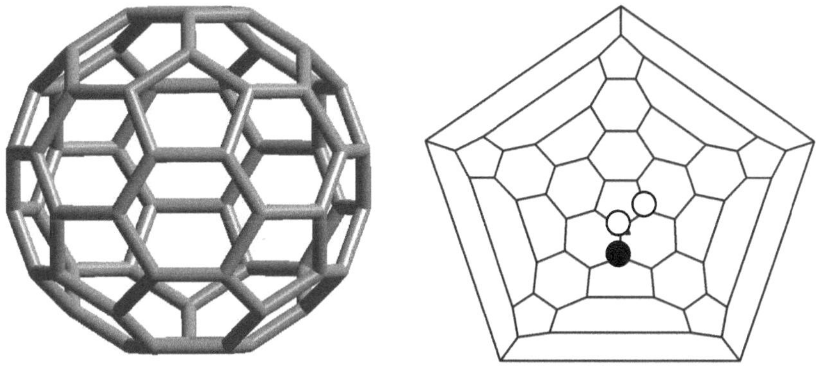

Figure 1: Left, view of the IPR C_{60} I_h ''buckyball'' with 60 symmetry-equivalent atoms. Bond lengths are 1.39 Å or 1.45 Å for bonds shared by two hexagons (black-white atoms) or by one pentagon and one hexagon (white-white atoms), respectively. Right, its planar Schlegel graphs evidencing 12 non-touching pentagons *p*.

Fig. **(1)** shows the tridimensional view of the IPR C_{60} I_h isomer and its planar representation (the *Schlegel graph*) with $n = 60$ vertices and $h = 20$ hexagonal faces, the 12th-pentagon coinciding with the external perimeter of the graph.

TM simulations are typically founded on the calculations and minimization of the topological potential Ξ proportional to a certain distance-based *topological invariant e.g.* a quantity that does not vary under an arbitrary re-labeling of graph nodes. In a chemical graph the concept of *distance* d_{ij} between two nodes i and j (the *chemical distance*) is purely topological, equaling the number of bonds along the *shortest* path connecting these two atoms; clearly the following relations hold: $d_{ij} = d_{ji}$ and $d_{ii} = 0$. When $d_{ij} = k$, the atom j lies in the k-th coordination shell of i and *vice versa*. Representing by M the length of the longest path of the graph - the integer M corresponds to the graph *diameter* - and with b_{ik} the number of k-neighbors of atom i, the effect of molecular connectivity on this atom is summarized by the topological invariant w_i:

$$w_i = \tfrac{1}{2} \textstyle\sum_k k b_{ik} \quad k=1,2,\ldots,M\text{-}1,M \tag{1}$$

where $n\text{-}1 = \Sigma_k\, b_{ik}$ and $b_{i1} = 3$ for any fullerene atom i. On a general molecular graph it is useful indicating with $\underline{w}$ the minimum of the w_i quantities *e.g.* $\underline{w}$ =

min$\{w_i\}$, assigning to the related atoms i the role of *minimal vertices* of the structure.

The *ordered* set $\{b_{ik}\}$ identifies the Wiener-weights (WW) of vertex i. The high symmetry of the C_{60} I_h molecule forces all atoms to share the WW sequence $\{b_k\}$ = {3,6,8,10,10,10,8,3,1} and $\underline{w}$ = 139 which derives from the direct application of equation (1) with M = 9. WW computation does not represent a difficult task. Algorithmic details may be found in various articles like [2] and directly implemented in computer codes working on the list of graph adjacencies like the one for the C_{60} I_h cage presented in Table **1** for an arbitrary node numbering.

Table 1: Adjacency list for the 60 nodes of the C_{60} I_h fullerene chemical graph

Atom	First Neighbors	Atom	First Neighbors	Atom	First Neighbors
1	2 5 46	21	22 25 53	41	42 45 47
2	1 3 15	22	21 23 32	42	41 43 59
3	2 4 20	23	22 24 26	43	39 42 44
4	3 5 6	24	17 23 25	44	10 43 45
5	1 4 45	25	13 21 24	45	5 41 44
6	4 7 10	26	23 27 30	46	1 47 50
7	6 8 19	27	26 28 33	47	41 46 48
8	7 9 29	28	27 29 37	48	47 49 60
9	8 10 38	29	8 28 30	49	48 50 51
10	6 9 44	30	18 26 29	50	11 46 49
11	12 15 50	31	32 35 54	51	49 52 55
12	11 13 52	32	22 31 33	52	12 51 53
13	12 14 25	33	27 32 34	53	21 52 54
14	13 15 16	34	33 35 36	54	31 53 55
15	2 11 14	35	31 34 57	55	51 54 56
16	14 17 20	36	34 37 40	56	55 57 60
17	16 18 24	37	28 36 38	57	35 56 58
18	17 19 30	38	9 37 39	58	40 57 59
19	18 20 7	39	38 40 43	59	42 58 60
20	3 16 19	40	36 39 58	60	48 56 59

In spite of its simplicity, the topological invariant w_i is of peculiar importance since it expresses the long-range effects induced by graph connectivity on the node i and therefore on the overall molecular structure. First of all, *topology fixes the upper-limit to the molecular symmetry* that the chemical structure assumes in the ordinary geometrical space once it is converted from a graph in a real tridimensional molecule, in the sense that two atoms i and j may be symmetry-equivalent only if they share the same ordered sets $\{b_{ik}\} = \{b_{jk}\}$. The last equality does not assure the geometrical equivalence but, on the contrary, one may surely affirm that atoms with different WW strings will never be symmetry-equivalent in the real space. The *deep* relationship between adjacency properties and molecular symmetry reverberates in one of the principal theoretical achievements of TM, *e.g.* the quick theoretical determination of the molecular fingerprint (number of resonance lines and relative intensities) in ^{13}C NMR experiments. Topological determination of resonance data has been proposed in the early article [14] on C_{76} fullerene isomers, by studying the properties of their *colored* graphs without any geometrical "ingredient", an approach which has a general validity as shown by recent literature [14, 16].

The descriptor w_i naturally conveys the concept of *molecular compactness*, assuming lower values for molecular sites that are centrally, deeply nested in the structure. For open molecules, like PAH's, it is straightforward to verify that atoms placed along the molecular border present higher w_i values than atoms close to molecular core, suggesting a possible "proportional" relationship between w_i value and atomic relative reactivity. The topological index which measures the molecular *compactness* is the Wiener index W, the oldest among the molecular invariants [2], expressed as the half-summation of the chemical distances d_{ij} of the graph with n nodes:

$$W(n) = \tfrac{1}{2} \sum_{ij} d_{ij} = \sum_i w_i \quad i,j=1,2,\ldots,n-1,n \tag{2}$$

The first basal assumption of the TM method is represented by the following topological minimum principle imposed on the *topological potential* $\Xi^W = W$: *chemically stable molecular structures tend to minimize their Wiener index.* For a molecular structure with n atoms, quantity $W(n)$ given by Equation (2) assumes the lowest values for the most compact isomers with the shortest distances d_{ij}. Within TM approximation, the most compact isomers correspond to chemical structures with higher chemical stability.

Moreover, topological minimum principle on Ξ^W implies that the minimal vertices of a graph are very good candidates to be stable sites of the structure.

TM simulations show in effect that a general molecule tries to preserve the coordination shells around its minimal vertices in order to preserve the lowest contributions w_i to the overall Wiener index according to Equation (2). Heuristically, imposing the minimum principle on Ξ^W may be justified thinking to the physical behavior of a network of harmonic oscillators that reaches the equilibrium configuration by minimizing the sum of squared distances d^2_{ij}, a condition that assures high structural compactness to any stable configurations. $W(n)$ works instead as the potential energy of a system of interacting nodes under the effect of the long-range topological interaction with of a *linear* dependence from chemical distances d_{ij}.

It is important to remember again that TM approximated methods are aimed to compare chemical stability of structures made with similar building units.

As we mentioned above, current studies on fullerenes evidence the effective role of the invariant W in ranking the molecular stability of various C_n isomers. The case of C_{60} fullerene is a relevant example of a chemical structure that selects, among 1,812 inequivalent isomers, the most compact icosahedral cage, minimizing in such a way the Wiener index with W=8,340.

As explained in reference [12], Wiener index represents a natural choice for the correlation *topology vs. relative stability* of fullerene isomers, the reason for that being the following: molecular compactness (measured by W) relates to the surface curvature of the molecule. Buckminsterfullerene itself exhibits in fact a spherical shape with constant curvature and the lowest W value among C_{60} isomers. In this way the Wiener index eliminates the most distorted and oblong isomers (characterized by higher W numbers) selecting relatively stable isomers with *quasispheroidal shapes*. Non IPR molecules with touching pentagons are likely to show high distortions and elongated carbon cages, confirming the validity of the choice $\Xi^W = W$.

It is very interesting to see in which way topology influences and describes structural and electronic properties of carbon systems with sp^2 hybridization.

Bond distortion can be evaluated by simple computational means, *e.g.* by setting the topological potential of the system as $\Xi^W = W$ and minimizing it. Stable configurations resulting from Ξ^W minimization may be then correlated to molecular energy and stability data derived from *ab-initio* simulations or experimental outcomes.

In [12] the Wiener numbers of the symmetry-distinct C_{80} IPR isomers, seven rather diversified cages, are computed and compared with the potential-energy coming from semi-empirical model calculations. The ground state structure has a D_{5d} symmetry and it is closely followed by a D_2 isomer. Table **2** reports the computed Wiener invariant for each cage and the corresponding potential-energy order (isomer labels are taken from [13]).

Table 2: Energy rank of IPR C_{80} fullerenes and their Wiener numbers

Isomer	ΔE_{rel} (Kcal/mol)	Symmetry	W
1	0	D_{5d}	17340
2	17	D_2	17352
3	75	D_{2v}	17412
5	115	D_S	17454
4	138	D_2	17368
6	154	D_{2v}	17500
7	314	D_2	17600

The fairly good $\Delta E_{rel}/W$ correlation given in Table **2** confirms that the Wiener index provides quick and valuable indications on fullerene stability. For C_{80} this topological invariant works even better than Hückel HOMO-LUMO gap [12]. However, the relative potential energies plotted against the Wiener index for all 24 IPR isomers of C_{84} results in a low degree of correlation. Topological potential $\Xi^W = W$ also fails in ranking the relative stability order of the 19 different IPR isomers of C_{86}.

These limitations point out that "with higher fullerenes there is a strong need for a simple *tool* to predict isomeric stabilities" and "also geometrical terms not truly topologically invariant, such as the measure of non-sphericality" have to be included in the simulations [12].

Before presenting the results of the original C_{50} study, the main computational steps of the topological method based on Ξ^{ρ} calculation and minimization are detailed in the following.

Connectivity lists, as the one given in Table **3** for the C_{50} D_3 fullerene, include the basic information necessary to compute topological sphericity of its molecular graph. This D_3 isomer has a graph diameter $M = 8$ matching the length of the longest path that connects, for example, nodes v_1 and v_{29} in its chemical graph, see Fig. (**2**). In such a way, the Wiener-weight w_i of any vertex v_i in C_{50} D_3 derives from the general definition (1) with $M = 8$, and considering $b_{i1} = 3$ and $b_{i2} = 6$ for all nodes:

$$w_i = \tfrac{1}{2}\,(15 + 3b_{i3} + 4b_{i4} + 5b_{i5} + 6b_{i6} + 7b_{i7} + 8b_{i8}) \quad (6)$$

Calculation shows that the group of six atoms v_4, v_{10}, v_{18}, v_{32}, v_{40}, v_{48} do not present atoms ($b_{i8} = 0$) in the 8th-coordination shell, coordination string and

Table 3: Adjacency list for the C_{50} D_3 fullerene chemical graph

Atom	First neighbors	Atom	First neighbors	Atom	First neighbors
1	2 3 7	19	21 18 24	37	33 36 15
2	1 5 16	20	17 21 42	38	26 35 39
3	1 4 6	21	20 19 22	39	38 40 24
4	49 3 5	22	21 46 23	40	34 39 49
5	4 2 33	23	22 24 25	41	42 45 8
6	3 8 17	24	19 23 39	42	41 43 20
7	1 8 12	25	23 26 29	43	42 44 46
8	6 7 41	26	38 25 27	44	48 43 45
9	12 10 45	27	28 30 26	45	9 41 44
10	50 9 11	28	29 27 32	46	22 43 47
11	10 13 14	29	25 28 47	47	29 46 48
12	13 7 9	30	31 36 27	48	44 50 47
13	11 12 16	31	30 32 14	49	4 18 40
14	11 15 31	32	31 28 50	50	32 10 48
15	14 16 37	33	34 37 5		
16	2 15 13	34	33 35 40		
17	6 20 18	35	38 36 34		
18	17 19 49	36	30 35 37		

Wiener-weight being $\{b_{ik}\} = \{3,6,8,10,9,7,6,0\}$ and $w_i = 104$. Similar conditions hold for carbon atoms v_2, v_{12}, v_{20}, v_{30}, v_{38}, v_{46}. For a general fullerene geometrically different from the highly-symmetrical ''buckyball'', the lack of spherical shape may force some atoms having null entries $b_{ik} = 0$ for high-order k, corresponding to the largest coordination shells.

From Fig. (**2**), the graphic check shows six nodes in the 7th-shell of vertex v_4 ($b_{4,7} = 6$), namely v_{28}, v_{29}, v_{32}, v_{47}, v_{48}, v_{50}, with $d_{4,28} = 7$ and analogous.

C_{50} D_3 cage presents 48 atoms distributed along 6 pairs of fused pentagons, a structural feature which increases strain energy, introducing instability regions in the fullerene. The two remaining atoms v_{49}, v_{50} stay at the center of three touching hexagonal rings and should correspond to relatively stable portions of the fullerene surface.

The topological measure of C_{50} D_3 sphericality is based on the set of Wiener-weights summarized in Table **4**, where topologically equivalent atoms are sorted by coordination strings $\{b_{ik}\}$. Table **4** shows nine sets of inequivalent atoms with relative cardinality (1:3:3:3:3:3:3:3:3) that originate in principle nine ^{13}C NMR lines experimentally detectable. This very fast method to predict resonance data of a chemical structure produces useful theoretical indications over the maximum possible symmetry a molecule may have, coherently with the topological connectivity and the permutation properties of the nodes of its molecular chemical graph, as documented in [14-16].

Table 4: Connectivity strings and Wiener-weights for C_{50} D_3 fullerene

Set/Line	Cardinality/ Intensity	$\{b_{ik}\}$	Equivalent Atoms	w_i
1	2	3 6 9 9 9 9 3 1	49 50	103 w
2	6	3 6 8 10 9 7 6 0	4 10 18 32 40 48	104
3	6	3 6 8 9 9 8 6 0	2 12 20 30 38 46	105
4	6	3 6 8 9 9 8 5 1	5 9 17 31 39 47	105.5
5	6	3 6 7 10 10 7 4 2	3 11 19 28 34 44	106
6	6	3 6 8 9 9 8 4 2	7 16 22 26 36 42	106
7	6	3 6 8 9 9 7 5 2	6 14 24 29 33 45	106.5
8	6	3 6 7 10 9 7 5 2	1 13 21 27 35 43	107
9	6	3 6 8 9 8 7 6 2	8 15 23 25 37 41	107.5 w

From Table **4** it is important to notice that "hexagonal" nodes v_{49}, v_{50} are the *minimal vertices* of the C_{50} D_3 isomer with $\underline{w} = 103$. These vertices represent the most-compactly embedded atoms in the molecular graph, resulting therefore in the most stable atoms of this isomer; on the other hand, nodes v_8, v_{15}, v_{23}, v_{25}, v_{37}, v_{41} provide the highest contribution $\underline{\underline{w}} = 107.5$ to the Wiener index corresponding to the topologically unstable (reactive) atoms of the C_{50} D_3 molecule. Remarkably, these topologically reactive nodes all belong to pentagonal rings although they do not participate in any of the pentagon-pentagon bonds, which are instead centered on the 12 shaded atoms represented in Fig. (**2**). These 12 nodes are grouped, see Table **4**, in two topologically distinct sets, set 5 with $w = 106$ and set 8 with $w = 107$ larger value, evidencing the peculiar instability of the 8th set of nodes v_1, v_{13}, v_{21}, v_{27}, v_{35}, v_{43} shared by fused pentagons of Fig. (**2**). By adopting the arbitrary color code given in Table **4**, the overall effect of the topological ranking on C_{50} D_3 nodes is reported in Fig. (**3**).

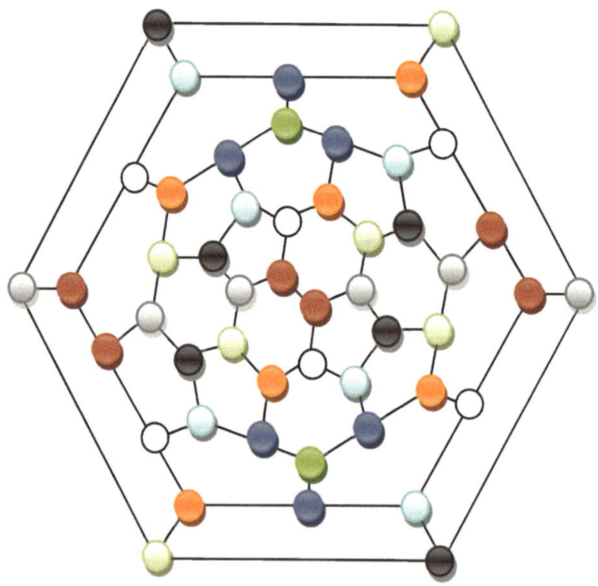

Figure 3: C_{50} D_3 has 9 sets of topologically equivalent atoms grouped by their coordination strings $\{b_{ik}\}$. Each group is colored with a different color, the most stable (unstable) carbon atoms being in green (red). Black nodes represent the most unstable atoms in pentagon-pentagon bonds.

Wiener-weights given in Table **4** allow computing the Wiener index (2) and topological efficiency invariants (4) and (5). Isomer C_{50} D_3 has therefore W = 5,291; ρ = 1.0274; ρ^E = 1.0437 reflecting long-range connectivity properties of this fullerenic cage. These topological descriptors play an important role in comparing relative stability of similar chemical structures, like the C_{50} isomers investigated in the following paragraph. For the C_{50} cages, the numbering sequence adopted in the authoritative theoretical investigation [17] will be applied, thus the molecule studied so far will be referred as isomer $C_{50}(1)$ D_3.

TOPOLOGICAL EFFICIENCY OF C_{50} ISOMERS

The interest for small non-IPR fullerenes C_n with $n < 60$ arises from their highly strained surface due to adjacent pentagonal faces affecting the fullerene stability and may lead to various adducts, oligomers, polymers, and solids with "unusual bonding and electronic properties" [18]. A certain set of *magic numbers* has been predicted by Kroto [19] for fullerenes C_n with $n = 24, 28, 32, 36, 50, 60$ and 70 that should present enhanced stability relative to other sizes. Mass spectrometric studies on carbon ions produced in hydrocarbon flames confirm the magic number model. Experimental observations point out that although C_{60} is the most diffuse fullerene, both C_{50} and C_{70} are also produced with relatively large abundance. As an important member of the fullerene class, C_{50} has 271 isomers and possibly represents the most stable carbon cage in the region $n < 60$ and it has been subject of several theoretical investigations in the past decade, see selected literature [17, 18, 20, 21] and related bibliography.

Authors of article [18] correctly consider the interest for such a hollow molecule made by 12 pentagons and 15 hexagons as a true scientific "passion" boosted by experimental investigations, including the synthesis of macroscopic quantity of $C_{50}Cl_{10}$ with high purity of about 99.5%.

In [20] the most stable structures of C_{50} isomers have been systematized by optimizing molecular geometry using Hartree-Fock methods and density functional theory (DFT). The electronic structure of the two molecules $C_{50}(1)$ D_3 and $C_{50}(2)$ D_{5h} provides those cages with an outstanding stability among the other isomers. In detail, the D_3 fullerene results more stable than D_{5h} but the difference in energy is small and these two structures are considered "as nearly isoenergetic isomers". The remaining C_{50} isomers present more than two fused pentagons, with considerable strain and relative instability.

$C_{50}(2)$ D_{5h} fullerene features two equivalent caps built around isolated pentagons surrounded by five hexagons and it has four types of symmetry-distinct carbon atoms [18, 20], labeled with C1, C2, C3, C4, see Fig. (**4**). The two poles, are bridged by five pairs of fused pentagons, intercalated by hexagons. In the case of $C_{50}(1)$ D_3 structure, Figs. (**2**) and (**3**), the 12 pentagons are grouped in six pairs surrounded by hexagons. The relative stability of D_3 and D_{5h} carbon cages arises from a competition of two structural elements [20].

→ First, the strain-related instability due to the presence of fused pentagons pairs implies that the D_3 isomer is less stable than D_{5h}.

The shape-effects largely influence the fullerene stability by increasing, for spherical isomers, the electronic delocalization which lowers the total molecular energy. Spherical cages exhibit therefore an enhanced molecular stability:

→ The stabilizing effects due to the *spherical distribution of pentagonal rings* have also to be taken in consideration.

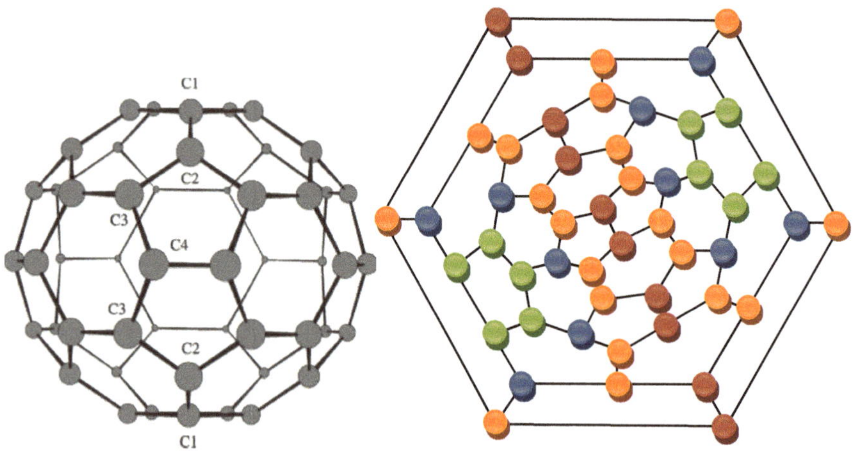

Figure 4: C_{50} D_{5h} planar graph shows 5 pairs of fused pentagons, connecting atoms C4 are in red; the 2 polar pentagons have green atoms C1. The molecule has 75 bonds represented by the same number of graph edges.

According to the *ab-initio* calculations reported in [20] and confirmed in [17], the relaxed carbon network of the D_3 molecule has a geometry more similar to a sphere than the other isomer and, consequently, should be more stable. Within the present graph-theoretical model however the task of measuring fullerene sphericality is assigned to the recently introduced topological efficiency invariants (4) and (5) *and both isomers are now suitable for a fast comparison in terms of topological stability.*

Once again, molecular connectivity provides the basic information to calculate topological descriptors W, ρ, ρ^E of $C_{50}(2)$ D_{5h} fullerene. By applying an arbitrary node numbering to the Schlegel network represented in Fig. (**4**), the coordination strings $\{b_{ik}\}$ are promptly computed for all atoms. Table **5** describes four topologically inequivalent sets of atoms, their $\{b_{ik}\}$ strings and the related color scheme shown in the $C_{50}(2)$ D_{5h} graph of Fig. (**4**).

Table 5: Connectivity strings and Wiener-weights for the atoms of the C_{50} D_{5h} fullerene

Set/Line	**Cardinality/ Intensity**	$\{b_{ik}\}$	w_i
1	10	3 6 8 10 9 7 4 2	105 $\underline{w}$
2	10	3 6 8 9 9 8 6 0	105
3	20	3 6 8 9 9 8 5 1	105.5
4	10	3 6 7 10 9 7 6 1	106.5 $\overline{w}$

Topological descriptors of $C_{50}(2)$ D_{5h} fullerene are easily obtainable: $W = 5275$, $\rho = 1.0048$, $\rho^E = 1.0143$, see Table **5**.

Moreover, four groups of topologically inequivalent atoms with cardinality (1:1:1:2) have been found by our fast TM computation, confirming for the D_{5h} isomer the expected ^{13}C NMR spectrum made by 4 resonance peaks, one with double intensity in respect to the remaining ones.

It is also very interesting to verify that the D_{5h} graph presents a set of ten *minimal vertices* with $\underline{w} = 105$, which includes the nodes of the two isolated polar pentagons, topologically the most stable regions of the cage. On the contrary, "equatorial" atoms that have the highest contribution $\overline{w} = 106.5$ to the molecular Wiener index, individuate – from the topological point of view – a belt of reactive sites in the C_{50} D_{5h} molecule. This TM prediction is confirmed by previous *ab-initio* theoretical works and experimental data [17, 18, 20, 21]. According to literature in fact, resonance measurements report a characteristic "Saturn-shaped structure" with D_{5h} symmetry for the $C_{50}Cl_{10}$ derivative. It is indeed quite intriguing that topology alone is able to reach such a precise prediction on the distribution of reactive sites in a complex molecule like a fullerene, present results pointing out that electronic properties of sp^2 systems are deeply rooted in topology.

The relative reactivity of C_{50}(2) D_{5h} equatorial sites, circled in red in Fig. (**4**), is chemically justified by the fact that these atoms are shared by two fused pentagons and one hexagon, whereas the other atoms, placed on one pentagon and two hexagons, should have higher stabilities. For this isomer the topological predictions fully match with previous theoretical outcomes [20]; for the C_{50}(1) D_3 molecule however, unstable nodes determined by TM methods, *e.g.* the red atoms in Fig. (**3**), do not correspond to reactive atoms normally reported in *ab-initio* studies, *e.g.* the black atoms shared in Fig. (**3**) by fused pentagons with $w = 107$ belonging in fact to the group of the second-most reactive atoms of Table **4**. This small discrepancy being compensated by the greater simplicity offered by TM calculations in respect to *ab-initio* methods. The two most-stable C_{50} isomers may be now analyzed in respect to the distance-based topological descriptors selected for the present TM study. Table **6** compares the Wiener numbers W and the sphericality related invariants ρ and ρ^E.

Table 6: Topological descriptors of the two most stable isomers of C_{50} fullerene. Starred minimum values reflect the higher stability of the D_{5h} cage

Isomer	W **compactness**	ρ **sphericity**	ρ^E **sphericity**
C_{50}(1) D_3	5291	1.0274	1.0437
C_{50}(2) D_{5h}	5275*	1.0048*	1.0143*

Topological invariant W reflects the highest topological compactness featured by C_{50}(2) D_{5h} molecule. This carbon cluster characterized by two opposite nanocones placed at "poles", shows a structure with two cups formed by one pentagon completely surrounded with five hexagons, typical in the icosahedral C_{60} molecule, whose structure may be in fact described as a fullerene made by 12 interconnected nanocones.

Under the effect of the topological potential $\Xi^W = W$ given in Equation (2), D_{5h} strongly competes with the D_3 cage, having a lower Wiener index $W(2) = 5275 < W(1) = 5291$. This result largely depends on the fact that the connectivity of the ten "green" nodes, increase the *average* value of $C_{50}(2)$ D_{5h} compactness.

TM algorithms also measure the *topological sphericity*. The quest for spherical isomers is fulfilled by adopting the topological potential $\Xi^\rho = \rho$ (or $\Xi^{\rho E} = \rho^E$).

Computational results show how the long-range connectivity properties attribute – again – the role of most stable isomer to the D_{5h} isomer being both functions Ξ^ρ and $\Xi^{\rho E}$ minimized for that molecule, see Table **6**. In this case, TM favors the topological sphericality over geometrical sphericality due to the "symmetrising" role played in the $C_{50}(2)$ D_{5h} graph by the two nanocone sub-units. This fact may be easily verified by considering the narrowest distribution of w_i values that characterizes the D_{5h} structure over the D_3 one, see the Wiener-weights entries reported by Tables **4** and **5**.

In summary, the present application of TM methods analyzes the *topological performance* of the two most stable C_{50} isomers and a good agreement with DFT methods in the predictions of relative stabilities has been found. The main difference with *ab-initio* outcomes consists in the *topological sphericality* that is larger for the $C_{50}(2)$ D_{5h} molecule, whereas normal literature data appear to favor the *geometrical sphericality* of the $C_{50}(1)$ D_3 isomer. The relative stability of D_{5h} is derived from its long-range connectivity properties. *Topological sphericality* represents therefore an original and very useful complementary theoretical tool on the way to fully understand fullerene formation mechanisms.

CONCLUSION

The "strong demand" for fast methods to simulate fullerene chemical stability inspired this work and the graph theoretical tools herewith proposed. In particular, this chapter deals with topological modeling of fullerene isomerisation. Two C_{50} molecules have been analyzed in detail, mapping the effects of two kinds of topological potentials, namely the Wiener index Ξ^W and the topological efficiency index Ξ^ρ and $\Xi^{\rho E}$. Carbon structures minimize then their topological invariants according with the adopted TM method. Non-trivial answers on relative stability of C_{50} isomers have been derived, predicting a certain predominance of the $C_{50}(2)$ D_{5h} molecule leaving the determination of energetic and thermodynamic quantities to future specific studies. The influence of topological long-range interaction on the spherical properties of fullerene represents an original theoretical result suitable for further investigations, extending this approach to other C_n fullerenic networks.

ACKNOWLEDGEMENTS

Ali Reza Ashrafi is partially supported by the University of Kashan under grant number 159020/74.

CONFLICT OF INTEREST

The authors confirm that this chapter contents have no conflict of interest.

ABBREVIATIONS

IPR	=	Isolated Pentagon Rule
QSAR	=	Quantitative Structure-Activity Rclationship
QSPR	=	Quantitative Structure-Property Relationship
TM	=	Topological Modeling
WW	=	Wiener-weights

REFERENCES

[1] Balaban, A. T., Ed. *Chemical Applications of Graph Theory*; Academic Press, London, **1976.**

[2] Todeschini, R.; Consonni. V. *Handbook of Molecular Descriptors*; Wiley, Weinheim, **2000**.

[3] Putz, M. V., Ed. *QSAR and SPECTRAL-SAR in Computational Ecotoxicology*; Apple Academic Press Inc., Weinheim, **2012**.

[4] Bonchev D.; Mekenyan O. A topologycal approach to the calculation of the π-electron energy and energy gap of infinite conjugated polymers. *Z. Naturforsch.*, **1980**, 35(a), 739-747.

[5] Alipour, M. A.; Ashrafi, A. R. A numerical method for computing the Wiener index of one-heptagonal carbon nanocone. *J. Comput. Theor. Nanosci.*, **2009**, 6, 1204–1207.

[6] Cataldo, F.; Ori, O.; Graovac A. Graphene topological modifications. *Int. J. Chem. Modeling*, **2011**, 3(1-2), 45–63.

[7] Cataldo, F.; Ori, O.; Iglesias-Groth S. Topological lattice descriptors of graphene sheets with fullerene-like nanostructures. *Molecular Simulation*, **2010**, 36(5), 341–353.

[8] Ori, O.; Cataldo, F.; Putz, M.V. Topological anisotropy of Stone-Wales waves in graphenic fragments. *Int. J. Mol. Sci.* **2011**, 12, 7934-7949.

[9] Iranmanesh, A.; Ashrafi, A.R.; Graovac, A.; Cataldo, F.; Ori, O. Wiener Index Role in Topological Modeling of Hexagonal Systems - From Fullerenes to Graphene. In: *Distance in Molecular Graphs – Applications*; Gutman, I., Furtula B., Eds.; Univ. Kragujevac, Kragujevac, **2012**; pp. 135-155.

[10] Ori, O.; D'Mello, M. A topological study of the structure of the C_{76} fullerene. *Chem. Phys. Lett.*, **1992**, 197, 49–54.

[11] Ori, O.; Cataldo, F.; Graovac, A. Topological ranking of C_{28} fullerenes reactivity. *Fullerenes, Nanotubes and Carbon Nanostructures*, **2009**, 17(3), 308 – 323.

[12] Slanina, Z.; Chao, M.-C.; Lee, S.-L.; Gutman, I., On applicability of the Wiener index to estimate relative stabilities of the higher-fullerene IPR isomers, *J. Serb. Chem. Soc.*, **1997**, 62(3), 211-217.

[13] Manolopoulos, D. E.; Fowler, P. W., Molecular graphs, point groups, and fullerenes, *J. Chem. Phys.,* **1992**, 96, 7603-7777.

[14] Ori, O.; D'Mello M., A topological study of the structure of the C_{76} fullerene, *Chemical Physics Letters,* **1992**, 197(1-2), 49-54.

[15] Ashrafi A.R.; Ahmadi M.R., Automorphism group and topological indices of the chemical graph of fullerenes, *Internet J Nanotechnol*, **2005,** 1(2), DOI: 10.5580/24c3.

[16] Ori, O.; Cataldo, F.; Vukicevic, D.; Graovac A., Topological Determination of ^{13}C–NMR Spectra of C_{66} Fullerenes, In: *The Mathematics and Topology of Fullerenes*; Cataldo, F.; Graovac, A.; Ori O., Eds., Carbon Materials: Chemistry and Physics, Springer Science+Business Media B.V., Dordrecht, 2011, pp. 205-216.

[17] Diaz-Tendero, S.; Alcami, M.; Martin, F., Fullerene C_{50}: Sphericity takes over, not strain, *Chemical Physics Letters*, **2005**, 407,153-158.

[18] Xu, Z.; Han, J.; Zhu, Z.; Zhang, W., Valence of D_{5h} C_{50} Fullerene, *J. Phys. Chem. A*, **2007**, 111, 656-665.

[19] Kroto, H.W., The stability of the fullerenes C_n, with n = 24, 28, 32, 36, 50, 60 and 70, *Nature*, **1987**, 329, 529-531.

[20] Xu, W.G.; Wang, Y.; Li, Q, S., Theoretical study of fullerene C_{50} and its derivatives, *Journal of Molecular Structure (Theochem)*, **2000,** 531, 119-125.

[21] Tian, W. Q.; Feng, J.-K.; Wang, Y.A.; Aoki, Y., Search for suitable approximation methods for fullerene structure and relative stability studies: Case study with C_{50}, *The Journal Of Chemical Physics*, **2006**, 125, 094105-10.

CHAPTER 2

Similarity in Chemical Reaction Networks: Categories, Concepts and Closures

Andrés Bernal[1,*], Eugenio Llanos[2], Wilmer Leal[3,4,5] and Guillermo Restrepo[5]

[1]Grupo de Química Teórica, Universidad Nacional de Colombia, Bogotá, Colombia; [2]Corporación Colombiana del Saber Científico SCIO, Bogotá, Colombia; [3]Fundación Instituto de Inmunología de Colombia FIDIC, Bogotá, Colombia; [4]Universidad del Rosario, Bogotá, Colombia and [5]Laboratorio de Química Teórica, Universidad de Pamplona, Pamplona, Colombia

Abstract: Similarity studies are important for chemistry and their applications range from the periodic table to the screening of large databases in the searching for new drugs. In this later case, it is assumed that similarity in molecular structure is related to similarity in reactivity. However, we state that structural formulas can be regarded as abstract representations emerging from the analysis of large amounts of data upon chemical reactivity. Hence, chemical formulas such as organic functions are not direct pictures of the atomic constitution of matter, but signs used to represent similarity in the reactivity of a class of substances. Therefore, reactivity, rather than molecular structure, becomes the fundamental feature of chemical substances. As reactivity is important, chemical identity is given by the relations substances establish with each other, giving place to a network of chemical reactions. We explore similarity in the network rather than in molecular structure. By characterising each substance in terms of the related ones, we show how Category Theory helps in this description. Afterwards, we study the similarity among substances using topological spaces, which leads us to concepts such as closure and neighbourhood, which formalise the intuition of things lying somewhere near around. The second focus of the chapter is the exploration of the potential of closure operators, and of topological closures in particular, as more general descriptors of chemical similarity. As we introduce the formalism, we develop a worked example, concerning the analysis of similarity among chemical elements regarding their ability to combine into binary compounds. The results show that several of the trends of chemical elements are found through the current approach.

Keywords: Binary compounds, category theory, chemical classification, chemical networks, closure, closure operators, directed hypergraphs, formal concept analysis, graph theory, network theory, order theory, periodic table, reaction networks, similarity, topology.

***Corresponding author Andrés Bernal:** Grupo de Química Teórica, Universidad Nacional de Colombia, Calle 44 N° 45-67 Bloque B9, Bogotá, Colombia; Tel: + 57 1 2227767; E-mail: afbernale@unal.edu.co

Subhash C. Basak, Guillermo Restrepo & José L. Villaveces (Eds)
Copyright © 2015 Bentham Science Publishers Ltd. Published by Elsevier Inc. All rights reserved.
10.1016/B978-1-68108-053-6.50002-8

INTRODUCTION

Similarity is a key concept at the grounds of the scientific enterprise, having produced many classification schemes ranging from archaeology and literature to chemistry [1]. Classification schemes separate object sets into classes of similar behaviour, allowing a reduction in the amount of data to handle. In this way, the total amount of information regarding the description of the whole set of objects, in terms of their attributes, is reduced to those trends found in classes. Objects sharing similar attributes are assumed to behave in a similar way, as long as the attributes used encapsulate their behaviour in a given domain. This assumption allows a generalization of the class behaviour to unknown instances, turning the classification scheme into a structuring predictive tool. In chemistry, the subject has a long tradition that may be traced back, at least, to the 17th century [2]. Among all existing examples, perhaps the most popular is that of the periodic law and the periodic table[1]. Elements were divided into classes of similar reactivity and arranged in a table representing similarity relationships. By assuming the periodic law, holes within this table were interpreted as missing elements and their properties were successfully estimated before their actual discovery.

In contemporary chemistry, similarity also plays a significant role for constructing predictive models in molecular design. QSAR models (Quantitative Structure-Activity Relationships) are motivated by the assumption that similarities in molecular structure, or in molecular fragments, produce similar activities of biochemical, physiological, or chemical nature. Within this approach, activity is regarded as determined by molecular structure. It is interesting to note that, in the historical development of structure theory, things happened exactly the other way round: compounds with similar reactivity were assumed to have similar structures, an assumption that led to the invention of chemical formulae as abstract representations of knowledge emerging from the analysis of large amounts of data regarding chemical change [3, 4]. From this perspective, key features of chemical formulae such as organic functions are not direct pictures of the atomic constitution of matter, but signs used to represent an element of similarity in the reactivity of a class of substances, that is, on the position they occupy within a potentially infinite reaction network [5].

[1]Some other examples of similarity in chemistry are discussed by Rouvray [1].

In recent years, this approach of studying similarity not on chemical structures but directly on the network of chemical reactions has been revisited by Klein and Ivanciuc [6] with promising results. However, much terrain remains to be explored regarding the mathematical foundations of reactivity similarity. This is a particularly pressing issue with arising interest in network modelling of biochemical systems, where proper understanding of the mathematics behind reactivity similarity would enrich the QSAR perspective by extending the notion of structure to metabolic networks, enhancing our capability to predict biological activities. In this chapter we delve into the unique challenge posed by similarity studies on reaction networks and propose the use of mathematical tools from category theory, topology and formal concept analysis to address it.

Similarity is defined with respect to some set of attributes of the objects under study. Attributes turn out to be n-ary relations, *i.e.* relations that involve n objects. When $n = 1$, attributes correspond to properties of each object, when $n = 2$ attributes correspond to couples of objects. In general, attributes in an n-ary relation correspond to n-tuples of objects.

Similarity studies usually take advantage of 1-ary relations by using attributes such as density, melting point, or refractive index, which can be described logically using single object predicates. The peculiarity of the study of similarity on a network comes from the fact that its elements have no quality whatsoever other than being related to each other. From a practical point of view, this means that the usual technique of identifying the objects to be compared with a vector of qualities cannot be applied straightforwardly. Therefore, the first step to speak of similarity on a chemical reaction network is to construct properties that refer to the relational identity of substances in the network, rather than to their isolated qualities when chemically disjoint from other substances. The first focus of the current chapter is the exploration of how the similarity of network properties can be modelled by using the concept of kernel of a relation. For the sake of clarity and generality, we develop the grounds of our framework upon Category Theory, which introduces, in a natural way, the notion of an adjacency-preserving mapping.

If we induce characteristic properties on chemical substances that describe their relational identity, we can start the study of similarity in chemical reaction networks. Presently, it is customary to quantitatively define chemical similarity regarding substances as elements of a metric space [7], which has brought significant advances in similarity studies. However, the concept of similarity is not necessarily restricted to its quantification. As we will argue, we may (and maybe should) think of similarity within a more general framework capable of grasping all its mathematical features and containing metric spaces as a particular case. We believe that a topological space provide a natural scenario wherein such framework can be developed.

Metric spaces are part of topological spaces, but not all topological spaces are metric ones. In a finite topological space the notion of distance can be disregarded, resorting instead to concepts such as closure and neighbourhood to formalise the intuition of things lying 'somewhere near around', even if we cannot tell exactly how near. Thus, the second focus of the current chapter is the exploration of the potential of closure operators, and of topological closures in particular, as more general descriptors of chemical similarity.

As we introduce the formalism, we will develop a working example concerning the analysis of similarity among chemical elements regarding their ability to combine into binary compounds, which will be contrasted against the periodic classification of elements. Along the chapter we assume the reader is familiar with the fundamental definitions of graph theory. Previous chapters of the present volume or references [8] and [9] may be looked up for a more or less detailed introduction to the topic. Also, some basic knowledge on topology can be of help, but it is not assumed.

Substance Properties from Chemical Relations

The Network Perspective in Chemistry

Several areas of science deal with systems governed by selective relations between their constituents, sociology being the classical example. In chemistry such examples comprise metabolism of living organisms, governed by chains of chemical reactions responsible for biomolecule synthesis, energy production and regulation; atmospheric reaction networks, where native and human-made molecules interact, consuming or accumulating products with potential damaging impact; and reaction mechanisms, *i.e.* collections of elementary reactions determining the kinetics of a chemical transformation. The term *Network Analysis* has been coined to denote a general approach to characterise the global and local structure of this kind of systems by means of the analysis of patterns of relations between their components.

Network analysis has been applied in chemistry to the study of chemical kinetics [10], atmospheric phenomena [11], astrochemistry [12], and the metabolism of living organisms [13, 14]. The common feature of these works is the representation of a system of interest as a network of substances interacting in chemical reactions; that is, as a *Chemical Reaction Network.*

Graph Theory has achieved a dominant status as the preferred mathematical framework for the analysis of chemical reaction networks. Several graph models have been proposed:

- An undirected graph, with one node for each substance, and an edge between two nodes if there is a precursor-product relation. This is known as the substrate graph. Fig. **(1b)** depicts the *substrate graph* of the reactions of Fig. **(1a)**.

- A directed graph, with one node for each substance and arcs pointing from reactants to products. This is known as the *directed substrate graph* (Fig. **(1c)**).

- A bipartite digraph, where nodes of one class represent substances, nodes of the other represent chemical reactions and arcs point from reactants to reactions and from reactions to products (Fig. **(1d)**).

These models are just approximations to the exact structure of a reaction network: chemical reactions in a reaction network are binary relations among *sets of substances*, which cannot be identified with graphs edges [14]. However, chemical reactions induce reactant-reactant, reactant-product, product-product, reactant-reaction, and reaction-product relations that can be represented by graph edges. Different graph models of reaction networks study distinct aspects of the structure of the full reaction network by focusing on distinct kinds of relations (*e.g.* reactant-product relations in the case of substrate graphs, reactant-reaction and reaction-product relations in the case of bipartite digraphs). Therefore, though graphs cannot achieve a full description of reaction networks in general, in specific cases a graph model gives a sufficiently good description of the particular aspects of the network being researched.

Example 1 (Periodic Table). A significant amount of information consigned in the periodic table concerns similarity in the behaviour of chemical elements as they combine to produce binary compounds. To illustrate the potential of the model proposed we will reconstruct a small piece of this knowledge in the formal mathematical language introduced above and further bellow.

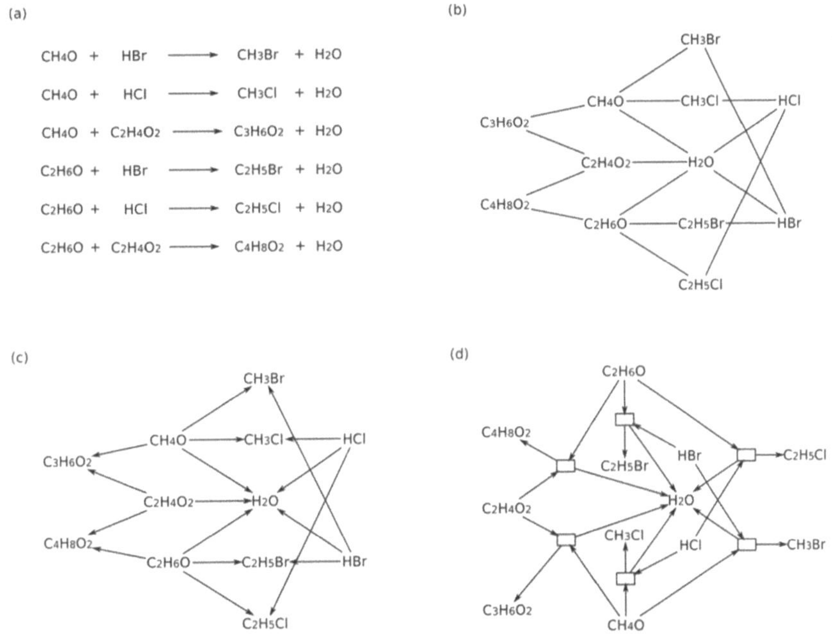

Figure 1: A network of six reactions (a) modelled as a simple substrate graph (b), as a directed substrate graph (c) and as a bipartite reactions-substances digraph (d). In the latter, reactions are depicted as unlabelled rectangular vertices.

We start by taking a reaction network of chemical elements and binary compounds, along with the reactions of elementary synthesis of binary compounds from their constituting elements. The first step comports the choice of a mathematical representation of the network, which is entirely composed of reactions of the form.

$$mu + nv \rightarrow u_m v_n \qquad (1)$$

u and v being chemical elements. We focus on similarity among chemical elements, meaning that binary compounds themselves are disregarded[2]. Furthermore, to keep things simple we avoid stoichiometric concerns. Hence, we are looking only at reactant-reactant relations existing between pairs of chemical elements that combine to produce at least one binary compound. We may represent this network by a graph $\mathcal{E}$, whose vertices are chemical elements, with edges (u, v) whenever elements u and v combine to produce a binary compound.

To be strict, the data set for constructing this graph should comprise a set of elementary synthesis reactions producing binary compounds that are known to be spontaneous in a thermodynamic sense. To do so we would require extensive data on free energies of formation of binary compounds. Since not enough consolidated data was found, we chose a simpler approach: we assumed that each compound reported in three consulted sources [15-17] can be spontaneously synthesized from its elementary constituents. This is a dubious supposition, yet we have found that the resulting network works reasonably well as a proxy for the real situation. Hence, we introduced an edge (u, v) for each binary compound constituted by u and v that was reported in the sources, obtaining the network of Fig. **(2)**.

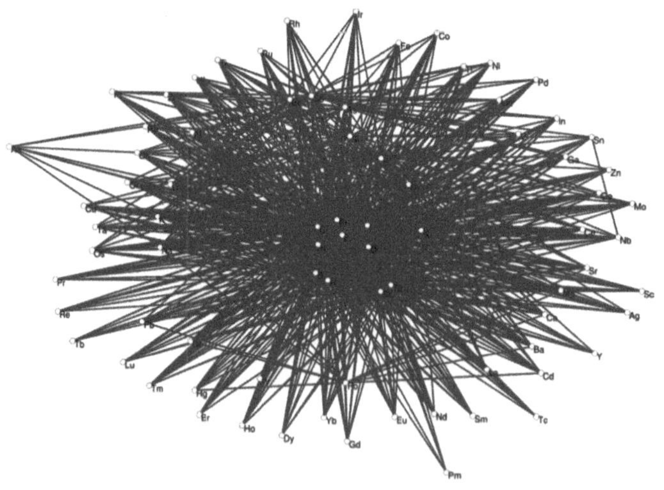

Figure 2: Image of the network of 84 chemical elements studied. An edge exists between two elements if they form at least one binary compound. Vertices with the most connections represent the most reactive elements and are drawn towards the centre of the picture.

---

[2]In the terms introduced in the forthcoming sections, this means that binary compounds are to be regarded as members of a single class.

Studies in philosophy of chemistry recognize the relevance of the network perspective in chemistry [5], revealing its status as a central epistemic approach long before the first mathematical models of chemical reaction networks [18, 19]. Even if research in material systems comprising hundreds of chemical reactions occurring in a shared environment is a recent phenomenon, the whole logical structure of chemistry can be understood as comprising an increasingly huge network of substances related by known chemical reactions [5]. This fact has important consequences in the study of chemical similarity.

Our first goal is to derive, starting from chemical relations, a set of characteristic properties of chemical substances with respect to which similarity is to be analysed. From the network perspective, chemical properties have the same logical structure of chemical operations [5]. By this we mean that just as substances are characterized in chemical operations as the terms of chemical reactions, their properties appear as the terms of archetypical chemical reactions. For instance, the 'alcohol' property can be understood as a pseudo-reactant/product appearing in certain classes of characteristic reactions of alcohols (*e.g.* esterification, halogenation, dehydration). The link between the network of substances and reactions and the network of properties and reaction classes is given by a structural invariance, intuited in the fact that we can map the terms of chemical reactions (the relations in the original network) into the terms of reaction classes (the relations of the abstract network of classes). Chemical properties thus emerge by mapping experimental reactions into abstract reaction types (Fig. **(3)**), which involves a transformation of an empirical reaction network into a reduced representation of the same kind, where the transformation preserves the essential logical structure of the original network. Category theory allows formalising this intuition by giving necessary and sufficient conditions for stating that a transformation between reaction networks is 'structure-preserving'.

Elements of Category Theory

The essential identity of mathematical objects is not determined by what the object 'is by itself', but by what it 'is related to'. For instance, $\{a, b, c\}$ and $\{\alpha, \beta, \omega\}$ are two different sets, yet they are essentially identical in the sense that they have exactly the same set-theoretical properties. Or, it does not matter what labels we use to denote the vertices of a graph: all what matters is the connectivity pattern among those vertices (Fig. **(4a)**), since that is what determines all graph-theoretical properties such as valences, branching, colourability, *etc.* In both cases, it is possible to transform one object into the other by point-wise transforming the elements/vertices of the original set/graph into the elements/vertices of the other.

Category Theory puts this idea into the light spot, characterizing mathematical entities exclusively in terms of their viable transformations. In the most restricted sense, this means that the transformation only changes the arbitrary labels used in a mathematical construct, as shown above. In such cases, the transformed object is ultimately identical to the original. But it is also possible to imagine transformations changing the identity of a mathematical object while still preserving certain features of its structure that we intuitively consider 'essential'.

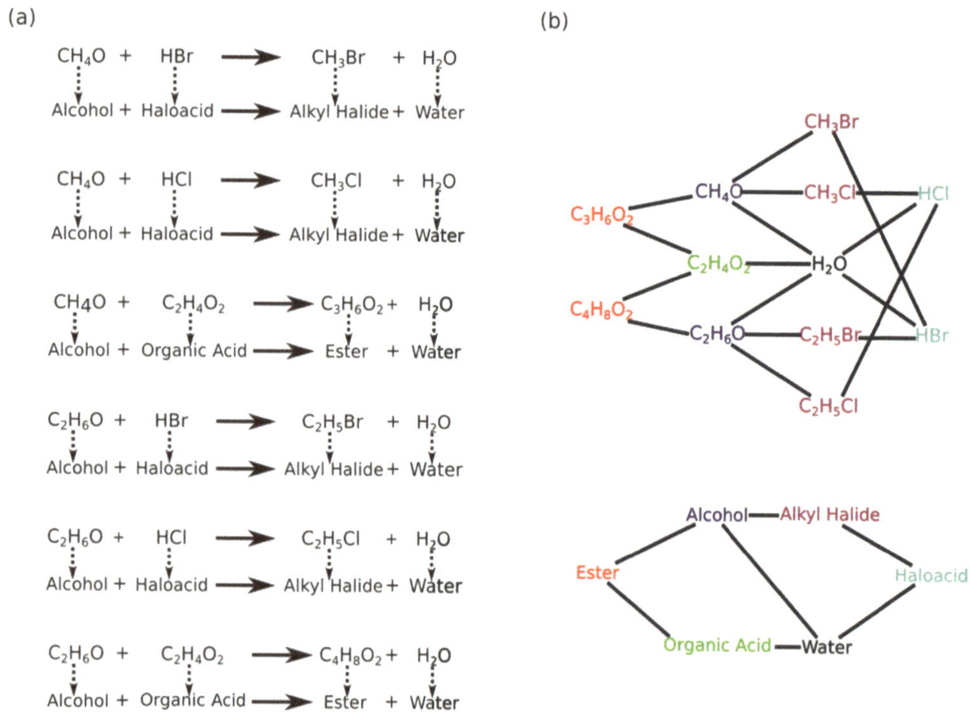

Figure 3: (a) Reactions 1, 2, 4 and 5 are mapped into the "halogenation" reaction class, and reactions 3 and 6 into the "esterification" reaction class. (b) This transformation corresponds to a unique morphism $\boldsymbol{f} \in \boldsymbol{Graphs_1}$ between the corresponding substrate graphs. In the picture, font colours match substances with their corresponding images under $\boldsymbol{f}$.

For example, in group theory, group homomorphisms map one group into another while keeping its algebraic structure; in topology preimages of open sets under continuous functions are open sets, which preserves the core topological structure of the space; in graph theory adjacency-preserving mappings transform graphs while preserving adjacencies among their vertices, and so on (Fig. **(4b)**). The core of the formalism of category theory is the generalization of these different criteria of structure-preserving transformation by means of the concept of *morphisms in a category*.

Definition 1 (Category). A *category* C consists of a collection C_0 of *objects* and a collection C_1 of *morphisms* such that

i). For each $f \in C_1$ there exist a unique object $\mathrm{Dom}(f) \in C_0$ called the *domain of f* and a unique object $\mathrm{Cod}(f) \in C_0$ called the *codomain of* f. A morphism f with $\mathrm{Dom}(f) = A$ and $\mathrm{Cod}(f) = B$ is noted as$f : A \to B$.

ii). For each $A, B, C \in C_0, f : A \to B$, and $f : B \to C$, there exists a unique morphism $g \circ f : A \to C$ in C_1 called the *composite of f and* g.

iii). For each $f, g, h \in C_1, h \circ (g \circ f) = (h \circ g) \circ f$.

iv). For each $A \in C_0$ there exists a morphism $1_A : A \to A$ in C_1, called the identity morphism of A, such that for all $B \in C_0, f : A \to B$, and $g : B \to A, f \circ 1A = f$ and $1B \circ f = f$.

A morphism $f : A \to B$ is an *isomorphism* if there is a $g : B \to A$ such that $f \circ g = 1_B$ and $g \circ f = 1_A$. In such a case it is said that g *is the inverse of f* (from now on noted f^{-1}), and that A and B are *isomorphic*, which is noted as $A \simeq B$. The set $\mathrm{Hom}_C(A, B)$ is defined as the set of all morphisms $f : A \to B$ in C_1.

In the following, some examples of categories are shown.

Example 2.

i) The category **Sets** of sets and functions, with composition defined in the usual way. Isomorphisms in this category are bijective functions and the identity morphism on a set A is the identity mapping on A.

ii) The category **Top** of topological spaces and continuous functions. Function composition and identities are defined as in **Sets,** which is correct because identity functions are continuous, and the composition of two continuous functions is continuous. Isomorphisms in this category are homeomorphisms.

iii) The category Graphs of **graphs** and adjacency-preserving mappings. Function composition and identities are defined as in **Sets**, which is correct since the identity mapping on the vertex set of a graph preserves adjacency, and so does the composition of adjacency-preserving mappings. Graph isomorphisms provide the isomorphisms of this category (Fig. **(3)**). The reaction network of **Example 1** may be regarded as an object in this category.

Most known mathematical constructions can be fully characterized by a category, as seen in the examples above. More specifically, we are dealing with concrete categories, which may be informally understood as categories whose objects are sets possibly endowed with an additional structure. For example, a metric space is a set structured by a metric function, a topological space is a set structured by a topology, and a graph may also be seen as a set of vertices structured by an edge set. The point to be noted is that *the category-theoretical definition of a structured set as an object X in a concrete category avoids introducing its internal structure explicitly: the collections of morphisms with domain and codomain in X are enough to characterize it.* This means, for instance, that as long as $\mathrm{Hom}_{\mathbf{Met}}(X, Y)$ is known for any two metric spaces X, Y, all relevant properties of any metric space can be determined without explicit reference to its actual metric. Category theory thus gives a description of mathematical objects in terms of what they can be transformed into, which is equivalent to the customary description in terms of what they are 'made of'. In this way, it provides the missing link between chemical substances and chemical properties: they conform two distinct networks that are related by a morphism in a suitable concrete category.

Emergence of Properties through Classification

The information accumulated through a series of chemical experiments can be represented as a structured set, whose elements represent the subjects of study, and whose structure embodies their relations. By regarding such a set as an object in a concrete category, we open the way for a description of its structure in terms of its structure-preserving transformations. As we will show, each of these transformations induces a classification on the set's domain, which at the same time endows each element of the set with a collection of properties consistent with the relational structure of the whole set.

Consider, for instance, the substrate graph model of Fig. **(1)**. From the perspective of category theory, a substrate graph G is an object in the concrete category **Graphs**, characterized by the collection $\bigcup \mathrm{Hom}_{\mathrm{Graphs}}(G, G')_{G' \in \mathbf{Graphs}_0}$ of adjacency-preserving mappings f with domain in G. Each of these morphisms describes a structural feature of G, by linking the structure of this set with that of the codomain set. The binary relation,

$$\ker f = \{(x, x') \in \mathrm{Dom}(f) \times \mathrm{Dom}(f) \colon f(x) = f(x')\} \tag{2}$$

known as the *kernel* of f is an equivalence relation that induces a partition of the domain of f in a set of equivalence classes parametrised on its codomain:

$$\ker f(y) = \{x \in \mathrm{Dom}(f) \colon y = f(x)\}_{y \in \mathrm{Cod}(f)} \tag{3}$$

Kernels of morphisms in a category can be thought of as 'natural' or 'viable' classifications on their corresponding domains, in the sense that they classify the elements of a structured set in accordance with its proper structure. Furthermore, it is clear that different functions cannot have the same kernel; that is, $\ker f$ is characteristic of each function. We see then that the category-theoretical perspective leads to a characterisation of chemical reaction networks in terms of the collections of viable classifications of their elements. This characterisation of the global structure of the network can be extended into a characterisation of its elements, by considering the effect of the transformation on each individual vertex: given a morphism $f : G \to G'$, each $x \in G$ is identified by f with an element $f(x) \in G'$ that parametrises the equivalence class of x. It is thus natural to regard G' as a (structured) set of *properties* attributable to the elements of G, in the sense that they can be ascribed to each substance in accordance to the relational structure of the reaction network.

In this way, morphisms in a category of reaction networks simultaneously induce characteristic properties on chemical substances and classify them according to those properties; so we may say equivalently that x belongs to the class $f(x)$, or that x *has the property* $f(x)$.

This classification methodology follows a well-known principle of network analysis, perhaps most explicitly stated in social sciences: 'actors of the same role relate in the same way to actors of the same given roles'. For instance, we say that a 'parent' is someone who has a 'son'; and conversely, a 'son' is someone who descends from a 'parent'. The apparent circularity of this definition is broken when we recognize that we are just stating the existence of a characteristic relation between social actors that is shared by several pairs of individuals. Fig. **(4b)** illustrates the connection between this principle and morphisms in a concrete category; in this particular case in **Graphs**. The graph to the left, a small part of a genealogical tree, is transformed by an adjacency-preserving mapping into the graph to the right. The kernel of this mapping naturally relates each individual in the genealogy with either the role of 'grandparent', 'parent', or 'son'. In other words, this particular morphism unveils the existence of a sociological invariant in this particular network –the parent-son relation–, and classifies the actors involved accordingly.

Chemical classifications follow an analogous principle: actors are chemical substances, and roles are chemical properties or classes such as acid, alcohol, halide, and so on. The principle behind chemical classifications is that 'substances of a given class react with substances of the same other classes to produce substances of other further classes'. For example, we say that a substance is an alcohol (equivalently, it *belongs* to the alcohol class or *has* the alcohol property) if it reacts with substances that are carboxylic acids to produce substances that are esters; reacts with substances that are haloacids to produced halides, and so on. Fig. **(3)** illustrates how such a classification corresponds with a morphism in **Graphs** with domain in the substrate graph of a reaction network.

Summarising: a chemical reaction network is described by an object G in a concrete category; each morphism with domain in G gives a viable classification of substances in the network into a set of classes given by its codomain. Last, such codomain set can equivalently be seen as a set of properties characterizing elements of G. There is but one point that we must address before closing this section: that there is hardly any similarity involved in the classifications considered up to this point.

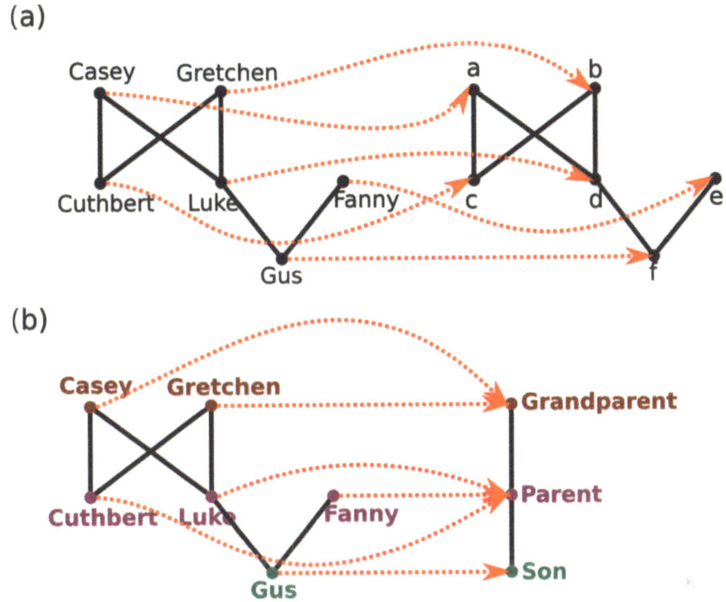

Figure 4: Structure-preserving transformations between graphs. (a) Two graphs differing only in their vertex labels are essentially identical. In such a case, they are related by an adjacency-preserving bijection (dotted arrows) $\boldsymbol{i}: \mathbf{G} \rightarrow \mathbf{G'}$ such that for any edge $(\boldsymbol{u}, \boldsymbol{v})$ in the graph's edge set $\boldsymbol{E_G}$, $(\boldsymbol{i}(\boldsymbol{u}), \boldsymbol{i}(\boldsymbol{v})) \in \boldsymbol{E'_G}$. These transformations are the isomorphisms of the category Graphs. (b) Two graphs differing in their connectivity pattern may still be related by a non-bijective adjacency preserving mapping, showing that it is possible to transform one graph into the other while preserving some essential features of its connectivity. These transformations are the morphisms of Graphs.

The Domain of Similarity

Though adjacency-preserving mappings solve the problem of attaching individual properties to each substance that are determined by the relational structure of a reaction network, this solution is inadequate for our original purpose of describing similarity among chemical substances. Indeed, the kernel of a mapping characterizes the elements of its domain as *equivalent*, rather than as *similar*: it induces a classification in mutually disjoint classes, meaning that each substance is identified with a unique property. Therefore, substances characterised by the same property would be equivalent rather than similar. For instance, one such classification may describe F and Cl as having the 'halogen' property, and oxygen as having the 'chalcogen' property –the former would be considered identical, as they are described by the same property, and completely unrelated to the latter, which is characterised by another property. Thus, no similarities between chlorine and oxygen would be unveiled.

This issue is a consequence of the kernel of a *function* being an equivalence relation, which is reflexive, symmetric, and *transitive*. The last property implies that, for any equivalence relation $\asymp, x_1 \asymp x_2 \asymp \cdots x_n$ implies $x_1 \asymp x_n$; yet our intuition tells that if $\asymp$ is a similarity relation it is entirely possible that in this case $x_1 \not\asymp x_n$. Therefore, similarity must be described by reflexive, symmetric, but possibly non-transitive relations. It is worth noting that these kinds of relations, nowadays mostly known as tolerance relations, were originally named *similarities* by Carnap and Poincaré [20].

In the following we show how the problem can be solved by using a category that has binary relations as morphisms, rather than functions. First, consider two graphs G, G' and a binary relation $r \subseteq G \times G'$ such that for any $(u, v) \in E_G$, there are (u, x) and (v, y) in r such that $(x, y) \in E_{G'}$. This binary relation is intuitively adjacency-preserving in the same sense of morphisms in **Graphs**; in fact, it is straightforward that such morphisms are particular instances of the binary relations just defined. Even more, we prove that the set of all graphs along with these 'adjacency-preserving relations' is a category:

i) For each adjacency-preserving relation $r \subseteq G \times G'$ we define $\mathrm{Dom}(r) = G$ and $\mathrm{Cod}(r) = G_0$.

ii) We define composition in the standard way for binary relations.

iii) It is known that composition of binary relations is associative.

iv) It is immediate that the identity mapping $1_G: G \to G$ is an adjacency preserving relation and thus the identity morphism of G.

We refer to this category that extends **Graphs** as **rGraphs.**

The usefulness of a category of binary relations is put in evidence when we consider the natural extension of the definition of kernel (of a function) to arbitrary binary relations: in analogy to (2), we define the *kernel of a binary relation* $r \subset X \times Y$ as,

$$\ker r = \{(x, x') \in X \times X : \exists y \in Y \text{ such that } (x, y), (x', y) \in r\} \qquad (4)$$

which reduces to the original definition, if and only if r is a mapping. Note that in any other case $\ker r$ is *non-transitive*, but it is nonetheless *symmetric* and *reflexive*! In short, $\ker r$ *is a dependency relation*. In this way, binary relations achieve the desired goal of capturing what we identified as the key features of similarity relationships.

Note that the classification induced by $\ker r$,

$$\{ker\ r(y) = \{x \in X : (x, y) \in r\}\}_{y \in Y} \qquad (5)$$

now comprises possibly non-disjoint classes parametrised in Y. That is, while previously each $x \in G$ was characterised under a function $f : G \to G'$ by exactly one property $f(x)$, in the new construction it can be characterised under a binary relation $r \subset G \times G'$ by arbitrarily many properties,

$$r(x) = \{y \in G' : (x, y) \in r\} \qquad (6)$$

In this situation, similarity between two substances u, v can be related to the presence of common and uncommon properties in their respective images under r. Going back to the previous example, Cl could be described by the 'electronegative' and 'monovalent' properties, and oxygen by the 'electronegative' and 'bivalent' properties. Under this description they share one property but differ in another, so they are described as similar but non-identical substances.

Example 3 (Periodic Table (continued)). By regarding the reaction network of **Example 1** as an object in **rGraphs**, new viable transformations appear. For instance, the relation $r = E_{\mathcal{E}}$ is a morphism $r : \mathcal{E} \to \mathcal{E}$ in this category:

i) $r \subset V_{\mathcal{E}} \times V_{\mathcal{E}}$

ii) Let $(u, v) \in E_{\mathcal{E}}$. Since $r = E_{\mathcal{E}}$ is symmetric, in that case $(u, v) \in r, (v, u) \in r$, and $(v, u) \in E_{\mathcal{E}} = \text{Cod}(r)$, so r is adjacency-preserving.

Note that $r(v)$, the set of properties assigned to v by r, is in this case the set of neighbours of v in $\mathcal{E}$, which fully characterizes v in the context of that graph. This means that r is a maximally discriminant classification on $\mathcal{E}$ in the sense that it can distinguish any pair of elements up to chemical equivalence in the particular reaction network considered. Thus, by moving towards the broader context of **rGraphs**, we achieve a thorough description of the elements of the network that is not possible in **Graphs**. Then, this classification has the disadvantage of being overly-discriminant, possibly hiding relevant similarity patterns due to excessive detail. Ideally we should look for a consensus between minimum and maximum discrimination.

Similarity Through Closure Operators

Now that we have dealt with the problem of inducing chemical properties from chemical relations, we can study similarity in chemical reaction networks. Previous works on chemical similarity have relied mostly on metric functions, assigning a real value to the degree of similarity between each pair of substances [21]. In the present contribution we use closure operators as the basic descriptors of similarity.

Definition 2 (Closure operator). A *closure operator* on a set X is a function $\text{cl} : P(X) \to P(X)$, where $P(X)$ is the power set of X, such that for any $A, B \subseteq X$:

i) $A \subseteq \text{cl}(X)$,

ii) $A \subseteq B$ implies $\text{cl}(A) \subseteq \text{cl}(B)$,

iii) $\text{cl}(\text{cl}(A)) = \text{cl}(A)$.

Closure operators are able to capture one of the most basic characteristics of similarity without introducing additional assumptions, namely that similarity is evaluated relative to the set of entities compared. For instance, any two humans are very similar to each other when compared with individuals from different species, but they are most likely quite dissimilar when compared with a pair of identical twins. Suitably defined closure operators on a substance set capture this idea, by adding to the closure of a subset of substances S all those substances that are 'as similar to elements of S as elements of S are among themselves'. The first two properties of closure operators are indispensable to follow this idea faithfully. Indeed, if we were to remove property i), a substance in S could be dissimilar to elements of S when considered in the context of S, which is absurd; and if we remove property ii), we could have $\mathrm{cl}(A) \nsubseteq \mathrm{cl}(B)$ for $A \subseteq B$, so that there exists at least one $x \in \mathrm{cl}(A) \setminus \mathrm{cl}(B)$, that is, a substance that is at the same time similar and not similar to those of A, which is absurd as well.

Only the third property looks somehow foreign to the fundamental principle invoked. It further demands that the closure of a set cannot grow by adding more similar substances in the new context of $\mathrm{cl}(A)$; *i.e.* we only consider a set to be closed with respect to similarity when we have found a group of substances that can be reasonably isolated from the rest of the substance space in accordance to its dissimilarity from all remaining substances. This is a reasonable demand, as it ensures that our closures produce robust similarity classes.

According to the principle that similarity is defined with respect to some properties, it is clear that the definition of an appropriate closure operator for the description of chemical similarity in reaction networks should start from, and be determined by, the choice of a morphism in a category of reaction networks. Formal Concept Analysis (FCA) provides us with a closure operator that fulfils this requirement.

Formal Concepts and Lattice Closure

Formal Concept Analysis (FCA) can be understood as a mathematical framework for deriving knowledge represented in the relations among a set of concepts with domain on given sets of elements and attributes. FCA starts by considering a *context*, defined as a relation linking elements with their attributes. Concepts are defined as pairs consisting of a subset of elements and a subset of attributes that uniquely determine each other, *i.e.* such that both, knowledge of concept's attributes determines its elements and knowledge of concept's elements determines its attributes.

Definition 3 (Formal Context). A *formal context* (*context* for short) consists of a set $\mathcal{G}$ of *elements*, a set $\mathcal{M}$of *attributes*, and a binary relation $r \subseteq \mathcal{G} \times \mathcal{M}$. It is said that $x \in \mathcal{G}$ has the attribute $y \in \mathcal{M}$ if and only if $(x, y) \in r$.

Definition 4. Let $\mathcal{G}, \mathcal{M}, r$ be a context. For any $O \subseteq \mathcal{G}$ and $A \subseteq \mathcal{M}$,

$$O^r = \{y \in \mathcal{M} : (x, y) \in r \text{ for all } x \in O\} \tag{7}$$

$$A^r = \{x \in \mathcal{G} : (x, y) \in r \text{ for all } y \in A\}$$

In other words, O^r is the set of common attributes to all elements in O, and A^r is the set of elements in $\mathcal{G}$ characterised by all attributes in A. For the sake of simplicity, we will note $\{x\}^r$as x^r.

Definition 5 (Formal Concept). Let $(\mathcal{G}, \mathcal{M}, r)$ be a context. A *formal concept (concept* for short) in $(\mathcal{G}, \mathcal{M}, r)$ is a pair (O, A), with $O \subseteq G$ and $A \subseteq \mathcal{M}$, such that $O^r = A$ and $A^r = O$. In other words,

i) For each $x \in O$ and for each $y \in A, (x, y) \in r$;

ii) For each $x \notin O$ there is a $y \in A$ such that $(x, y) \notin r$;

iii) For each $a \notin A$ there is an $x \in O$ such that $(x, y) \notin r$.

For any given concept (O, A), O and A are known as the *extent* and the *intent* of the concept, respectively.

The idea behind this definition is to make concepts closed, both regarding their intent and their extent. Indeed, the extent of a concept is a maximal subset of $\mathcal{G}$ characterised by the attributes in its intent, and the intent of a concept is a maximal set of attributes in $\mathcal{M}$ characteristic of all elements in its extent. Formally, this means that $O \subseteq O^{rr}$ and $A \subseteq A^{rr}$, and $O^{rrrr} = O^{rr}$ and $A^{rrrr} = A^{rr}$ for any $O \in \mathcal{G}, A \in \mathcal{M}$, so that the operator ^{rr}is a closure both in $\mathcal{G}$ and M. A formal concept can be then constructed either by closing a subset $O \subseteq \mathcal{G}$ with this operator, which produces the concept (O^{rr}, O^{r}), or by closing a subset $A \subseteq \mathcal{M}$, producing the concept (A^{r}, A^{rr}). Also, note that $O_1^{rr} \subset O_2^{rr}$ if and only if $O_2^{r} \subset O_1^{r}$.

Note that for any given category C of reaction networks, *e.g.* **rGraphs**, each morphism $r : G \to G' \in C_1$ defines a context with $\mathcal{G} = G$ and attribute set $\mathcal{M} = G'$, induced by the category-theoretical identity of those sets.

Concepts on this context are defined with respect to the properties induced on G by $\ker r$. Thus, for any $V \subset G$, the closure V^{rr} gives us the extent of the minimum concept containing V, that is, the set of substances that share the maximum set of attributes common to all elements of V. Intuitively, this is the complete set of substances that are as similar to elements of V as elements of V are among themselves. We refer to V^{rr} as the *lattice closure* of V.

Topological Closure

Previous work by Restrepo and collaborators shows that topological spaces form a rich framework for the study of chemical similarity. Topological invariants such as closure, interior, exterior, and boundary have been found to provide valuable information on similarity among chemical elements [23-25]. However, a meaningful definition of these invariants requires a concept of closure that is slightly more restricted than the one introduced previously. Indeed, topological spaces are determined by a special class of closure operators, where property ii) of **Definition 2** is replaced by the strongest condition

$$\mathrm{cl}(A \cup B) = \mathrm{cl}(A) \cup \mathrm{cl}(B) \quad (8)$$

Lattice closure does not fulfil this condition, as the union of concept extents is not necessarily a concept extent. Topological invariants, then, cannot be introduced directly in the formalism presented on the previous section. We want to modify that formalism just enough to allow for the definition of a topological space associated with a classification $r : G \to G'$ while keeping some consistency with its original motivation. Hence, we could exploit the potential of different topological invariants as similarity descriptors.

Note that, unlike union, intersection of concept extents *is* a concept extent. This means that, though they do not immediately conform a topology themselves, concept extents are a basis for a topology on G. By construction, this is the smallest topology where concept extents in (G, G', r) are open sets, which we call the r-topology on G and note it as τ_r. Now, consider the *dual context* of (G, G', r), that is, the context (G, G', R) with $R = G \times G' \setminus r$. We shall prove that, for the finite case, τ_r is the smallest topology on G where concept extents in (G, G', r) are closed sets. Before we do so, we require some preliminary results.

Proposition 1. Let (X, Y, r) be a context. $O \subset X$ is a concept extent in (X, Y, r) if and only if $O = \bigcap_{y \in A} \ker r(y)$ for some $A \subset Y$.

Proof. Note that $\bigcap_{y \in A} \ker r(y)$ is the set of objects that have all properties in A, that is, A^r. In consequence,

$\rightarrow$ for a given set $O = \bigcap_{y \in A} \ker r(y)$, $(O, O^r) = (A^r, A^{rr})$, and since $A^{rrr} = A^r$, this pair is a concept with extent O.

$\leftarrow$ for a given concept (O, A) in (X, Y, r), since $O = A^r$, then $O = \bigcap_{y \in A} \ker r(y)$

Interestingly, this proposition gives a direct connection between τ_r and the classification given by r:

Corollary 1. τ_r is the topology induced by the subbase of classes $\{\ker r(y)\}_{y \in Y}$.

From now on we will refer to a concept extent just as a 'concept' as long as there is no room for confusion.

Proposition 2. Let (X, Y, r) be a context. $O \subset X$ is a concept in (X, Y, R) if and only if $O = \bigcap_{Y \in A} X \setminus \ker r(y)$ for some $A \subset Y$.

Proof. By **Proposition 1** O is a concept in (X, Y, R) if and only if $O = \bigcap_{y \in A} \ker R(y)$ for some $A \subset Y$; and since $\ker R(y) = \{x \in X : (x, y) \in R\} = \{x \in X : (x, y) \notin r\} = X \setminus \ker r(y)$, we conclude $O = \bigcap_{y \in A} X \setminus \ker r(y)$.

Proposition 3. Let (X, Y, r) be a finite context. τ_r is the smallest topology on G where concepts in (X, Y, r) are closed sets.

Proof. Since concepts in (X, Y, R) are a basis for τ_r, we may express any open set O in τ_r as,

$$O = \bigcup_{A \in \mathcal{A}} \bigcap_{y \in A} X \setminus \ker r(y) \tag{9}$$

where $\mathcal{A} \subset P(Y)$. Since closed sets are the complements of open sets, any closed set $K \subset X$ may be expressed as,

$$\begin{aligned} K &= X \setminus \left(\bigcup_{A \in \mathcal{A}} \bigcap_{y \in A} X \setminus \ker r(y) \right) \\ &= \bigcap_{A \in \mathcal{A}} \left(X \setminus \bigcap_{y \in A} X \setminus \ker r(y) \right) \\ &= \bigcap_{A \in \mathcal{A}} \bigcup_{y \in A} X \setminus (X \setminus \ker r(y)) \\ &= \bigcap_{A \in \mathcal{A}} \bigcup_{y \in A} \ker r(y) \end{aligned} \tag{10}$$

which, since Y is finite, may also be expressed as,

$$K = \bigcup_{A' \in \mathcal{A}'} \bigcap_{y \in A'} \ker r(y) \tag{11}$$

for some $\mathcal{A}' \subset P(Y)$. By **Proposition 2** we know that the intersection of sets in this expression is a concept in (X, Y, r). Thus, closed sets in τ_r are unions of concepts in this context, which trivially implies that concepts are closed sets themselves. Last, suppose that there is a smaller topology τ satisfying this property. Then there is some $K \in \tau_r$ such that $K \notin \tau_r$; but since all closed sets in τ_r can be expressed as unions of concepts in (X, Y, r), this would imply that there would be a union of closed sets in τ that is not a closed set, which is absurd. Thus, τ_r is the smallest topology where concepts in (X, Y, r) are closed sets.

Corollary 2. τ_r is the topology induced by the closed sets subbase of classes $\{ker\ r(y)\}_{y \in Y}$.

In conclusion, τ_r is the closest topological analogue of concept closure that is directly induced by the classification associated with a morphism $r : G \to G' \in$ **Graphs$_1$**. Closure in τ_R is slightly but significatively different from closure in (G, G', r). The latter gives the smallest concept containing a given subset $V \subset X$. On the other hand, closure in τ_r (noted $\bar{V}$) gives the smallest *union* of concepts containing V. Hence, $\bar{V} \subseteq V^{rr}$ for any $V \subset G$. Also, since x^{rr} is the smallest concept containing x, it follows immediately that $\bar{V}$ is given by,

$$\bar{V} = \bigcup_{x \in V} x^{rr} \tag{12}$$

In this sense, we may say that if ^{rr} closes V 'as a whole', closes V, 'element by element'. The following proposition gives additional insight into this intuition.

Proposition 4. Let (X, Y, r) be a formal context and $V \in (X, \tau_R)$. For any $u \in X, u \in \bar{V}$ if and only if there is a $v \in V$ such that $v^r \subset u^r$.

Proof. $\to$ From (12) we know that $u \in \bar{V}$ implies $u \in v^{rr}$ for some $u \in V$. Since u^{rr} is the smallest concept containing u, this means that $u^{rr} \subset v^{rr}$, which in turn implies $v^r \subset u^r$.

The proof in the opposite direction just follows the inverse argument.

The closure of a set is also characterised as the set of its *adherent points*. A point x is adherent to a subset V of a topological space if all neighbourhoods of x contain points of V. Intuitively, this means that any region around x intersects V, so x is 'adhered' to that set. The previous proposition shows how points in the closure of a subset V of an R-topological space are adhered to it: for any $u \in \bar{V}$ there is a point $v \in V$ such that u is in all concepts containing v, which intuitively means that u is 'closely attached to v' in terms of their similarity with respect to the properties induced by r.

Example 4 (Periodic table (continued)). Figs. **5**, **6** and **7** depict the closures of some chemical groups in the periodic table. Naturally, lattice closures are always larger than or equal to topological closures; yet both of them share some interesting trends. First, in Fig. **5** we show that closures of groups at the extremes of the periodic table are sharper than those of groups towards the middle, a result that matches chemical knowledge: alkali metals, halogens and (in a lesser extent) chalcogens are known to comprise strongly similar elements, nearly identical in their chemical behaviour and clearly distinguished from elements outside their families; on the other hand, similarities among transition metals are much more diffuse and less defined.

Also, note that closures of families of metallic/non-metallic elements, located to the lower-left/upper-right side of the periodic table, comprise almost exclusively metallic/non-metallic elements. On this regard, groups located in the limit between metallicity and non-metallicity show an interesting behaviour (Fig. (**6**)). Naturally, these families comprise both (predominantly) metallic and non-metallic elements, and in consequence their closures include both metallic elements, located to the left of the group, and non-metallic elements, located to its right. Then, if we cut the group by the horizontal line separating metallic from non-metallic elements, the upper set 'closes to the right', while the lower set 'closes to the left', emulating the idea that the separation between metallic and non-metallic elements is operating at a fundamental level that goes across different families in the periodic system. It is also worth noting that alkali and alkaline-earth metals appear repeatedly in the closures of transition metal groups. This is reasonable, since our data set concerns only the synthesis of binary compounds and ignores stoichiometric data. Within this context, the highly reactive groups 1 and 2 appear as the ultimate realization of the idea of metallic behaviour, that are thus 'adhered' to all families of less-ideal metallic elements.

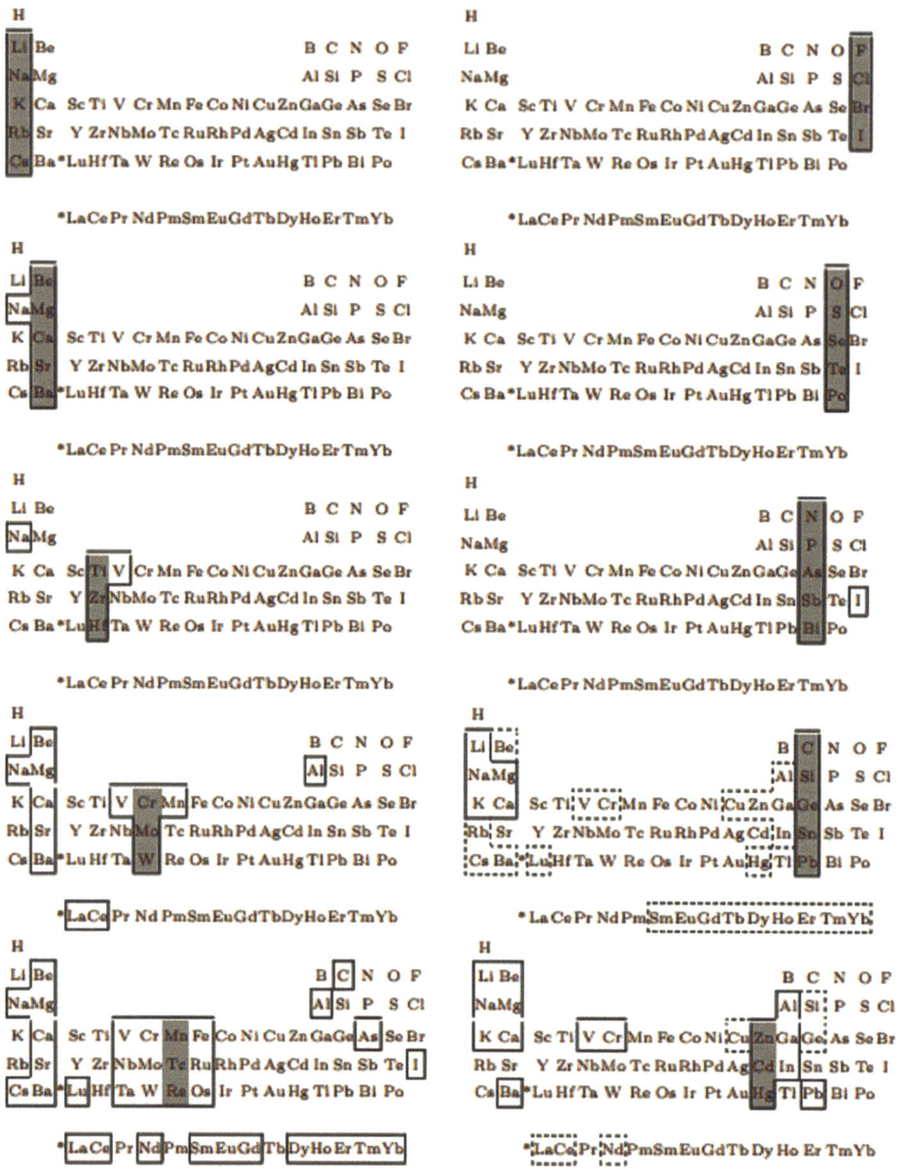

Figure 5: Topological closures (continuous squares) and lattice closures (topological closure + dashed squares) of some chemical families (grey squares).

Fig. (**6**) analyses three triads of transition metals that are customarily considered to comprise a single chemical family -namely, those of groups 8, 9, and 10. We computed the closure of all triads together, and those of each triad separately. The closure of the first triad (Fe, Ru, Os) is the sharpest, and does not include any element of the other two. The second triad (Co, Rh, Ir) has the most diffuse closure, including elements of the other two triads, and the third triad (Ni, Pd, Pt) stands somewhere in the middle, including an element of the second triad. This result gives an interesting illustration of the relativistic character of similarity, materialised in this case in some sort of asymmetry. In fact, while the second triad is arguably similar to the first (the closure of the second triad includes elements of the first), the opposite does not hold: the first triad is clearly not similar to the second (the closure of the first triad does not capture any element of the second). Closure operators allow us to capture this subtle facet of similarity, a feat that is not achieved when we just regard all three groups as conforming a single equivalence class.

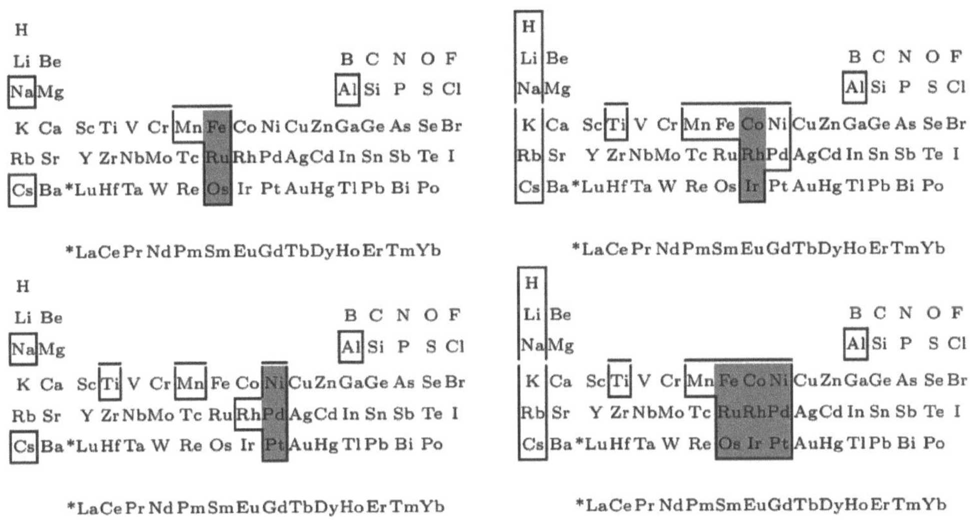

Figure 6: Topological closure (continuous squares) of groups 8 (upper-left), 9 (upper-right) and 10 (lower- left) of the periodic table, and of all three groups together (lower-right).

Before concluding, we analyse the meaning of another topological invariant in τ_R: the interior of a set.

Interior

The interior A° of a set A is the largest open set that is a subset of A. Of course, this means that A° is the union of all concepts in (X, Y, R) contained in it. This observation immediately produces the following simple, yet useful result:

Proposition 5. Let (X, Y, r) be a context. For any $V \subset (X, \tau_R)$ and $x \in X, x \in V^\circ$ if and only if $x^{RR} \subset V$.

Proof. This is a direct consequence of the fact that x^{RR} is the smallest open set containing x.

Furthermore, in the following we introduce a proposition that characterizes V° in terms of *concepts in* (X, Y, r).

Proposition 6. Let (X, Y, r) be a context. For any $\subset (X, \tau_R)$ and $v \in X, v \in V^\circ$ if and only if for any $x \notin V, x^r \setminus v^r \neq \emptyset$.

Proof. $\rightarrow$ By **Proposition 5** we now that $\emptyset$ implies $v^{RR} \in V^\circ$. In that case, $X \setminus v^{RR}$ is a closed set containing any $x \notin A$, which according to equation (12) means that $x^{rr} \subset X \setminus v^{RR}$. In turn, this implies $v \notin x^{rr}$ and then $v^{rr} \not\subset x^{rr}$, which implies $x^r \not\subset v^r$, or equivalently, $x^r \setminus v^r \neq \emptyset$.

$\leftarrow$ Let $v \notin X$ such that $x \notin V$ implies $x^r \setminus v^r \neq \emptyset$;. Then $v^R \setminus x^R \neq \emptyset$; or equivalently, $v^R \not\subset x^R$, which implies $x^{RR} \not\subset v^{RR}$, and then $x \notin v^{RR}$. Since x is an arbitrary element of $\setminus V$, we conclude that $v^{RR} \subset V$, and by **Proposition 5** $v \in V^\circ$.

The previous proposition shows that it is possible to distinguish any $v \in V^\circ$ from any x in its complement by a property that x has and v does not. We may then say that the interior of a set conforms a strong core of that set, comprising those elements that are 'separated' from its complement by significant dissimilarity in their properties. This may be easier to understand if we note that when we compute the interior of *V* we remove any element of *V* that is in the closure of its complement, and is thus 'adhered' to elements of V. The interior of V then consists of all remaining points, which are not adherent to those outside of V and can be 'separated' from the rest of the space.

The interior of a concept turns out to be particularly interesting. When defining a concept, we are closing a subset of X that is described by a collection of properties that exclusively characterise its elements. Then, when computing its interior, we are separating a strong core of the concept that is further characterised by the properties that distinguish it from the rest of the space. This observation suggests a connection between the lack of those properties absent from the interior and the presence of those properties characteristic of the concept. In the following we introduce a proposition formalizing this intuition in a useful principle of inference.

Proposition 7. Let $O \in X$ be a concept in (X, Y, r). Then $u \in O^{\circ}$ if and only if for any $x \in X, u^R \subset x^R$ implies $O^r \subset x^r$.

Proof. → By **Proposition 5** we now that $u \in O^{\circ}$ implies $u^{RR} \subset O$. Now let $x \in X$, and suppose $u^R \subset x^R$, which implies $x^{RR} \subset u^{RR}$, and then $x \in u^{RR}$. Since $u^{RR} \subset O$, this also implies $x \in O$. Now, O is a concept and thus a closed set, so according to equation (12) $x \in O$ implies $x^{rr} \subset O$, which implies $O^r \subset x^r$. We conclude then that $u^R \subset x^R$, implies $O^r \subset x^r$.

← Let $u \in X$ such that $u^R \subset x^R$implies $O^r \subset x^r$ for any $x \in X$. Since $u^R \subset x^R$ means $x^{RR} \subset u^{RR}$, then $x \in u^{RR}$. By hypothesis, in that case $O^r \subset x^r$, which means $x^{rr} \subset O^{rr} = O$, and thus $x \in O$. Since x is an arbitrary element of u^{RR} we conclude $u^{RR} \subset O$, which by **Proposition 5** implies $u \in O^{\circ}$.

In other words: for any $u \in O^{\circ}$, if x lacks all the properties that u lacks, then x has the properties O^r characteristic of the concept O.

The exterior of a set $V \subset (X, \tau_R)$ is the interior of its complement. Mirroring the same analysis carried on V°, we can say that $\overline{X / A} \subset (X, T_R)$ comprises elements that are significantly dissimilar from those of V, being 'separable' from its adherent points.

Example 5 (Periodic Table (concluded)). Concept interiors single a core of highly similar elements on each of the diffuse lattice closures found for transition metals groups. In principle and according to the periodic law, one would expect this core to comprise the elements of the group being closed. Such is not the case in general. For example, as seen in Fig. **(7)**, the interior of the lattice closure of the manganese group (Mn, Tc, Re) comprises two elements of that group (Tc, Re) and two elements of the adjacent chromium group (Mo, W). Though it is unadvisable to extrapolate too far from results obtained with such a small data set, the suggestion that stronger similarities may exist between members of different columns in the periodic table is nonetheless worth of further exploration. This can be seen for example in the case of the groups 8, 9, and 10 (headed by Fe, Co, and Ni, respectively), which were regarded as a single group (VIIIB) in the former group nomenclature, as the result of their strong similarities. In fact, due to the strong similarity of Ru, Rh, Pd, Os, Ir and Pt, these elements are called platinum metals.

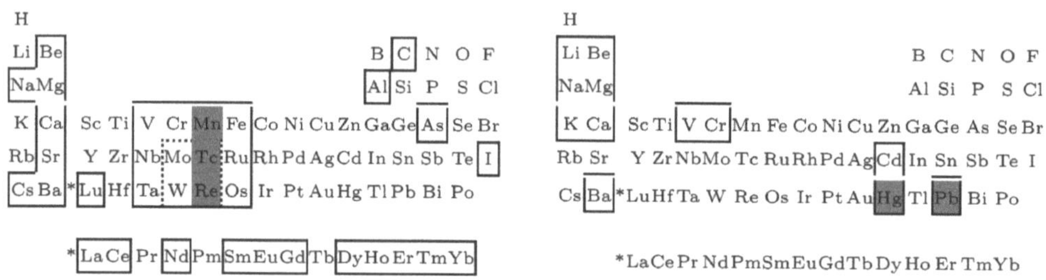

Figure 7: Left: Closure (continuous squares) of the group 7 (grayed squares) of the periodic table and interior of the closure (dotted squares). Right: Closure (continuous squares) of the interior of the Zinc group (grayed squares). Compare this closure with that of the group itself in Fig.(5).

On this regard, it is interesting to find that the interior of the closure of the Zinc group (Fig. **(5)**) comprises two lone elements, Hg and Pb, furthermore, the closure of this pair of elements is much sharper than that of the Zinc group: other than the ubiquitous alkali and alkaline-earth metals, it includes one element from the zinc group (Cd) and two transition metals (V and Cr) that also tend to appear in the closures of many transition metal groups as seen in Fig. **(5)**, but it excludes the elements from the boron and carbon groups that were included in the closure of the zinc group. These observations suggest that, despite what the structure of the periodic table tells, it is lead and not zinc that conforms, along with cadmium and mercury, a sharp group that materializes one particular subtype of metallic behaviour. Of course, once more we must remember that this result must be taken with some reservations: valence plays a big role in determining the periodic classification and this factor cannot make its appearance in a model that overlooks the stoichiometry of chemical combinations.

FINAL REMARKS

Following Schummer's approach to study chemical similarity of substances through structuring their experimental chemical information in a network-like approach,

Schummer's suggests building up a network whose nodes are substances and their relations represent the ability to combine them to form new substances. Hence, we proposed in the current chapter using category theory as a formal framework for studying thc aforementioned objects and relations through the category ***r* Graph**. Thus, moving away from using distance functions that assign real values as the degree of similarity between objects, we studied similarity through adjacency preserving transformations of a graph. In particular, we showed the usefulness of tolerance relations, which we consider are better suited for dealing with similarity relationships than equivalence relations, this latter widely used in classification schemes. The use of tolerance relations give us the notion of neighbour of an object as those objects near the object in question, taking into account the relational nature of objects without metric constraints. We showed how tolerance relations induce a context in the sense of the Formal Concept Analysis methodology, where concepts are defined according to the properties induced by the kernel of a relation r on the set of objects. Moreover, we defined a topological space associated to the tolerance relation by using concept's extents as basis for a topology τ_R and we built up the topology τ_R associated to it by considering the complement of the relation. We also proved that this topology τ_R, made up of closed sets, is the analogue of the concept of closure that is induced by the classification given by the relation r. By applying the methodology here described to the case of the chemical elements, we found several of their trends, which shows that by taking 'distance' from metric constraints and using only relational properties coming from experiments, it is possible to build up a wealth amount of knowledge of one of the icons of chemistry.

ACKNOWLEDGEMENTS

Guillermo Restrepo thanks the Universidad de Pamplona for the financial support to conduct this research.

CONFLICT OF INTEREST

The authors confirm that this chapter contents have no conflict of interest.

REFERENCES

[1] Rouvray, D. H. J. Definition and role of similarity concepts in the chemical and physical sciences. *Chem. Inf. Comp. Sci.* **1992**, *32*, 580-586.

[2] Mi Gyung Kim Affinity, that elusive dream: a genealogy of the chemical revolution; MIT Press, **2003**.

[3] Rocke, A. Chemical atomism in the nineteenth century: from Dalton to Cannizzaro; Ohio University Press, **1984**.

[4] Rocke, A. The Quiet Revolution: Hermann Kolbe and the Science of Organic Chemistry; Berkeley: University of California Press, **1993**.

[5] Schummer, J. HYLE. The Chemical Core of Chemistry I: A Conceptual Approach. *Int* J *Philos Sci* **1998**, *4*, 129-162.

[6] Ivanciuc, T.; Ivanciuc, o.; Klein, D. Posetic Quantitative Superstructure/Activity Relationships (QSSARs) for Chlorobenzenes. *J. Chem. Inf. Model*. **2005**, 45, 870-879.

[7] Willett, P.; Barnard, P.; Downs, G. Chemical Similarity Searching. J *Chem Inf Comp Sci* **1998**, *38*, 983-996.

[8] Harary, F. Graph Theory. Addison-Wesley, **1969**.

[9] Gross, J.; Yellen, J. Graph Theory and its applications; Chapman & Hall, **2006**.

[10] Temkin, O.; Zeigarnik, A.; Bonchev, D. Chemical reaction networks: a graph-theoretical approach; CRC Press, **1996**.

[11] Gleiss, P.; Stadler, P.; Wagner, A.; Fell, D. Relevant cycles in chemical reaction networks. *ACS* **2001**, *4*, 207-226.

[12] Solé, R.; Munteanu, A. The Large-Scale Organization of Chemical Reaction Networks in Astrophysics. *EPL,* **2004**, 68, 170; Preprint.

[13] Klamt, S.; Haus, U.; Fabian. Effective but Costly, Evolved Mechanisms of Defense against a Virulent Opportunistic Pathogen in Drosophila melanogaster. *Theis PLos Biology* **2009**, 5, e1000385.

[14] Bernal, A.; Daza, E. Metabolic Networks: Beyond the Graph. C*urr Comput Aided Drug Des*. **2011**, *7*, 122-132.

[15] Greenwood, N.; Earnshaw, A. Chemistry of the elements; Elsevier, **1997**.

[16] NIST Chemistry webbook; http://webbook.nist.gov/chemistry/; **2009**.

[17] The chemical thesaurus; http://www.chemthes.com/; **2009**.

[18] Bernal, A.; Daza, E. HYLE. On the Epistemological and Ontological Status of Chemical Relations *Int* J *Philos Sci.* **2010**, *16*, 80-103.

[19] Vincent, B. B.; Jonathan Simon Chemistry: The impure science; Imperial College Press, **2008**.

[20] Bartol, W.; Pioro, K.; Rossello, F. On the covering generalized rough sets. *Inf Sci,* **2004**, *166*, 193-211.

[21] Willett, P. Chemoinformatics and computational chemical biology; Bajorath, J., Ed.; Humana Press, **2011**; Chapter Similarity searching using 2D structural finger- prints, pp 133-158.

[22] Carpineto, C.; Romano, G. Concept data analysis; John Wiley & sons, **2004**.

[23] Restrepo, G.; Llanos, E; Mesa, M.; Villaveces, J. L. Topological Study of the Periodic System. *J Chem. Inf. Comp. Sci.* **2004**, 44, *68*.75.

[24] Restrepo, G.; Llanos, E; Mesa, M. Topological space of the chemical elements and its properties. *J Math Chem* **2006**, *39*, 401-416.

[25] Leal, W.; Restrepo, G.; Bernal, A. A network study of chemical elements: from binary compounds to chemical trends. *MATCH Commun. Math. Comput. Chem.* **2012**, *68*, 417-442.

CHAPTER 3

Discrimination of Small Molecules Using Topological Molecular Descriptors

Chandan Raychaudhury* **and Debnath Pal**

Bioinformatics and Supercomputer Education Research Centre, Indian Institute of Science, Bangalore 560012, India

Abstract: One of the subjects of special interest in chemical structure handling is to be able to associate a unique quantitative value to each and every chemical compound. The job is not only huge from the stand point of the number of compounds known and to be known, but quite difficult as well from the angle of developing a suitable method. One of the common situations is to be able to discriminate isomeric structures where a large number of compounds having closely related structures for same number of atoms come into picture and this number grows very fast as the number of atoms increases. Getting quantitative descriptors having power of discriminating one compound from another is an important requirement for storage, retrieval and handling of chemical structures as well as for predicting molecular properties/activities. In this chapter, we review several molecular descriptors, mostly topological distance based, in the form of topological indices considering the connectivity aspect of molecular structures only, that have useful discriminative power.

Keywords: Chemical documentation, discrimination of graphs, distance based topological index, distance sum, graph distance complexity, graph vertex complexity, information content of a graph, information-theoretical topological index, molecular connectivity, molecular graph, paths in a graph, spectra of graph, structure property/activity analysis, topological distance, topological index.

INTRODUCTION

Quantitative analysis of the structural information of chemical compounds is of paramount importance from both the perspective of structure handling and prediction of their property/activity [1-4]. While for chemical documentation, uniqueness of structural descriptors is highly important, fair amount of discrimination of structures using molecular descriptors can serve a useful

***Corresponding author Chandan Raychaudhury:** Bioinformatics and Supercomputer Education Research Centre, Indian Institute of Science, Bangalore 560012, India; Tel: +91-80-2293-2901; Fax: +91-80-2360-2648; E-mail: chandanrc53@yahoo.co.in

Subhash C. Basak, Guillermo Restrepo & José L. Villaveces (Eds)
Copyright © 2015 Bentham Science Publishers Ltd. Published by Elsevier Inc. All rights reserved.
10.1016/B978-1-68108-053-6.50003-X

purpose in explaining molecular properties/activities [5]. Presumably, a quantitative descriptor would correlate well with the properties/activities of chemical compounds if its values for the compounds are consistent with the property/activity values under consideration. This will depend on the discriminating power of the descriptor(s). While some properties may need high discrimination, some may not. In fact, some properties/activities may exhibit very little or, no change in their values for different molecules and in that case a descriptor with lower discriminating power may correlate well and out perform more discriminating indices.

So far the handling of chemical structures is concerned, one of the highly intensive applications of molecular descriptors is in searching databases of chemical compounds. In those situations, it is often important that a database does not contain any molecule more than once *i.e.*, to avoid repetition of any structure in the database. A well known way of doing that is to see whether two molecules have the same connectivity, in other words whether the two (molecular) graphs are isomorphic [6]. This may be verified by doing row/column interchanges of the adjacency matrices of the graphs. If at any point of this exercise, the adjacency matrices of two structures coincide, the structures can be regarded as the same structure.

However, this process may require n! iterations where n are the number of atoms (vertices) in a molecule (molecular graph). Doing this for all or, most of the compounds in a large database is, presumably, a huge task. Perhaps, such problems can be handled in a more reasonable manner if highly discriminating molecular descriptors can be used in this purpose. Even, it has been proposed [7] that a `super index' formed by a number of indices instead of a single index may be used. In this approach, two molecules would be considered to be discriminated if at least one of the indices of the super index has different values for the two graphs representing them. However, the uniqueness of any super index in discriminating isomers has not yet been mathematically established.

A large number of topological indices are now available in the literature and they show varied discriminating power. However, in this chapter, we will restrict ourselves to a group of topological indices which have been explored by several researchers to investigate their discriminating power. In this connection, it may also be noted at this point that in this article, while dealing with discrimination, only carbon skeleton of structures will be considered. It is known that presence of other heavy atoms like N, O, halogens *etc.* and various types of chemical bonds in organic compounds is important for working with such molecules. Although such problems may be handled by using suitable weights on the vertices and/or, edges of molecular graphs, our present purpose is to look exclusively into the topological aspects of molecular structures like branching, cyclicity *etc.* and the discriminating power of some topological indices for these types of molecular connectivity based structural changes in the compounds will be investigated.

Molecular Graph

A molecular graph is essentially a connected graph [6] that is used to represent the structural formula of a chemical compound where the atoms are depicted by vertices and the bonds (commonly covalent bonds) by edges. The connectedness of the vertices by edges in a molecular graph, also referred to as molecular topology, is a unique representation of a molecule in that the change in connectivity changes the molecule. At the same time, it is also of interest to understand what kind of changes in the structure take place with the change in connectivity. Such characteristics are exclusively topological in nature and are known as branching, cyclic/acyclic aspects of chemical structure, neighborhood of atoms in a molecule *etc.* These structural properties are distinctly different from geometrical properties like bond length, bond angle *etc.*, and therefore give researchers a new direction of studies. Even books have been written on the fundamental aspects as well as applications of molecular graphs and descriptor development from molecular connectivity [8-11].

Molecular graphs can be drawn keeping hydrogen atoms present in the corresponding molecules and are commonly referred to as hydrogen-filled molecular graphs. However, some time hydrogen atoms are not considered in the graph models and such models are commonly referred to as hydrogen-suppressed graphs. In particular, for investigating discriminating power of connectivity based molecular descriptors, hydrogen-suppressed graphs of molecular skeletons are considered. Also, in many graph models, multiplicity of bonds is represented by multiple edges between pairs of vertices, which give a multigraph [6] model of a molecule.

To bring more closeness to chemical compounds, vertices may also be labeled with atomic symbols such as 'C' for carbon, `O' for oxygen, `H' for hydrogen *etc.* Furthermore, suitable vertex and edge weighted graphs [6] may also be considered as models to derive molecular descriptors that may be of use for chemical documentation and structure-activity/structure property relationship studies.

Topological Index

One of the most widely used applications of molecular topology is in terms of topological indices. To our knowledge, the term 'Topological Index' was first coined by Hosoya [12]. Topological indices are a kind of molecular descriptors that translate the connectivity of the atoms (vertices) in a molecule by bonds (edges) and are essentially graph invariants [6]. Therefore, two isomorphic (molecular) graphs would have the same value for any kind of topological molecular descriptors. However, the converse may not be true which means two different graphs may have the same value for a topological descriptor. This, in a way, generates the necessity of developing topological indices which can give different values for different molecules. In other words, developing topological indices of high/very high power of discrimination is a requirement for identifying molecular graphs as uniquely as possible in terms of such indices. This, in turn, could be considerably helpful in chemical structure/database handling and chemical documentation.

Distance Based Topological Indices

Although a number of topological indices have been developed so far; those proposed on the basis of topological distances between pairs of vertices have gained much attention in discriminating chemical structures. Historically, an index W developed by Wiener [13] is believed to be the first topological index that was used to find correlation between physical properties and the index W for a series of paraffin. The index is computed by adding all the distances between pairs of vertices in a molecular graph. Hosoya [12] showed that this can be obtained by adding all the entries of the upper/lower off-diagonal elements of the distance matrix associated with the molecular graph. It is, therefore, computed on the magnitude of the inter-atomic topological distances. However, much later a number of more discriminating distance based topological indices were proposed. Balaban proposed an index J [14] that showed very high discriminating power. Using information-theoretical formalism, Bonchev and Trinajstic [15] and later on Raychaudhury *et al.* [16] proposed distance based information theoretical topological indices which also showed high discriminating power.

Distance based topological indices which are computed on the basis of topological distances between pairs of vertices in a graph, can be developed in various ways. As mentioned earlier, Wiener index W can be obtained by summing up all the entries of the upper/lower triangle off the main diagonal (whose entries are zero). For example, in graph G_1 (Fig. (**1**)) the distance matrix $D(G_1)$ may be given as:

G_1

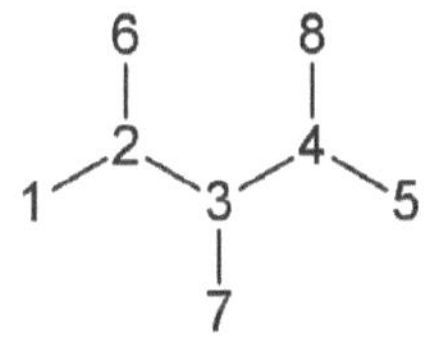

Figure 1: Carbon skeleton (hydrogen-suppressed graph) of one of the octane isomers.

$$D(G_1) = \begin{array}{c} \\ 1 \\ 2 \\ 3 \\ 4 \\ 5 \\ 6 \\ 7 \\ 8 \end{array} \begin{array}{c} \begin{array}{cccccccc} 1 & 2 & 3 & 4 & 5 & 6 & 7 & 8 \end{array} \\ \begin{bmatrix} 0 & 1 & 2 & 3 & 4 & 2 & 3 & 4 \\ 1 & 0 & 1 & 2 & 3 & 1 & 2 & 3 \\ 2 & 1 & 0 & 1 & 2 & 2 & 1 & 2 \\ 3 & 2 & 1 & 0 & 1 & 3 & 2 & 1 \\ 4 & 3 & 2 & 1 & 0 & 4 & 3 & 2 \\ 2 & 1 & 2 & 3 & 4 & 0 & 3 & 4 \\ 3 & 2 & 1 & 2 & 3 & 3 & 0 & 3 \\ 4 & 3 & 2 & 1 & 2 & 4 & 3 & 0 \end{bmatrix} \end{array}$$

Therefore, Wiener index W and Balaban index J may be computed using the distance matrix entries in the following equations:

$$W = \sum_{i,j;i>j} d_{ij} \tag{1}$$

$$J = \frac{m}{\mu + 1} \sum_{i,j} (D_i D_j)^{-\frac{1}{2}} \quad (2)$$

where, m is the number of edges in a graph; μ is the cyclomatic number of the graph *i.e.*, the minimum number of edges required to be removed such that a (poly)cyclic graph becomes acyclic and D_i is the sum of all the entries in i^{th} row/column and D_j is that in j^{th} row/column.

Now, the values of W and J for G_1 may be computed using (1) and (2) respectively and the values are:

W = 65; J = 3.4629

Information-Theoretical Topological Indices

Use of Shannon's information theoretical formalism [17] has been found to be useful in developing information theoretical topological indices. In this approach, an information theoretical measure, called information content of graph, is computed taking into consideration different graph elements, such as, number of vertices/edges in a graph, topological distances between pairs of vertices in a graph. The elements of a graph are considered to form a finite discrete system and are partitioned into disjoint classes on the basis of an equivalence relation defined on the elements of the system.

So, if there are n vertices in a (molecular) graph and they are partitioned into k disjoint classes C_1, C_2, …, C_k, having n_1, n_2, …, n_k elements in the respective partitioned classes, then a probability scheme can be associated with the partition as:

$$C_1 \; C_2 \; \ldots\ldots\ldots \; C_k$$

$$p_1 \; p_2 \; \ldots\ldots\ldots \; p_k$$

where,

$$p_i = \frac{n_i}{n}; \; p_i > 0 \text{ and } \sum_{i=1}^{k} p_i = 1$$

On the basis of this partition scheme, an information theoretical measure, Information Content, for the system (I_s) can be obtained using Shannon's formula [17]:

$$I_s = -\sum_{i=1}^{k} p_i \log p_i = \sum_{i=1}^{k} \frac{n_i}{n} \log \frac{n_i}{n} \tag{3}$$

It may be noted that this measure is done taking logarithm to the base 2 ($\log_2$) and is expressed in bits.

Shannon's information was first used by Rashevsky [18] by partitioning the vertices of a graph on the basis of the equality of vertex degrees of the first-neighbor vertices. Subsequently, Trucco [19] formalized it on the basis of the automorphism group of graphs and vertices belonging to the same orbit were considered as equivalent. Later on, Sarkar *et al.* [20] proposed a method based on the neighborhood of the vertices of multi-graph and gave a formalism for computing first-order neighborhood information content (IC) from the equivalence of the vertices on the basis of the chemical nature of the vertices (atoms), type of connecting bonds (single bond, double bond *etc.*) with the first neighbor vertices and the degrees of those vertices. Subsequently, two related measures, structural information content (SIC) by Basak *et al.* [21] and complementary information content (CIC) by Raychaudhury and Ghosh [22] were also proposed. Very soon Basak and Magnuson [23] found useful application of CIC in Quantitative Structure-Activity Relationship (QSAR) studies with a series of alcohols. By considering chemical nature and bond types, the multi-graph models were closer to actual chemical structures.

Distance Based Information-Theoretical Topological Indices

The information theoretical measures mentioned in the previous section were based on the partition of vertices. So, it was quite natural that researchers working in this area would look for developing information-theoretical measures on various other elements of graphs in addressing their problems in hand. In this regard, so far molecular structural discrimination is concerned, the power of discriminating molecular graphs by information-theoretical topological indices improved significantly by considering the topological distances between pairs of vertices. Bonchev and Trinajstic [15] first adopted a distance based approach for developing information theoretical topological indices and such indices showed fair amount of discriminating power. In their work, they considered partition of the lower/upper off-diagonal distance elements of the distance matrix of a graph as well as the actual distance values (magnitude of topological distances) between pairs of vertices in developing their indices.

Later on, Raychaudhury *et al.* [16] proposed few more distance based indices which also showed high discriminating power. Such distance based information-theoretical topological indices may be computed as follows:

Let there be n_j vertices at a distance d_j from a vertex v in a graph G having n vertices, j = 1,.....,e where e is the eccentricity of v, the maximum distance from v. Evidently, v is at a zero distance from itself. This may be called a distance distribution associated with v in G and can be expressed in the following manner for convenience:

$$v\colon 1;\ 1,n_1;\ 2,n_2;\ \ldots,\ e,n_e;\ \text{and}\ n = 1 + \sum_{j=1}^{e} n_j$$

Now, if d(v) be the sum of the distances of all the vertices in G from v, then d(v), termed as distance sum [15] may be given by:

$$d(v) = \sum_{j=1}^{e} n_j d_j \tag{4}$$

Bonchev *et al.* [15] proposed two indices $\bar{I}^{E}{}_{D}$, from the partition of the distances between pairs of vertices on the equality of distance and $\bar{I}^{W}{}_{D}$, from the partition of distance sum d(v) on the equivalence of distance magnitude.

This may be illustrated by taking distance matrix $D(G_1)$ of graph G_1 (Fig. (**1**)) as an example:

In $D(G_1)$, considering the upper/lower triangular elements above/below the main diagonal one gets 7 entries of distance 1, 9 entries of distance 2, 8 entries of distance 3 and 4 entries of distance 4. Now, it may be said that the 28 entries are partitioned into 4 disjoint classes on the basis of the equality of the distance values as 28(7, 9, 8, 4). Hence, a measure of information content for G_1 from this partition can be computed using (1) as:

$$\bar{I}^E{}_D = (7/28 \log 28/7 + 9/28 \log 28/9 + 8/28 \log 28/8 + 4/28 \log 28/4) = 1.9433.$$

Furthermore, a partition of the distance magnitude is also possible in the form of a partition of the upper/lower off-diagonal elements *i.e.*, of W, the Wiener index value. Bonchev proposed an index $\bar{I}^W{}_D$ from this partition and this measure for G_1 from $D(G_1)$ may be done in the following manner:

The sum of all the off-diagonal entries is (7*1 + 9*2 + 8*3 + 4*4) = 65.

Hence, a partition of 65 may be given as 65 (7*1 +9*2 + 8*3 + 4*4) with the distance distribution in G_1 and the index $\bar{I}^W{}_D$ may be computed using (1) as:

$$\bar{I}^W{}_D = 7*1/65 \log 65/1 + 9*2/65 \log 65/2 + 8*3/65 \log 65/3 + 4*4/65 \log 65/4 = 4.6680$$

However, it was felt that information-theoretical indices for the vertices could be developed and used them to develop an index for graph and that was believed to show more discriminating power since the graph distances would be considered in a greater detail in this process.

In doing that, Raychaudhury *et al.* [16] proposed some distance based information-theoretical vertex indices and such indices were further used to develop information-theoretical graph indices. The indices were developed in the following manner:

By considering equation (4), one gets a distance distribution associated with a vertex in graph. In the distance matrix of the graph, each row/column gives the distance distribution associated with the corresponding vertex and the sum of all the entries in that row/column gives the distance sum d(v). So far the number of vertices in the graph is considered, the vertex under consideration which is at a distance zero with itself has to be taken into consideration. Therefore, for computing information content on the partition of vertices, this has to be included and the number of vertices would be $1 + n_1 + n_2 + ... + n_e$ (from eq. 4).

However, for computing vertex index on the basis of magnitude of the distances, d(v) can be used as it is given in (4).

Now, a measure of information content of a vertex v from the partition of vertices in a graph G having n vertices, called Vertex Complexity (V^c) [16], on the basis of its distances from other vertices in G may be defined using (4) as:

$$V^c(v) = \frac{1}{n}\log n + \sum_{j=1}^{e} \frac{n_j}{n} \log \frac{n}{n_{j\,i}} \tag{5}$$

Using (5), one can compute the V^c value for vertex 1, v_1, of G_1 (Fig. (**1**)) from the partition of the entries of the first row/column of the distance matrix $D(G_1)$ from the equality of the distances. Here, the distance distribution of the eight vertices of G associated with v_1 will be:

v_1: 1; 1,1; 2,2; 3,2; 4,2

and hence

$$V^c(v_1) = 2^* (1/8) \log (8/1) + 3^* (2/8) \log (8/2) = 2.2499.$$

Proceeding in a similar way, another measure of vertex information content, called Vertex Distance Complexity (V^d) [16], may also be defined in the following way by considering distance magnitude as given in (4):

$$V^d(v) = \sum_{j=1}^{e} n_j \{\frac{d_j}{d(v)}\} \log \{\frac{d(v)}{d_j}\} \tag{6}$$

Therefore, the V^d value of v_1 in G may be computed using (6) as:

$$d(v_1) = 1 + 2^*2 + 2^*3 + 2^*4 = 19.$$

And, hence,

$$V^d(v_1)=(1/19)\log(19/1)+2^*(2/19)\log(19/2)+2^*(3/19)\log(19/3)+2^*(4/19)\log(19/4)=2.6948$$

Having vertex indices V^c and V^d defined, one can now have measures of information content of a graph using these indices. This can be done by defining average information measures [17] of a graph G using V^c and V^d.

Let V^c (v_1), V^c (v_2), …, V^c (v_n) be the vertex complexities of the n vertices of G. An index called Graph Complexity (H^V) [16] is defined as:

$$H^v(G) = \frac{1}{n}\sum_{i=1}^{n} V^c(v_1) \tag{7}$$

Therefore, from the V^c values of the eight vertices of G_1, one can compute $H^V(G_1)$ as:

$$H^V(G_1)=(1/8)[0.6773+0.5736+0.4231+0.5736+0.6773+0.6773+0.5268+0.6773]=1.9958$$

Again, let V^d (v_1), V^d (v_2), …, V^d (v_n) be the Vertex Distance Complexities of the n vertices of G.

Also, let,

$$R = \sum_{i=1}^{n} d(v_i) \tag{8}$$

Let R be partitioned into n number of d(v) values for the n vertices in G and let

$$r_i = \frac{d(v_i)}{R}; r_i > 0 \text{ and } \sum_{i=1}^{n} r_i = 1 \tag{9}$$

Then, a measure of information content of a graph G, called Graph Distance Complexity (H^D) [16] may be computed using:

$$H^D(G) = \sum_{i=1}^{n} r_i * V^d(v_i) \tag{10}$$

Therefore, the H^D value of G_1 may be computed as:

$$R = 19 + 13 + 11 + 13 + 19 + 19 + 17 + 19 = 130.$$

$$H^D(G_1) = (1/130)\,[4*19*0.8112 + 2*13*0.8011 + 11*0.8224 + 17*0.8229] = 2.6967$$

It is evident that R value is the sum of all the entries in the distance of G and this value is twice the value of Wiener number W. However, as mentioned earlier, since in the process of computing vertex indices, particularly vertex distance index V^d, for all the vertices in G one considers the distance distribution in a graph in greater details, one can expect such indices to be significantly discriminating. This is discussed in the following sections.

Discrimination of Small Molecules by Distance Based Indices

For finding the discriminating power of a topological index, discrimination of structural isomers comes first in one's mind. Such isomers are a class of compounds which have very close structural architecture. For example, in alkane isomers (C_nH_{2n+2}), where n is the number of carbon atoms, this closeness becomes very clear, particularly for structures of higher n values. Therefore, if any topological index can discriminate such structures reasonably well, one can expect that the index would be able to discriminate a large number of small molecules quite effectively and in a useful manner. Presumably, a topological index that is derived considering greater details of the connectivity in a molecule may be expected to have higher discriminating power. Distance based topological indices have been found to be quite effective in this regard. It was reported [16] that H^D and J indices could discriminate carbon skeletons (as shown in Fig. (**1**)) of all the alkane isomers having four to eight carbon atoms and a number of monocyclic graphs having four to six vertices while some other indices, considered there, failed to do that.

As we have seen earlier, Wiener index W [13] and Balaban index J [14] are based on the magnitude of the distance and therefore get effect not only of the number of different topological distances present in a graph but their values as well. However, while proposing a new highly discriminating index, molecular identification number (this will be dealt with later in this chapter) Randić [24] demonstrated with examples where these indices failed to discriminate structures. Later on, in a comparative study with distance based topological indices, Mihalic *et al*. [25] also discussed with examples the relative discriminating power of some of such indices.

The information theoretical approach in defining $\bar{I}^{W}{}_{D}$ by Bonchev *et al.* [15] and H^D by Raychaudhury *et al.* [16] also take care of such distance magnitudes quite thoroughly and therefore could be expected to show high discriminating power. In fact, it was shown [16] that these information theoretical indices could discriminate the graphs given in Fig. (**2**):

$$H^D = (1{:}2.702,\ 2{:}2.686);\ \bar{I}^{W}{}_{D} = (1{:}4.673,\ 2{:}4.657)$$

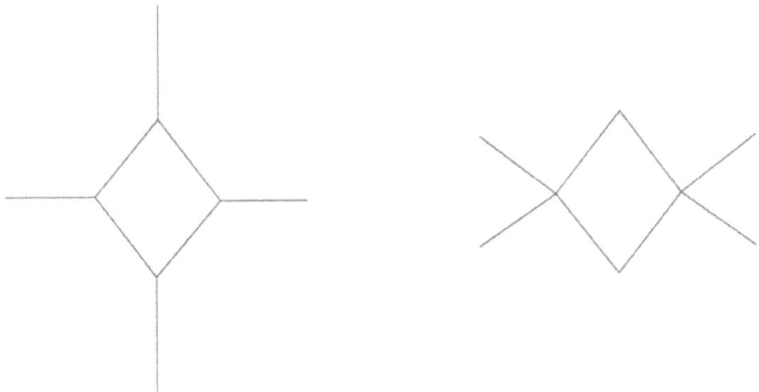

Figure 2: Two graphs of eight vertices and one cycle, (1) vertex attachment at four vertices of the cycle, (2) vertex attachment at two vertices of the cycle.

However, for two graphs of five vertices, shown in Fig. (**3**), containing two and three cycles, H^D and $\bar{I}^{W}{}_{D}$ failed to discriminate the structures.

$$H^D = (3{:}1.920,\ 4{:}1.920);\ \bar{I}^{W}{}_{D} = (3{:}3.236,\ 4{:}3.236)$$

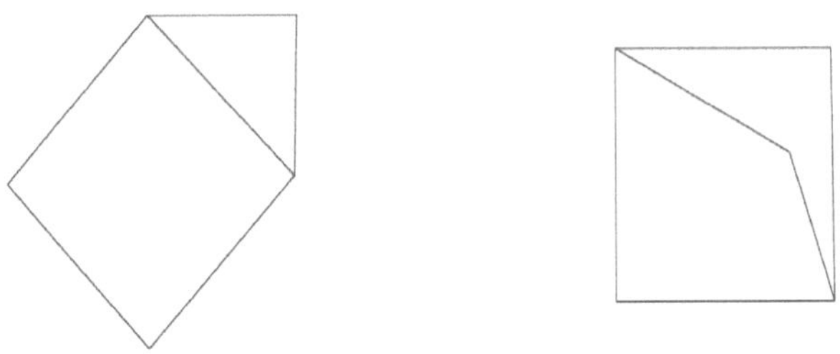

Figure 3: Two graphs of five vertices, (3) one has two cycles and (4) the other has three cycles.

Looking into these interesting findings, it becomes tempting to expect that such distance based indices may be considered in forming a super index along with other indices of high discriminating power. In this process, a super index consisting of fewer indices could possibly be formed.

Molecular Identification Number of Randić

During the same period when Raychaudhury *et al.* [16] proposed distance based information-theoretical indices, Randić [24] proposed a Molecular Identification Number (denoted by ID) which had elements from the connectivity index developed by him earlier [26] together with the consideration of all paths in a molecular graph. The scheme consists of the use of weighted bond type (this is how connectivity index of Randić [26] is computed which he put forward for characterizing molecular branching) and enumeration of all bond-weighted paths in a graph. Bond weights are given by $(d_i * d_j)^{-1/2}$, where d_i and d_j are the degrees of vertices [6] i and j connected by the edge under consideration in the graph.

Furthermore, acknowledging that finding the longest path in a graph is a known NP-Complete problem [27], Randić explored the discriminating power of this ID number for some 400 graphs which might be expected to represent some of the commonly used compounds and found no degeneracy. He also furnished data [24] to show the superiority of ID numbers over a number of well known topological indices in discriminating several graphs. However, he pointed out [24] that since such numbers are condensed form of structural details (in fact this is true for any topological index), some degeneracy may surface in due course.

Very soon Szymanski *et al.* [28] demonstrated it with counter examples and found that ID numbers had degenerate values for alkane isomers of 15 vertices and this degeneracy increased to 88 for alkane isomers having 20 vertices.

Spectra of Graph

Another set of values associated with graphs is the spectra of a graph [6]. The eigenvalues [6] of the characteristic polynomial of a graph constitute the spectra. Two graphs which have same spectra are said to be isospectral graphs. Balasubramanium and Basak [29] have shown that the first few principal components obtained from principal component analysis (PCA) on a number of topological indices could discriminate different pairs of isospectral graphs quite effectively. They [29] also furnished data to show the superiority of this approach in discriminating isospectral graphs in comparison to some other topological indices.

In another work, Mihalic *et al.* [30] considered distance polynomial and distance spectra of graph in evaluating their discriminating power. In the measure, instead of adjacency, topological distances between pairs of vertices in a graph were considered. They observed [30] that such distance based approaches could show higher discriminating power compared to what could be obtained from the characteristics polynomial/adjacency matrix and the corresponding spectra. They showed that two mono-cyclic isospectral graphs of 10 vertices (Fig. (**4**)) had different distance polynomials and distance spectra.

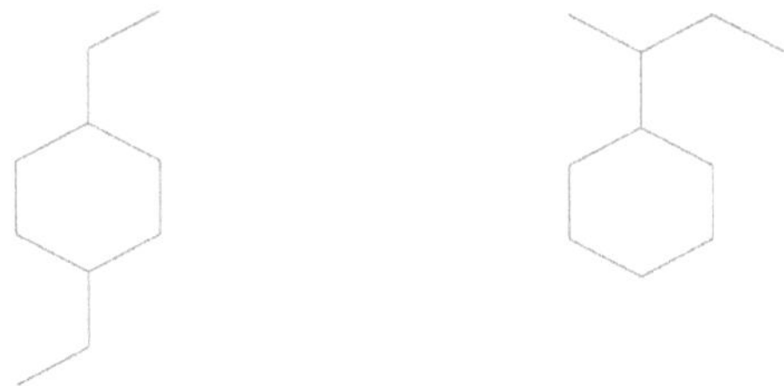

Figure 4: Carbon skeletons of two mono-cyclic isospectral graphs of 10 vertices; (5) 1,4-divinylbenzene, (6) 2-phenylbutadiene.

It is interesting to note that the distance based topological indices W, $\bar{I}^W{}_D$, H^V H^D, considered in this article, can also discriminate these two graphs:

$$W = (5{:}125, 6{:}121);\ \bar{I}^W{}_D = (5{:}5.287, 6{:}5.313);\ H^V = (5{:}2.471, 6{:}2.382);\ H^D = (5{:}2.999, 6{:}3.017)$$

At the same time, they also demonstrated that two graphs representing carbon skeleton of 17-carbon alkanes (Fig. (**5**)) had degeneracies in distance polynomial and distance spectrum.

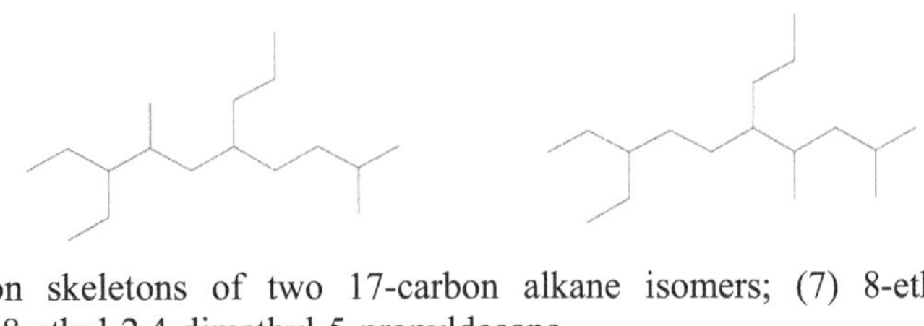

Figure 5: Carbon skeletons of two 17-carbon alkane isomers; (7) 8-ethyl-2,7-dimethyl-5-propyldecane, (8) 8-ethyl-2,4-dimethyl-5-propyldecane.

However, among the above mentioned four distance based indices too, only H^V could discriminate these two graphs whereas the indices W, $\bar{I}^W{}_D$ and H^D failed to do that:

$W = (7{:}584, 8{:}584)$; $\bar{I}^{W}_{D} = (7{:}6.883, 8{:}6.883)$; $H^{V} = (7{:}2.923, 8{:}2.929)$; $H^{D} = (7{:}3.827, 8{:}3.827)$

It is, therefore, apparent that one distance based topological indices or the other are capable of discriminating molecular graphs quite effectively and the indices considered here can discriminate reasonably large molecular graphs which could be of much use for practical purposes.

CONCLUSION

In this chapter, we discussed the role of some topological indices, derived from molecular graphs, in discriminating small molecules and their usefulness in some areas of application. Discrimination of small molecules is very important for handling chemical structures in a suitable manner. It may be useful to find whether one molecule is structurally different from another one or, to eliminate duplicate structures in a database. It had been discussed that distance based topological indices like Balaban's J index and some information theoretical indices like I^{W}_{D} and H^{D}, Distance polynomial/distance spectra of graph, Randić's molecular ID number showed high discriminating power due to the methods adopted in computing them by considering structural connectivity in greater detail. In dealing with discrimination studies in this chapter, only connectivity aspect of the molecules was considered. However, researchers have worked on three dimensional structures of molecules and have taken 3-D distance matrix to compute different 3-D distance based indices [25] which is, however, beyond the scope of the present article.

It may be noted that many of the topological indices considered here have found significant application in molecular similarity studies too [31]. This can have important role to play in drug discovery based on the belief that similar structures may possess similar activities. It is also noteworthy that distance based vertex indices, like V^{d}, which was the basis for defining highly discriminating index H^{D} and the then newly proposed distance exponent index [34] have found useful applications in a rule based method for structure-activity/structure-toxicity analysis [32-35] from identification of activity related substructures. This is believed to be due to the ability of these vertex indices to discriminate different substructures effectively helping search the substructural patterns which could be responsible for the pharmacological and toxicological activities of chemical compounds. It has also been discussed that while some highly discriminating indices may be required for chemical documentation and/or, database handling purposes, some not so discriminating indices may find suitable application in structure-activity/structure-property analysis. Therefore, what seems important is to be able to understand the discriminating power and characteristics of individual topological indices and use them suitably for a problem in hand.

ACKNOWLEDGEMENTS

The authors are thankful to Bioinformatics Centre, Indian Institute of Science, Bangalore, India for extending the facilities required for this work. Financial support received by Chandan Raychaudhury from Department of Science & Technology, Government of India, during this work is hereby thankfully acknowledged. The authors are also thankful to Dr. S. C. Basak for his help in computing values of some topological indices using POLLY software.

CONFLICT OF INTEREST

The authors confirm that this chapter contents have no conflict of interest.

REFERENCES

[1] Stuper, A. J.; Brugger, W. E.; Jurs, P. C., *Computer Assisted Studies of Chemical Structure and Biological Function*, Wiley-Interscience, New York, **1979**.

[2] i) Basak, S. C., In: Topological indices and related descriptors in QSAR and QSPR, J. Devillers and A. T. Balaban Eds., Gordon and Breach Science Publishers, The Netherlands, 563-593, **1999**; ii) Ivanciuc, O; Ivanciuc, T., In: Topological indices and related Descriptors in QSAR and QSPR, J. Devillers and A. T. Balaban Eds., Gordon and Breach Science Publishers, The Netherlands, 563-593, **1999**;

[3] Roy, K., Topological descriptors in drug design and modeling studies, Mol. Divers. **2004**, 8, 321-457.

[4] Galvez, J. and Garcia-Domenech, R. On the contribution of molecular topology to drug design and discovery, Current Computer-Aided Drug Design **2010**, 6, 252-268.

[5] Balaban, A. T., In: Steric Fit in QSAR, Lecture Notes in Chemistry, Springer, Berlin, **1980**.

[6] Harary, F., Graph Theory, Addison-Wesley, Reading, MA, **1972**.

[7] Bonchev, D.; Mekenyan, O.; Trinajstic, N. J. Comput. Chem. **1981**, 2, 127.

[8] Kier, L. B.; Hall, L. H., Molecular Connectivity in Chemistry and Drug Research, Academic Press, New York, **1976**.

[9] Trinajstic, N., Chemical Graph Theory, Vol. 2, Chapter 4, CRC Press, Boca Raton, Florida, **1992.**

[10] Bonchev, D., Information Theoretic Indices for Characterization of Chemical Structures, Wiley-Research Studies Press, Chichester, U.K., **1983.**

[11] Kier, L. B.; Hall, L. H., Molecular Connectivity in Structure Activity Analysis, Wiley-Research Studies Press, Chichester, **1986**.

[12] Hosoya, H., Topological index. A newly proposed quantity characterizing the topological nature of structural isomers of saturated hydrocarbons, Bull. Chem. Soc. Jpn. **1971**, 44, 2332-2339.

[13] Wiener, H., Structural Determination of Paraffin Boiling Point., J. Am. Chem. Soc. **1947**, 69, 17-20.

[14] Balaban, A., T., Highly Discriminating Distance-Based Topological Index, Chem. Phys. Lett. **1982**, 89, 399-404.

[15] Bonchev, D.; Trinajstic, N., Information Theory, Distance Matrix and Molecular Branching, J. Chem. Phys. **1977**, 67, 4517-4533.

[16] Raychaudhury, C.; Ray, S. K.; Ghosh, J. J.; Roy, A. B.; Basak, S. C., Discrimination of Isomeric Structures Using Information Theoretic Topological Indices, J. Comput. Chem, **1984**, 5, 581-588.

[17] Shannon, C.; Weaver, W., Mathematical Theory of Communication, University of Illinois, Urbana, **1949**.

[18] Rashevsky, N., Life, Information Theory and Topology, Bull. Math. Biophys. **1955**, 17, 229-235.

[19] Trucco, E., A Note on Rashevsky's Theorem about Point Bases in Topological Biology, Bull. Math. Biophys. **1956**, 18, 65-85.

[20] Sarkar, R.; Roy, A. B.; Sarkar, P. K., Topological Information Content of Genetic Molecules-I, Math. Biosci. **1978**, 39, 299-312.

[21] Basak, S. C.; Roy, A. B.; Ghosh, J. J., in Proceedings of the 2nd International Conference on Mathematical Modeling, University of Missouri, Rolla, Vol 2, **1979**.

[22] Raychaudhury, C,; Ghosh, J. J., Proceedings of the Third Annual Conference of the Indian Society for Theory of Probability and Its Applications, Aug.22-24, 1981, Wiley Eastern Limited, New Delhi, **1984**.

[23] Basak, S. C.; Magnuson, V. R., Molecular topology and narcosis: A Quantitative Structure-Activity Relationship (QSAR) study of alcohols using complementary information content (CIC), Arzneim. -Forsch./Drug Res., **1983**, 33, 501-503.

[24] Randić, M. On molecular identification number, J. Chem. Inf. Comput. Sci. **1984**, 24, 164-175.

[25] Mihalic, Z.; Nikolic, S.; Trinajstic, N., Comparative study of molecular descriptors derived from distance matrix, J. Chem. Inf. Comput. Sci. **1992**, 32, 28-37.

[26] Randić, M. On characterization of molecular branching, J. Am. Chem. Soc. **1975**, 97, 6609-6615.

[27] Garey, M. R.; Johnson, D. S. Computers and Intractability – A Guide to the Theory of NP-Completeness, W. H. Freeman, San Francisco, **1979**.

[28] Szymanski, K.; Muller, W. R.; Knop, J. V.; Trinajstic, N. On Randić's molecular identification number. J. Chem Inf. Comput. Sci. **1985**, 25, 413-415.

[29] Balasubramanium, K.; Basak, S. C. Characterization of isospectral graphs using graph invariants and derived orthogonal parameters, J. Chem. Inf. Comput. Sci **1998**, 38, 367-373.

[30] Mihalic, Z; Veljan, D.; Amic, D.; Nikolic, S.; Plavsic, D.; Trinajstic, N. The distance matrix in chemistry. J. Math. Chem. **1992**, 11, 223-258.

[31] Basak, S. C.; Magnuson, V. R.; Niemi, G. J.; Regal, R. R., Determining Structural Similarity of Chemicals using Graph- Theoretic Indices, Discrete Appl. Math. **1988**, 19, 17-44.

[32] Klopman, G.; Raychaudhury, C., A Novel Approach to the use of Graph Theory in Structure-Activity Relationship Studies: Application to the Qualitative Evaluation of Mutagenicity in a Series of Nonfused Ring Aromatic Compounds, J. Comput. Chem. **1988**, 9, 232-243.

[33] Klopman, G.; Raychaudhury, C., Vertex Indices of Molecular Graphs in Structure-Activity Relationships: A Study of the Convulsant-Anticonvulsant Activity of Barbiturates and the Carcinogenicity of Unsubstituted Polycyclic Aromatic Hydrocarbons. J. Chem. Inf. Comput. Sci. **1990**, 30, 12-19.

[34] Raychaudhury; C.; Klopman, G., New Vertex Indices and Their Applications in Evaluating Antileukemic Activity of 9-Anilino- Acridines and the activity of 2',3'-Dideoxy-Nucleosides against HIV. Bull. Soc. Chim. Belg. **1990**, 99, 255-264.

[35] Raychaudhury, C.; Dey, I,; Bag, P.; Biswas, G.; Das, B. N.; Roy, P.K.; Banerjee, A., Use of a Rule Based Graph-Theoretical System in Evaluating the Activity of a Class of Nucleoside Analogues against Human Immunodeficiency Virus. Arzneim.- Forsch./Drug Research, **1993**, 43(II), 1122-1125.

CHAPTER 4

The Periodicity of Molecules

Fanao Kong*, Weiqiang Wu, Na Ji and C. L. Calson

The Institute of Chemistry, Chinese Academy of Sciences, Beijing, 100190, China

Abstract: Mendeleev periodic table of atoms is one of the most important principles in natural science. However, there is not such a thing for molecules. Here we propose three periodic tables for diatomic molecules, triatomic molecules and AH_3 tetratomic molecules, respectively. The tables not only contain isolated molecules, but also the "virtual" diatomic molecules in polyatomic molecules. The form of these molecular periodic tables is analogous to that of Mendeleev periodic table. In the table, molecules are classified and arranged by their group number G, which is the number of valence electrons, and the periodic number P, which represents the size of molecules. Basic molecular properties, including bond length, binding energy, force constant, ionization potential, spin multiplicity, chemical reactivity, or bond angle, among others. change periodically with the tables. This periodicity originates from the shell-like electronic configurations of molecules. The periodic tables can be used to predict unknown properties of molecules, to understand the role of virtual molecules in polyatomic molecules, and to initiate new research fields such as the periodicity of aromatic compounds, clusters, or nanoparticles.

Keywords: Bond angle, bond energy, chemical reactivity, diatomic molecules, dissociation energy, ionization potential energy, molecular classification, molecular periodic table, molecular properties, multiplicity, periodic change, periodic table, periodicity, spectral term, triatomic molecules.

INTRODUCTION

Periodicity has been known as the most significant rule in chemistry [1]. Atoms are classified into different groups in a periodic table according to their electronic configurations. In the periodic table, atomic properties change periodically with the number of electrons. Subsequent interesting questions have then been arisen. Does periodicity of molecules exist? Is there a periodic table correlating molecular properties and electronic configuration of molecules?

***Corresponding author Fanao Kong:** Institute of Chemistry, Chinese Academy of Sciences, Beijing 100190, China; Tel: +8610-6255 7842; Fax: +8610-6256 3167; E-mail: fkong4407@163.com

Subhash C. Basak, Guillermo Restrepo & José L. Villaveces (Eds)
Copyright © 2015 Bentham Science Publishers Ltd. Published by Elsevier Inc. All rights reserved.
10.1016/B978-1-68108-053-6.50004-1

In the 1930s, Clark catalogued diatomic molecules into different groups and periods on the basis of a "period matrix" of the composed atoms [2]. However, the matrices are simple combinations of element symbols lacking of physical insight, for at the time the Molecular Orbital (MO) theory had not yet come into being. Moreover, the correctness of the matrices could not be inspected by molecular properties, as both experimental and theoretical data were very scarce at that time [3].

Since the 1980s, Hefferlin has made some small pieces of periodic tables, including the table of homologous nuclear diatomic molecules and the tables of oxides and halides. Some properties of these molecules, as transition moment, internuclear separation, ionization potential and dissociation potential are studied [4, 5]. However, the regularity in the diagrams is obscure, making vague the molecular periodicity. Kong proposed a simple periodic table for diatomic molecules in 1982 [6], and for triatomic molecules in 1989 [7]. With limited experimental data, he indicated that molecular properties change periodically with the table. Hefferlin *et al.* classified molecules using group theory and correlated molecular properties through group dynamics [8, 9]. However, the casual relationship between molecular structure and molecular properties is not explicit.

Many efforts have been made following the "atomic electronic configuration" in molecules. For example, Sidgwick [10] proposed a rule of 18 electrons to explain the periodicity in organometallic compounds. Haas suggested a principle of element displacement, correlating the pseudohalogens with perfluoroorgano-element groups [11]. Recently, Castleman *et al.* correlated the photoelectron spectroscopy of the molecular ions of TiO^-, ZrO^-, and WC^- with those of the isoelectronic superatoms of O or C by shifting six positions in the atomic periodic table [12]. The above correlations refer to the electronic configuration of the central atom, but not to the whole molecular orbitals. These studies therefore relate merely to the periodicity of atoms, but not to molecular periodicity, which we discuss in this paper.

Here we propose three periodic tables referring to diatomic, triatomic and AH_3 type tetratomic molecules, respectively. The tables are built up in the light of simple molecular orbital (MO) theory. It is amazing to see that molecular properties change periodically with these tables. To represent a well-defined electronic state of a molecule it is required to take into account configuration interactions (CI) of electronic orbitals. However, in a rough approximation, the simple MO theory considers that each electron occupies a single orbital, neglecting the interaction between electronic configurations. Following the Aufbau principle, the electrons fill sequentially the simple molecular orbitals in the energetic sequence; molecules increase their size shell by shell. The electronic configuration (EC) has a shell-like feature, changing periodically with the electron number. Furthermore, since molecular properties are determined by the EC, molecule properties thus exhibit periodicity with the total electron numbers.

Molecules of same point group have the same EC series. The geometric frame of a molecule can be expressed by the related irreducible representatives of their point group. Diatomic molecules and linear triatomic molecules are catalogued by the point group $C_{\infty v}$. Nonlinear triatomic molecules can be classified by C_{2v} or D_{2h} groups. And the AH_3 type tetra-atomic molecules are catalogued by the C_{3v} or D_{3h} point groups. To reveal molecular periodicity, suitable forms of periodic tables are needed. We have made the periodic tables of diatomic, tri-atomic and AH_3 type tetra-atomic molecules, respectively. All these tables keep resemblance with Mendeleev's periodic table.

The periodic tables are built up using two parameters, vertically (G) and horizontally (P). Here G is the number valence electrons, called as group number. The molecules of the same G have similar valence EC, therefore they have analogous properties. P is the number of the electronic shells, called as periodic number. Increasing P means that the electron shell increases, or the molecular size grows up.

DIATOMIC MOLECULES

The Isolated Diatomic Molecules

According to the simple MO theory, valence electrons in homogeneous nuclear diatomic molecules fill molecular orbitals in the order of $(1s\sigma)^2$, $(1s\sigma^*)^2$, $(2s\sigma)^2$, $(2s\sigma^*)^2$, $(2p\sigma)^2$, $(2p\pi)^4$, $(2p\pi^*)^4$ and $(2p\sigma^*)^2$….; the EC of a molecule is thus determined. For example, N_2 has 14 electrons, with EC $(1s\sigma)^2$, $(1s\sigma^*)^2$, $(2s\sigma)^2$,

$(2s\sigma^*)^2$, $(2p\sigma)^2$, $(2p\pi)^4$. Four electrons occupy the inner shell of the principal quantum number n = 1. The remaining ten electrons are the valence electrons, filling the valence shell of n = 2.

The symmetry of g (gerade) and u (ungerade) is broken in the heterogeneous nuclear diatomic molecules. For example, CO and N_2 have an analogous electron configuration of $1s\sigma^2$, $1s\sigma^{*2}$, $2s\sigma^2$, $2s\sigma^{*2}$, $2p\pi^4$, $2p\sigma^2$. The properties of two analogous molecules are thus similar. Since the outmost shell or the valence shell dominates molecular properties, it is interesting to systematically investigate the relationship of molecular properties and EC. The dependence can be manifested through a table similar to the Periodic Table of atoms.

Classification of Diatomic Molecules by G and P

Molecules are classified firstly into different groups G. G is defined as the number of valence electrons in a molecule, therefore G equals to $g_1 + g_2$ for diatomic molecules, where g_1 and g_2 are the group numbers of the composed atoms in the atomic periodic table. For example, N_2 and CO are classified into the G = 10 group, since each molecule has 5 + 5 = 10 valence electrons.

On the other hand, molecular properties slightly depend upon molecular size. Here molecular size can be expressed by the periodic number P, which is the sum of the atomic periods of the composed atoms, $P = p_1 + p_2$, where p_1 and p_2 are the periods of the composed atoms in the atomic periodic table. Small diatomic molecules as CN, N_2, NO and O_2, have P = 2 + 2 = 4, since each molecule is consisted of two small atoms of the second period in the atomic table. Similarly, larger diatomic molecules as CP, P_2 and PSe belong to P = 5, 6, and 7, respectively.

Periodic Table of Diatomic Molecules

Having classified diatomic molecules, a concise diatomic periodic table is shown in Fig. (**1**). Each column is characterized by a group number G. The molecules of the same period P are arranged in a row. The sequence of the grids in the table is numbered by N, which is called molecular number.

The style of the molecular periodic tables is analogous to that of the atomic periodic table. In the atomic periodic table, there are eight main groups of atoms ordered in eight columns. Similarly, there are 16 groups of diatomic molecules ordered in 16 columns for the molecular diatomic periodic table.

Free diatomic molecules	^{11}NO	^{11}: Molecular number
	NN-H	
	$NN\text{-}CH_3$	
Virtual molecules	$NN\text{-}NH_2$	
	NN-OH	
	NN-F	

P	Group Number (G)															
	1	2	3	4	5	6	7	8	9	10	11	12	13	14	15	16
4	$^{1}LiNe$	$^{2}Li_2$	$^{3}LiBe$	$^{4}Be_2$	^{5}BeB	$^{6}B_2$	^{7}BC	$^{8}C_2$	^{9}CN	$^{10}N_2$	^{11}NO	$^{12}O_2$	^{13}OF	$^{14}F_2$	^{15}FNe	$^{16}Ne_2$
5	$^{17}NaNe$	$^{18}NaLi$	$^{19}NaBe$	$^{20}MgBe$	^{21}MgB	^{22}AlB	^{23}AlC	^{24}SiC	^{25}SiN	^{26}PN	^{27}PO	^{28}OS	^{29}SF	^{30}ClF	$^{31}ClNe$	$^{32}NeAr$
6	^{33}KNe	$^{34}Na_2$	$^{35}NaMg$	$^{36}Mg_2$	$^{37}MgAl$	$^{38}Al_2$	$^{39}AlSi$	$^{40}Si_2$	^{41}SiP	$^{42}P_2$	^{43}PS	$^{44}S_2$	^{45}SCl	$^{46}Cl_2$	$^{47}ClAr$	$^{48}Ar_2$
7	$^{49}RbNe$	^{50}KNa	^{51}KMg	$^{52}MgCa$	$^{53}AlCa$	$^{54}AlGa$	$^{55}GaSi$	$^{56}GeSi$	^{57}GeP	^{58}AsP	^{59}AsS	^{60}SeS	$^{61}SeCl$	$^{62}BrCl$	$^{63}BrAr$	$^{64}ArKr$
8	$^{65}RbAr$	$^{66}K_2$	^{67}KCa	$^{68}Ca_2$	$^{69}GaCa$	$^{70}Ga_2$	$^{71}GaGe$	$^{72}Ge_2$	$^{73}GeAs$	$^{74}As_2$	$^{75}AsSe$	$^{76}Se_2$	$^{77}SeBr$	$^{78}Br_2$	$^{79}BrKr$	$^{80}Kr_2$
9	$^{81}RbKr$	^{82}RbK	$^{83}RbCa$	$^{84}SrCa$	$^{85}SrGa$	$^{86}InGa$	$^{87}InGe$	$^{88}GeSn$	$^{89}SnAs$	$^{90}AsSb$	$^{91}SbSe$	$^{92}TeSe$	$^{93}TeBr$	^{94}BrI	^{95}KrI	$^{96}KrXe$
10	$^{97}RbXe$	$^{98}Rb_2$	$^{99}RbSr$	$^{100}Sr_2$	$^{101}InSr$	$^{102}In_2$	$^{103}InSn$	$^{104}Sn_2$	$^{105}SnSb$	$^{106}Sb_2$	$^{107}TeSb$	$^{108}Te_2$	^{109}TeI	$^{110}I_2$	^{111}XeI	$^{112}Xe_2$

Figure 1: Periodic table of diatomic molecules.

Virtual Diatomic Molecules

The molecules listed in periodic table can be extended to the "virtual diatomic molecules". In a new concept of virtual molecule, two adjacent atoms linked by a chemical bond in a polyatomic molecule can be considered as a diatomic molecule, and called as virtual diatomic molecule. Virtual diatomic molecules possess both their own valence electrons and their "pseudo" valence electrons donated by their neighboring atoms.

N_2 is again taken as example, Fig. (**1**). One of the N atoms can be substituted by CH, becoming a molecule of HC-N. The C-N part in the molecule has nine valence electrons, four from C and five from N. In addition to the nine electrons, the molecular orbital in the CN vicinity is filled up by another "pseudo" valence electron, donated by the neighboring H atom. The total valence electrons around

CN vicinity is thus ten, the same number as N_2. In this sense, the CN part in HCN can be considered as a virtual diatomic molecule of the group 10. Similarly, the H atom can be further substituted by a group R, *e.g.* CH_3, NH_2, OH, or a larger group. Each group donates one "pseudo" valence electron to the CN part in the polyatomic molecules of R-CN (CH_3CN, NH_2CN or HOCN). As a virtual diatomic molecule, the CN part in the polyatomic molecules can thus be considered as the new member of N_2 family, labeled by G = 10 and P = 4. The members of the molecular periodic table are thus expanded.

Members of the molecules in the table can be further expanded, for isoelectronic molecules can be included. In the first row (P = 4) isoelectronic molecules of LiB, LiC, BeC, BeN, BN, BO, CO, CF, NF, NeN and NeO are listed together with their analogous molecules of Be_2, BeB, B_2, BC, C_2, CN, N_2, N_2, NO, O_2, OF and F_2, respectively.

Periodicity of Molecular Properties

Variation of Molecular Properties with N

It is verified that molecular properties change repeatedly with the molecular number N. Six basic physical and chemical properties of diatomic molecules are surveyed. The properties varying with the molecular number N are shown in Fig. (**2**). In each chart an analogous pattern repeats for seven rounds, corresponding to P = 4 to 10. The patterns are featured by a U-shape for bond length in the chart (a), Λ-shape for binding energy and force constant in charts (b) and (c), N-shape for ionization potential in (d), and different zigzags for spin multiplicity and chemical inertia in (e) and (f), respectively. The periodicity of the properties reveals an important rule: molecular properties are the periodic functions of the molecular number N. The experimental data in the figure are adapted from NIST table [13]. The other properties are obtained from quantum chemical calculations, which were carried out for both the free diatomic and the related polyatomic molecules shown in Fig. (**2**). For example, the polyatomic molecules of H-NN, CH_3-NN, NH_2-NN, HO-NN, and F-NN are calculated. Each molecule contains a "virtual" diatomic molecule of NN, which is analogous to the free molecule NO. All these molecules belong to the group 11. The calculating method of the first four periods, P = 4 - 7, is B3LYP/6-311G. The method for the rest three periods, P = 8 to P = 10, is B3LYP/LanL2DZ.

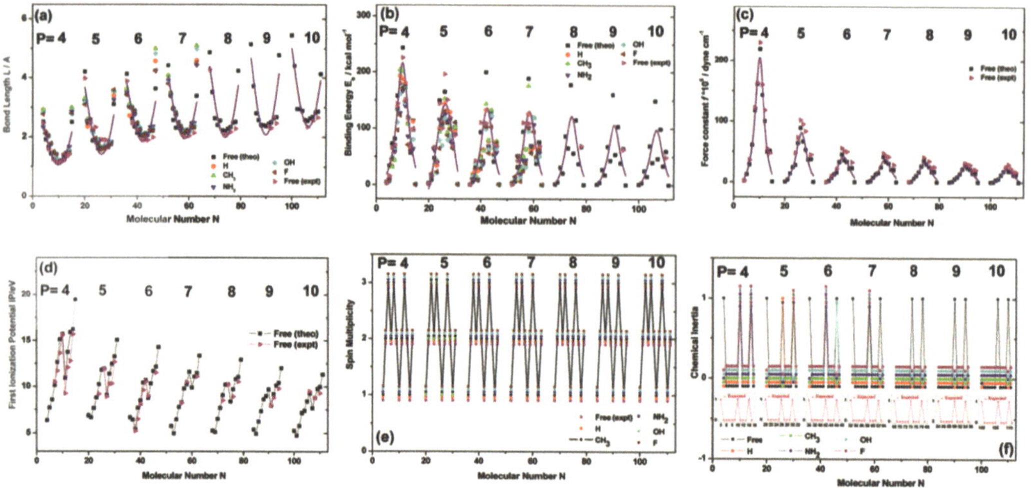

Figure 2: Molecular Properties *versus* Diatomic Molecular Number N. Six molecular properties varying with the molecular number N are displayed in the charts (a) to (f), individually. The curve in each chart repeats an analogous pattern for seven rounds, referring to seven periods of the table. The period number P is indicated on the top of the charts.

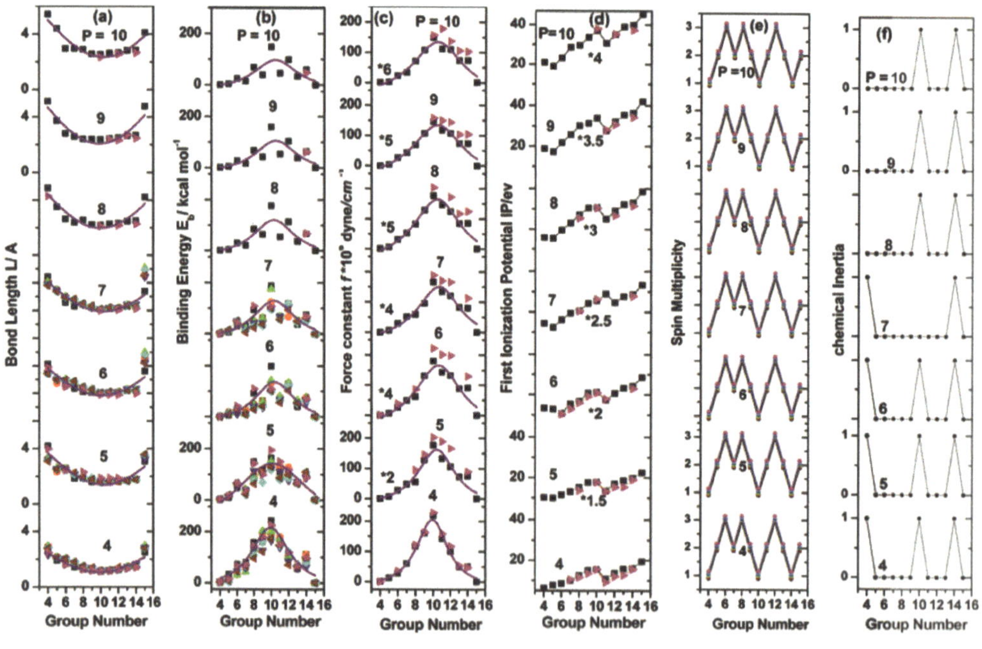

Figure 3: Diatomic molecular properties *versus* group number G. Variations of six molecular properties with group number G are shown in the charts (a) - (f). The molecular properties are mainly determined by the group number G. The symbols and colors in the figure are the same as those in Fig. (**1**). (a) Bond length, (b) Binding energy, (c) Force constant, (d) Ionization potential, (e) Spin multiplicity, (f) Chemical reactivity.

Variation of Molecular Properties with G

It is further found that the six important properties of diatomic molecules depend mainly upon the group number G. The dependence is shown in the charts (a)-(f) of Fig. (**3**). Each chart shows a property by seven analogous curves, corresponding to seven periods (P = 4 - 10). The similarity of the patterns in each chart verifies that the molecular properties are indeed determined by the group number G. The data can be fitted by some simple mathematic functions as shown in Fig. (**4**): (a) Bond length L: Each period performs a U shaped curve, which is fitted by a Lorentz function: $L = L_0 + A\omega/[4(G\text{-}G_c)^2+\omega^2]$. (b) Binding energy E_b: Each period performs a lambda type curve. Binding energy of a virtual diatomic molecule is defined as the dissociation energy of the diatomic molecule of interest. The data are fitted by a Lorenz function: $E_b = E_{b0} + A\omega/[4(G\text{-}G_c)^2+\omega^2]$. (c) Force constant k: The force constant of the fundamental vibration in the ground state; only the free diatomic molecules are shown. The data are fitted by a Lorentz function: $k = k_0 + A\omega/[4(G\text{-}G_c)^2+\omega^2]$. (d) Ionization Potential IP: The first ionization potential of the free diatomic molecules is shown; each period performs an N-shaped zigzag. (e) Spin multiplicity 2S + 1: Each period displays an analogous zigzag, showing the same odd/even alternation of multiplicity. The data

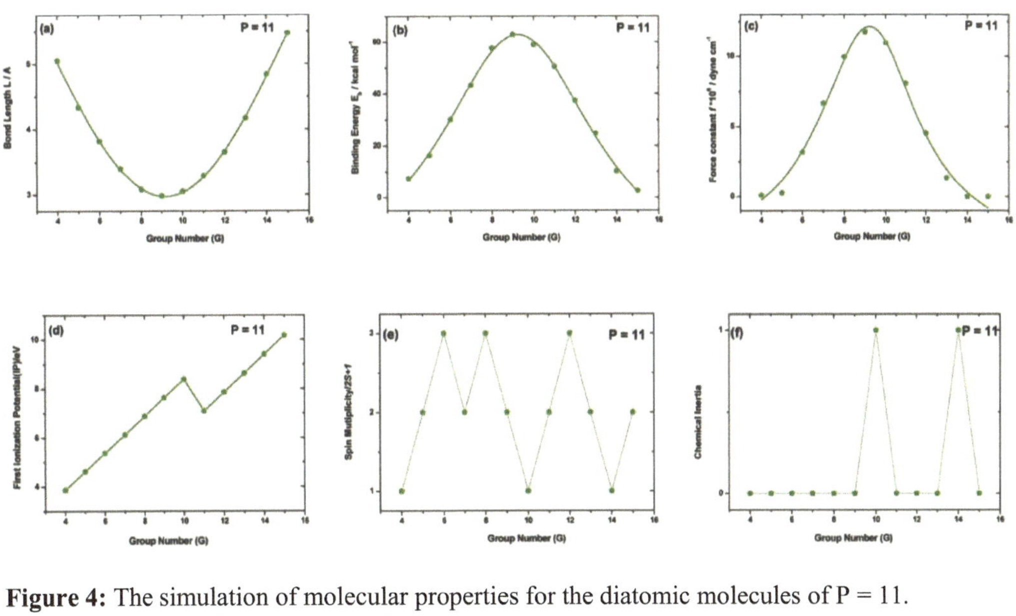

Figure 4: The simulation of molecular properties for the diatomic molecules of P = 11.

are adapted either from the spectral observation shown in NIST table [13] or from the input parameter required in the quantum calculations. (f) Chemical inertia CI: The matter labeled by 1 in the vertical axis means chemical inert so that existing in nature, and thus appearing in the conventional handbooks. On the contrary, the matter labeled by 0 means that the species is not inert, or it is the reactive species. The matter should not appear in the conventional handbooks. The chart shows the chemical inertia of free diatomic molecules. Two exceptions are NO and O_2, which are the reactive species while listing in the handbooks.

Origin of the Periodicity

The shapes of the curves in the G diagrams can be interpreted by the simple MO theory. For the molecules of P = 4, the highest occupied molecular orbitals (HOMO) are $(2s\sigma)^2$, $(2s\sigma^*)^2$, $(2p\sigma)^2$, $(2p\pi)^4$, $(2p\pi^*)^4$ and $(2p\sigma^*)^2$. For the molecules from G = 5 to 10, six valence electrons fill into the bonding orbitals of $(2p\sigma)^2$, and $(2p\pi)^4$ one after another. Bond strength is enhanced gradually by increasing the bonding electrons number. Adding the bonding electrons to the molecule would enhance the binding energy and the force constant, and shorten the bond length. On the contrary, for the molecules of G = 11 to 16, six more electrons fill up the anti-bonding orbitals of $(2p\pi^*)^4$ and $(2p\sigma^*)^2$ sequentially. The anti-bonding electrons would weaken the bond strength, causing a negative effect on the binding energy, force constant and bond length. The above two effects thus cause the U-shaped curve of bond length in Fig. (**3**) (a), the Λ-shaped curves of binding energy and force constant in (b) and (c), respectively. The turning point of the above curves is at G = 10, where the molecules own the maximum number of net bonding electrons.

The N-shaped pattern of the ionization potential in Fig. (**3**) (d) can also be explained. Removing an anti-bonding electron of $2p\pi^*$ from the molecules of G = 11 requires much less energy than that of the $2p\pi$ bonding electron in the molecules of G = 10. It causes a sudden drop of IP at G = 11 in Fig. (**3**) (d).

The number of unpaired electrons determines two other basic properties, spin multiplicity and chemical reactivity. The valence electrons in the molecules of even G number groups, except for G = 8 and G = 12, are spin paired. The representative molecules are Li_2, B_2, N_2, F_2, respectively in P = 4. The spin multiplicity of spectral terms for these molecules is singlet. On the other hand, molecules of odd G numbers have an unpaired valence electron. Their spin multiplicity is expected to be 2, or the doublet. The molecules of the group 8 and 12, represented by C_2 and O_2, have two unpaired valence electrons. The spin multiplicity is predicted as 3. These expected multiplicities are identified by the spectroscopic observations. The zig-zag in Fig. (**3**) (e) shows the experimentally observed multiplicities, which are adapted from NIST table [13]. It can be seen in the comparison that the observed data fit exactly the expected multiplicities.

Chemical reactivity of molecules is also determined by the electronic configuration, for the valence electrons are paired in the molecules of even G numbers, except for G = 8 and G= 12 again. These molecules are inert in bimolecular reactions. A concept of chemical inertia, which is the opposite of chemical reactivity, is introduced here. Chemical inertia can be denoted roughly by a delta function, manifesting the occurrence of the matter in nature or not. The inert molecule must occur in nature or can be synthesized in the laboratory, so that it should be listed in conventional handbooks. Fig. (**3**) (f) shows the chemical inertia of diatomic molecules *versus* N. The inserts in the bottom of Fig. (**3**) (f) express the expected chemical inertia of molecules. Status 1 in the vertical axis means that the matter is chemical inert, as the molecules of Li_2, Be_2, B_2, N_2, F_2, so that they appear in conventional handbooks. On the other hand, molecules of odd G numbers own one unpaired valence electron. The molecules are thus free radicals, very reactive in bimolecular reactions. The free radicals do not occur in nature, so that they should not appear in the conventional handbooks. Representative molecules are LiNe, LiBe, BeB, BC, CN, NO, OF, or NeF. In Fig. (**3**) (f) the status of chemical inertia for these molecules is expressed by 0. The molecules of the group 8 and 12, represented by C_2 and O_2, have two unpaired valence electrons. These triplet molecules are known as carbenes, the extremely reactive biradicals in chemistry. The status of the chemical inertia must be 0 also.

The above predictions are compared with the real chemical inertia of diatomic molecules. The top of Fig. (**3**) (f) displays the real status of chemical inertia, which is discriminated by the appearance in handbooks. Comparing the top and bottom of Fig. (**3**) (f) it can be seen that the real chemical inertia well agrees with the expected chemical inertia. Such coincidence identifies that periodicity of chemical reactivity also exists.

Triatomic Molecules

For triatomic molecules, there are two kinds of geometric shapes. The linear shape corresponds to the point group of $C_{\infty v}$. While the non-linear, or the bent shape, corresponds to D_h point group. However, according to Walsh diagrams, the geometric shape of triatomic molecules is determined by the number of their valence electrons. Those molecules possessing 16 to 19 valence electrons have a bent shape. While those molecules having less valence electrons than 16 or more than 19 are linear ones.

In Fig. (**5**) we show a periodic table of triatomic molecules. The table has 24 columns containing 24 groups of triatomic molecules. Group number G is defined as the total number of valence electrons, $G = g_1 + g_2 + g_3$. The period P of the table is defined as the sum of the periods of the composed atoms in the atomic periodic table, $P = p_1 + p_2 + p_3$. The sequential number N of the molecules in the table is called as molecular number N.

P	Group Number (G)																						
	1	2	3	4	5	6	7	8	9	10	11	12	13	14	15	16	17	18	19	20	21	22	23	24
4	1 LiH_2^{2+}	2 LiH_2^{+}	3 LiH_2	4 BeH_2	5 BH_2	6 CH_2	7 NH_2	8 H_2O	9 H_2F	10 H_2Ne	11 HHeNe	12 He_2Ne												
5	13 Li_2H^{2+}	14 Li_2H^{+}	15 Li_2H	16 HLiHe	17 HBe_2	18 HBeB	19 B_2H	20 HBC	21 C_2H	22 HCN	23 N_2H	24 HNO	25 HO_2	26 HOF	27 HF_2	28 HFNe	29 HNe_2	30 $HeNe_2$						
6	31 Li_3^{2+}	32 Li_3^{+}	33 Li_3	34 Li_2Be	35 $LiBe_2$	36 Be_3	37 Be_2B	38 BeB_2	39 B_3	40 B_2C	41 C_2B	42 C_3	43 C_2N	44 CN_2	45 N_3	46 N_2O	47 NO_2	48 O_3	49 FO_2	50 F_2O^2	51 F_3	52 NeF_2	53 Ne_2F	54 Ne_3
7	55 Li_2Na^{2+}	56 Li_2Na^{+}	57 Li_2Na	58 Li_2Mg	59 $NaBe_2$	60 Be_2Mg	61 NaB_2	62 B_2Mg	63 B_2Al	64 B_2Si	65 AlC_2	66 SiC_2	67 C_2P	68 SiN_2	69 N_2P	70 N_2P	71 PO_2	72 SO_2	73 ClO_2	74 SF_2	75 ClF_2	76 ArF_2	77 Ne_2Cl	78 Ne_2Ar
8	79 Na_2Li^{2+}	80 Na_2Li^{+}	81 Na_2Li	82 Na_2Be	83 $LiMg_2$	84 Mg_2Be	85 Mg_2B	86 $BeAl_2$	87 Al_2B	88 Al_2C	89 Si_2B	90 Si_2C	91 Si_2N	92 CP_2	93 P_2N	94 P_2O	95 NS_2	96 S_2O	97 S_2F	98 Cl_2O	99 Cl_2F	100 $NeCl_2$	101 Ar_2F	102 Ar_2Ne
9	103 Na_3^{2+}	104 Na_3^{+}	105 Na_3	106 Na_2Mg	107 $NaMg_2$	108 Mg_3	109 Mg_2Al	110 $MgAl_2$	111 Al_3	112 Al_2Si	113 $AlSi_2$	114 Si_3	115 Si_2P	116 SiP_2	117 P_3	118 P_2S	119 PS_2	120 S_3	121 S_2Cl	122 SCl_2	123 Cl_3	124 $ArCl_2$	125 Ar_2Cl	126 Ar_3
10	127 Na_2K^{2+}	128 Na_2K^{+}	129 Na_2K	130 Na_2Ca	131 KMg_2	132 Mg_2Ca	133 Mg_2Ga	134 $CaAl_2$	135 Al_2Ga	136 Al_2Ge	137 $GaSi_2$	138 Si_2Ge	139 Si_2As	140 GeP_2	141 P_2As	142 P_2Se	143 AsS_2	144 SeS_2	145 S_2Br	146 $SeCl_2$	147 $BrCl_2$	148 $KrCl_2$	149 Ar_2Br	150 Ar_2Kr
11	151 NaK_2^{2+}	152 NaK_2^{+}	153 NaK_2	154 K_2Mg	155 $NaCa_2$	156 $MgCa_2$	157 Ca_2Al	158 $MgGa_2$	159 $AlGa_2$	160 Ga_2Si	161 $AlGe_2$	162 $SiGe_2$	163 Ge_2P	164 $SiAs_2$	165 PAs_2	166 As_2S	167 Se_2P	168 Se_2S	169 Se_2Cl	170 SBr_2	171 Br_2Cl	172 $ArBr_2$	173 Kr_2Cl	174 $ArKr_2$
12	175 K_3^{2+}	176 K_3^{+}	177 K_3	178 K_2Ca	179 Ca_2K	180 Ca_3	181 Ca_2Ga	182 $CaGa_2$	183 Ga_3	184 Ga_2Ge	185 $GaGe_2$	186 Ge_3	187 Ge_2As	188 $GeAs_2$	189 As_3	190 As_2Se	191 $AsSe_2$	192 Se_3	193 Se_2Br	194 $SeBr_2$	195 Br_3	196 $KrBr_2$	197 Kr_2Br	198 Kr_3
13	199 K_2Rb^{2+}	200 K_2Rb^{+}	201 K_2Rb	202 K_2Sr	203 $RbCa_2$	204 Ca_2Sr	205 Ca_2In	206 $SrGa_2$	207 Ga_2In	208 Ga_2Sn	209 $InGe_2$	210 Se_2Sn	211 Ge_2Sb	212 $SnAs_2$	213 As_2Sb	214 As_2Te	215 $SbSe_2$	216 Se_2Te	217 Se_2I	218 $TeBr_2$	219 Br_2I	220 $XeBr_2$	221 Kr_2I	222 Kr_2Xe
14	223 KRb_2^{2+}	224 KRb_2^{+}	225 KRb_2	226 Rb_2Ca	227 KSr_2	228 $CaSr_2$	229 Sr_2Ga	230 $CaIn_2$	231 $GaIn_2$	232 In_2Ge	233 $GaSn_2$	234 $SeSn_2$	235 Sn_2As	236 $GeSb_2$	237 $AsSb_2$	238 Sb_2Se	239 $AsTe_2$	240 $SeTe_2$	241 Te_2Br	242 SeI_2	243 BrI_2	244 KrI_2	245 Xe_2Br	246 $KrXe_2$
15	247 Rb_3^{2+}	248 Rb_3^{+}	249 Rb_3	250 Rb_2Sr	251 $RbSr_2$	252 Sr_3	253 Sr_2In	254 $SrIn_2$	255 In_3	256 $SnIn_2$	257 Sn_2In	258 Sn_3	259 Sn_2Sb	260 $SnSb_2$	261 Sb_3	262 Sb_2Te	263 $SbTe_2$	264 Te_3	265 Te_2I	266 TeI_2	267 I_3	268 XeI_2	269 Xe_2I	270 Xe_3

Figure 5: The Periodic table of triatomic molecules.

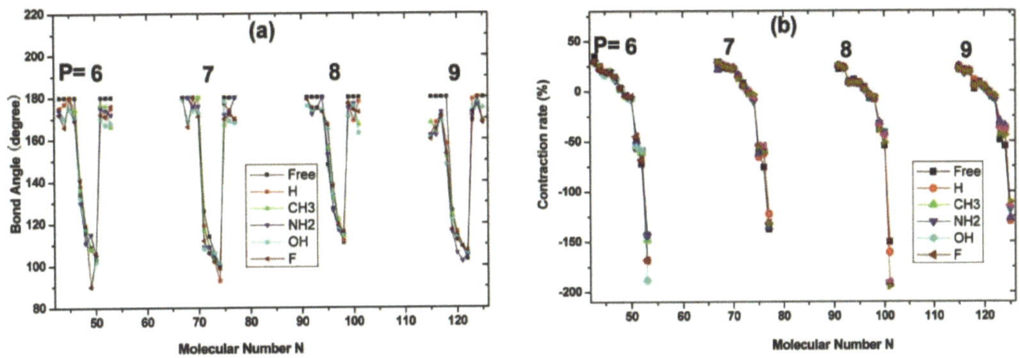

Figure 6: Triatomic molecular properties *versus* N. Charts (a) and (b) show the periodicities of bond angle and bond length with the molecular number N, respectively. The ν-shaped curves in Chart (a) indicate that the linear triatomic molecules become bent and stretch out again. Chart (b) shows the contraction rate of two bond lengths, $C = (R_{co} - R_{sum})/R_{co}$.

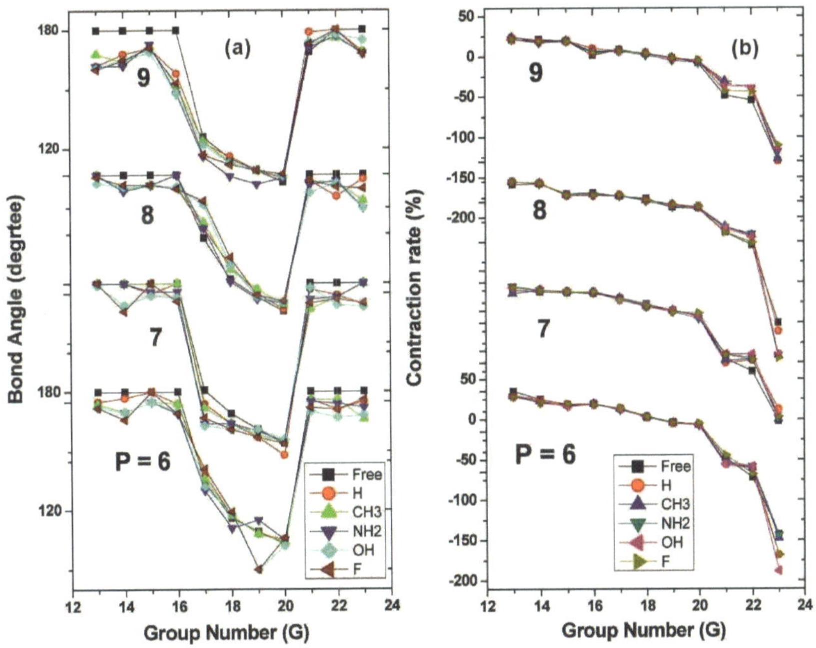

Figure 7: Triatomic molecular properties *versus* G. Charts (a) and (b) denote the bond angle and bond length *versus* group number G, respectively.

To discover the periodicity of triatomic molecules, some free molecules and virtual triatomic molecules are surveyed in polyatomic molecules. The surveyed molecules are of G = 13-23, and P = 4-7 in Fig. (7). Two molecular properties, bond angle and bond length, are investigated by quantum chemical calculations.

Fig. (**6**) (a) and (b) show the variations of the bond angle and the bond length with molecular number N, respectively. In each chart, an analogous pattern repeats for four rounds. The result clearly shows that periodicity occurs also among triatomic molecules.

Fig. (**7**) (a) shows the G dependence of bond angle. The curve is characterized by a ν-shaped pattern. Linear molecules become bent and stretch back to linear shape again when G increases. The triatomic molecules of $G < 17$ have linear shape. Molecules start to be bent at $G = 17$ and stretch out again at $G = 21$. The variation is due to the change of electronic configuration, and can be interpreted by Walsh diagrams [14].

The bond length in triatomic molecules is also investigated. Total contraction of two bond lengths is used to measure the average bond strength of the triatomic molecules. Sum of the calculated internuclear distances between atoms a and b, b and c, $R_{sum} = (R_{ab} + R_{bc})$, is compared to the normal covalent bond distances R_{co}, $R_{co} = (r_a + 2r_b + r_c)$. The terms of r_a, r_b and r_c are the standard covalent radii of the atoms a, b and c respectively. The relative contraction of bond length is defined as $C = (R_{co} - R_{sum})/R_{co}$. The more contracted the bond, the stronger the chemical binding. Both Fig. (**6**) (b) and Fig. (**7**) (b) show that the relative contraction C increases with the molecular number N and the group number G. The curves indicate that bond contraction, and thus bond strength of triatomic molecules, varies also periodically. The fact further identifies the existence of periodicity among triatomic molecules.

AH_3 TYPE TETRA-ATOMIC MOLECULES

There are different shapes of tetra-atomic molecules, including the AB_3 type molecules (C_{3v} or D_{3h}), the Y shape molecules of ABC_2 (C_{2v} or D_{2h}), *etc.* Each kind of molecules has a corresponding periodic table. Here we show the periodic table of AH_3 type molecules, which are the simplest tetra-atomic molecules.

The group number G is defined as the number of valence electrons of a molecule. Considering that three electrons from the hydrogen atoms should be counted, the total valence electron number of an AH_3 molecule is $G = g_A + 3$. Here g_A is the valence electron number of the central atom. The periodic number P of an AH_3 molecule is defined as the number of electron shells. $P = p_A + 1$, where p_A is the periodic number of the central atom A in the atomic periodic table. The periodic table of AH_3 molecules, Fig. (**8**), is compiled with the G and P numbers of the molecules.

P	Group Number (G)					
	4	5	6	7	8	9
5	BeH_3^+	BeH_3	BeH_3	CH_3	NH_3	H_3O
6	MgH_3^+	MgH_3	AlH_3	SiH_3	PH_3	H_3S
7			GaH_3	GeH_3	AsH_3	
8		InH_3^+	InH_3	SnH_3	SbH_3	H_3Te
9		TlH_3^+	TlH_3	PbH_3	BiH_3	

Figure 8: The Periodic table of AH_3 molecules.

Properties of the AH_3 molecules vary periodically with the number of valence electrons. The collected experimental data as well as *ab-initio* calculation results of these properties are listed in the appendix.

Bond Angle

The geometric shapes of AH_3 molecules can be depicted by the bond angle of HAH. The AH_3 molecules of D_{3h} type are flat molecules. All three bond angles are 120 degrees. On the other hand, those molecules with the vertex angles smaller than 120 degrees hold pyramid shape. The smaller the angle, the more acute the pyramid.

Fig. (**9**) (a) shows that the bond angle of HAH changes with the valence electron number G. The bond angle is only determined by G. The angles are 120 degrees or the AH_3 molecules are flat if their valence electrons are equal or less than six, while those molecules with more than six valence electrons have pyramid shape.

The regularity can be understood in the light of the Walsh Diagram [14] shown in Fig. (**10**). According to the simple MO theory, the first six valence electrons of an

AH_3 molecule occupy the lowest orbitals of a_1, e and e, which are the sp^2 hybrid orbitals. The bond angles tend to be 120 degrees. Therefore, molecules, as AlH_3 or GaH_3, are flat with D_{3h} symmetry. On the other hand, the seventh and eighth valence electrons of a molecule occupy the orbitals of a_2-a_1S_A. The electrons in these orbitals are the lone paired electrons. The electrons constrain hydrogen atoms out of the molecular plan, causing the bond angles bent. Each a_2 electron constrains the angle about 10 degrees. Therefore, the HAH bond angles of the molecules with six, seven and eight valence electrons are approximately 120, 110 and 100 degrees, respectively.

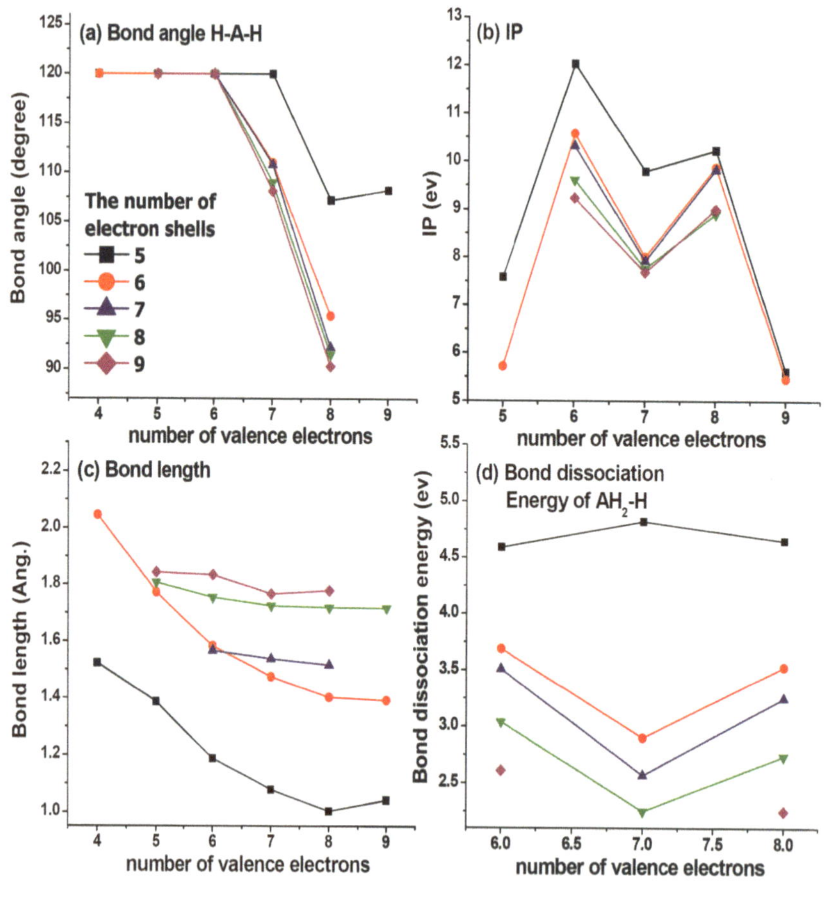

Figure 9: Properties of AH_3 molecules which vary periodically with G. (a) Bond angle between H-A-H; (b) Ionization potential; (c) Bond length of A-H; (d) Bond energy between AH_2-H.

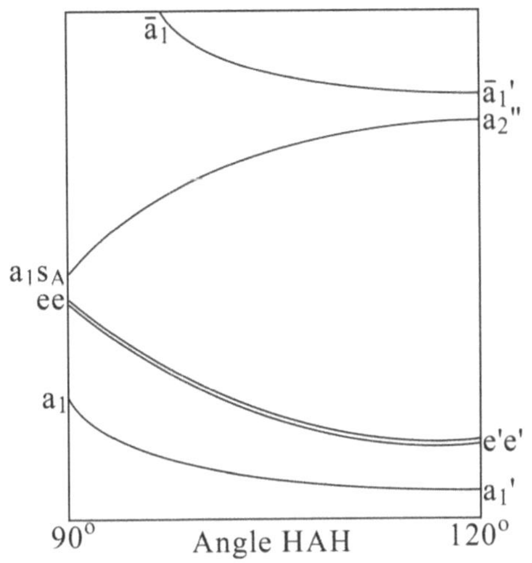

Figure 10: Walsh Diagram of the AH_3 Molecules.

Ionization Potential

Ionization Potential (IP) is determined by the energy of the HOMO electron. Fig. (**9**) (b) show the variation of IP with G. For each period, there is an M type line in the figure. The drop of IP at G = 7 is caused by the minimum ionization potential of the seventh valence electron, which is easier to remove.

Bond Length

Bond length reflects the strength of the A-H bonds. Since the electronic orbitals of a_1, a_1, e and a_2 have bonding or nonbonding characters, the molecules owning five to eight valence electrons are more stable. As a result, the A-H bond length of each period shrinks in the sequence as shown in Fig. (**9**) (c).

Bond Dissociation Energy

Bond dissociation energy of AH_2-H is another measurement of the stability of an AH_3 molecule. The periodicity can be clearly seen in Fig. (**9**) (d).

Vibrational Frequencies

For D_{3h} geometry, the four characteristic vibrational modes are $A_1'(\nu_1)$, $A_2''(\nu_2)$, $E'(\nu_3)$ and $E'(\nu_4)$. The corresponding four modes in C_{3v} geometry are $A_1(\nu_1)$, $A_1(\nu_2)$, $E(\nu_3)$ and $E(\nu_4)$ respectively, as shown in Fig. (**11**).

The data of these vibrational frequencies of the AH_3 molecules are shown in Fig. (**12**). We see again that the periodicity occurs with each vibration mode.

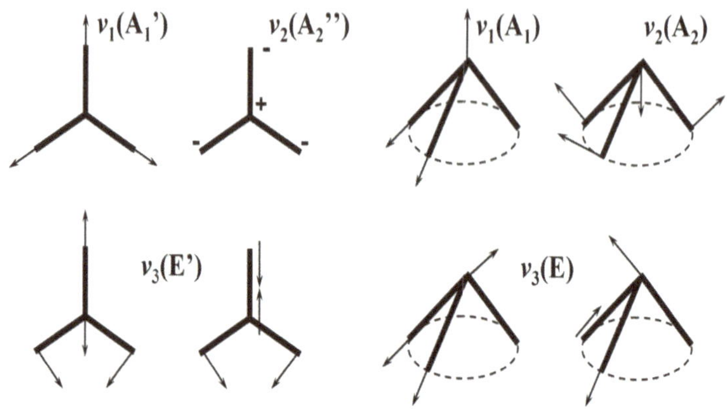

Figure 11: The characteristic vibrational modes for molecules with D_{3h} and C_{3v} geometries.

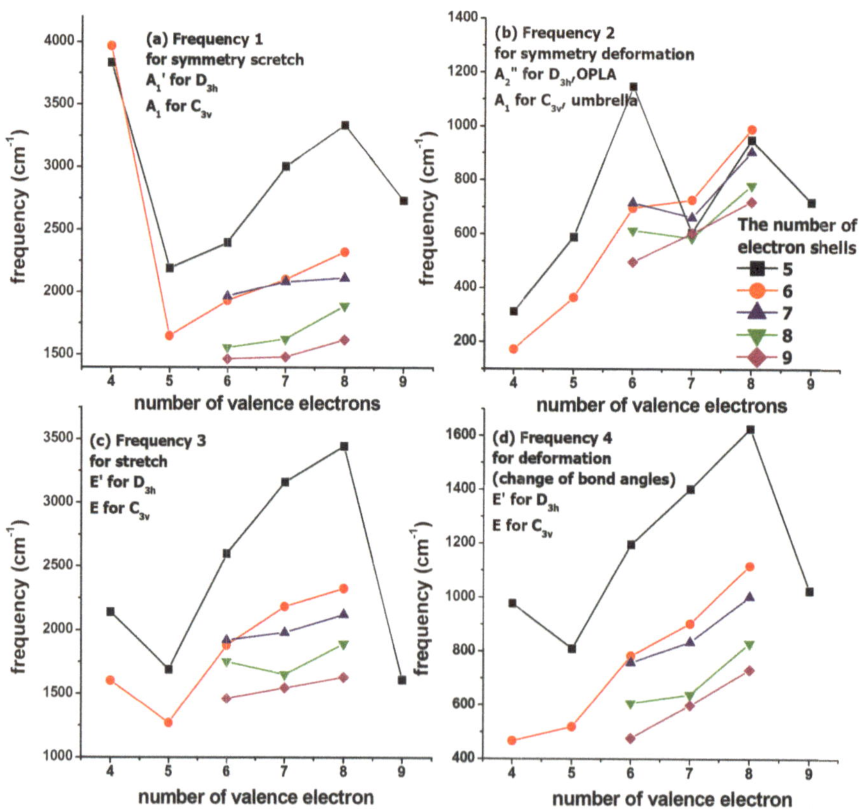

Figure 12: The vibrational frequencies for AH_3 molecules with both D_{3h} and C_{3v} geometries.

ACKNOWLEDGEMENTS

The authors would thank Professor Ray Hefferlin for his encouragement and appreciate Dr. Kunhui Liu and Dr. Xiaonan Ma for their technical assistances in drafting the electronic version.

CONFLICT OF INTEREST

The authors confirm that this chapter contents have no conflict of interest.

APPENDIX

Table 1: Some properties of AH_3 molecules

	Ionization Potential (ev)	**Dissociation Energy (eV) (AH_2-H)**	**Average Bond Length (Å)**	**Bond Angle (°)**
BeH_3^+			1.523[18]	120
BeH_3	7.58*		1.3864*	120
BH_3	12.03[15]	4.59[25]	1.188[43]	120[25]
CH_3	9.79[16]	4.82[25]	1.079[44]	120
NH_3	10.23[17]	4.65[25]	1.00249[25]	107.21[25]
H_3O	5.62*		1.0423*	108.2*
MgH_3^+			2.045[18]	120
MgH_3	5.72*		1.7725*	120
AlH_3	10.58[18]	3.69[25]	1.58407[25]	120[25]
SiH_3	8.01[19]	2.9[25]	1.47587[25]	111[27]
PH_3	9.87[20]	3.53[25]	1.40314[25]	95.41[25]
GaH_3	10.32	3.51(De)	1.567*	120*
GeH_3	7.92[21]	2.57[26]	1.539[28]	110.8[28]
AsH_3	9.82[22]	3.25[24]	1.517[24]	92.2[24]
InH_3^+			1.807[23]	120
InH_3	9.61[23]	3.04(De)[23]	1.754[23]	120
SnH_3	7.769**	2.25[26]	1.7248**	109**
SbH_3	8.9[24]	2.74[24]	1.719[24]	91.5[24]
TlH_3^+			1.843[23]	120
TlH_3	9.24[23]	2.61[23]	1.8337**	120
PbH_3	7.68**		1.766**	108.1**
BiH_3	9[24]	2.25[24]	1.7786**	90.3[24]

Table 2: Vibrational frequencies of AH_3 molecules (cm^{-1})

	ν_1^a	ν_1^b	ν_2^a	ν_2^b
BeH_3^+	4307[29]	$3833^{0.89**}$	350[19]	$311.5^{0.89**}$
BeH_3		2190*		589*
BH_3	2693[30]	$2397^{0.89**}$	1225[30]	1147.5[37]
CH_3	3285[30]	3004[31]	308[30]	606.5[38]
NH_3	3690[30]	3337[32]	1207[30]	950[32]
H_3O		2734*		719*
MgH_3^+	4460[29]	$3969^{0.89**}$	194[29]	$172.7^{0.89**}$
MgH_3		1651*		364*
AlH_3	2813[30]	$1934^{0.89**}$	760[30]	697.8[35]
SiH_3	2364[30]	$2104^{0.89**}$	875[30]	727.9[39]
PH_3	2606[30]	2323[32]	1140[30]	992[32]
GaH_3		1969*		717.4[35]
GeH_3	2212[28]	$2085^{0.89**}$	782[28]	663[40]
AsH_3	2360[28]	2116[32]	1021[28]	906[32]
InH_3^+				
InH_3	1749**	$1557^{0.89**}$		613.2[35]
SnH_3	1828**	$1627^{0.89**}$	658.5**	$586.1^{0.89**}$
SbH_3		1891[32]		782[32]
TlH_3^+				
TlH_3	1649**	$1467^{0.89**}$	558.5**	$497.1^{0.89**}$
PbH_3	1668**	$1484^{0.89**}$	679.0**	$604.3^{0.89**}$
BiH_3	1821**	$1621^{0.89**}$	810.6**	$721.5^{0.89**}$
	ν_3^a	ν_3^b	ν_4^a	ν_4^b
BeH_3^+	2403[29]	$2139^{0.89**}$	1098[29]	$977.2^{0.89**}$
BeH_3		1689*		809*
BH_3	2813[30]	2602[33]	1305[30]	1197[41]
CH_3	3461[30]	3161[34]	1540[30]	1403[42]
NH_3	3823[30]	3444[32]	1849[30]	1627[32]
H_3O		1612*		1025*
MgH_3^+	1802[29]	$1604^{0.89**}$	525[29]	$467.2^{0.89**}$
MgH_3		1270*		519*
AlH_3	2029[30]	1883[35]	846[30]	783.4[35]
SiH_3	2383[30]	2185[36]	1015[30]	$903.3^{0.89**}$
PH_3	2601[30]	2328[32]	1271[30]	1118[32]

GaH_3		1923[35]		758.7[35]
GeH_3	2227[28]	1982$^{0.89**}$	937[28]	833. 9$^{0.89**}$
AsH_3	2362[28]	2123[32]	1129[28]	1003[32]
InH_3^+				
InH_3		1754[35]		607.8[35]
SnH_3	1856**	1652$^{0.89**}$	719.1**	640.0$^{0.89**}$
SbH_3		1894[32]		831[32]
TlH_3^+				
TlH_3	1644**	1464$^{0.89**}$	536**	477.0$^{0.89**}$
PbH_3	1741**	1550$^{0.89**}$	674**	599.9$^{0.89**}$
BiH_3	1835**	1633$^{0.89**}$	822.3**	731.9$^{0.89**}$

[a]Calculated vibrational frequencies: multiplied by 0.89 as scaling.
[b]Experimentally measured vibrational frequencies, used in the figures. For molecules whose experimental data are not available, the scaled calculated values are used, with marked "0.89**".
*: Calculated by Gaussian 94; B3LYP/6-311G(3df,2p) as basis set.
**: Calculated by Gaussian 94; B3LYP/LanL2DZ as basis set.

REFERENCES

[1] Mendeleev, D. I. On the Relationship of the Properties of the Elements to their Atomic Weights. St. Petersburg, **1869** (in Russian).

[2] Clark, C. H. The periodic Groups of Non-Hydride Di-Atoms. *Trans. Faraday Soc.*, **1935**, *31*, 1017.

[3] Hefferlin, R. Systematics of Diatomic Molecular Transition Moments: Encouraging Progress -- II. *J. Quant. Spectrosc. Radiat. Transfer.*, **1976**, *16*, 1101.

[4] Hefferlin, R.; Kuzner, M. Systematics of Ground-State Potential Minima between two Main-Group Atoms of Ions. *J. Chem. Phys.*, **1981**, *75*, 1035.

[5] Hefferlin, R.; Innis, W. The Differential Coefficient (dp/dn)|n e for Properties of Diatomic Molecules and Atoms. *J. Quant. Spectrosc. Radiat. Transfer.*, **1983**, *29*, 97.

[6] Kong, F. The Periodicity of Diatomic Molecules. *J. Mol. Struct.*, **1982**, *90*, 17.

[7] Kong, F. An Alternative Periodic Table for Triatomic Molecules, In: *Periodic Systems and Their Pelation to the Systematic Analysis of Molecular data*; Hefferlin, R., Ed.; Edwin Mellen Press: Lewistin, **1989**.

[8] Carlson, C. M.; Cavanaugh, R. J.; Hefferlin, R.; Zhuvikin, G. V. Periodic Systems of Molecular States from the Boson Group Dynamics of SO(3)xSU(2)s. *J. Chem. Inf. Comput. Sci.*, **1996**, *36*, 396.

[9] Hefferlin, R.; Sepman, A.; Zhuvinkin, G. Molecular Multiplets of Alkaline Atoms. *J. Chem. Inf. Comput. Sci.*, **1996,** *36*, 399.

[10] Sidgwick, N. V. The nature of the non-polar link. *Trans. Faraday. Soc.*, **1923**, *19*, 469.

[11] Haas, A. Periodic system of functional groups: formalism only or heuristic principles. *Pure and Applied Chemistry*, **1991**, *63*, 1577.

[12] Peppernick, K.D.; Gunaratne, K. D.; Castleman, A.W. Superatom spectroscopy and the electronic state correlation between elements and isoelectronic molecular counterparts. *Proc. Natl. Acad. Sci. U. S. A.*, **2010,** *107*, 975.

[13] http://webbook.nist.gov/chemistry/

[14] Walsh, A. D. The electronic orbitals, shapes, and spectra of polyatomic molecules. Part IV. Tetratomic hydride molecules, AH3. *J. Chem. Soc.*, **1953**, *4*, 2296.

[15] Ruscic, B.; Mayhew, C. A.; Berkowitz, J. Photoionization studies of (BH3)_n (n=1,2). *J. Chem. Phys.*, **1988**, *88*, 5580.

[16] Pople, J. A.; Curtiss, L.A. Theoretical thermochemistry. 2. Ionization energies and proton affinities of AHn species (A = C to F and Si to Cl); heats of formation of their cations. *J. Phys. Chem.*, **1987**, *91*, 155.

[17] Pople, J. A.; Curtiss, L. A. Theoretical thermochemistry. 3. A modified procedure for ionization energies of AHn species. *J. Phys. Chem.*, **1987**, *91*, 3637.

[18] Curtiss, L. A.; Pople, J.A. Theoretical thermochemistry. 4. Ionization energies and proton affinities of AHn species (A = lithium to boron and sodium to aluminum): geometries and enthalpies of formation of their cations. *J. Phys. Chem.*, **1988**, *92*, 894. The bond lengths of BeH^{3+}, BH^{3+}, MgH^{3+} and AlH^{3+} are the result after average (C_{2v}).

[19] Berkowitz, J.; Greene, J. P.; Cho, H.; Ruscic, B. Photoionization mass spectrometric studies of SiHn (n=1–4). *J. Chem. Phys.*, **1987**, *86*, 1235.

[20] Ruede, R.; Troxler, H.; Beglinger, C.; Jungen, M. *The dissociation energies of the positive ions NH+3, NF+3, PH+3, PF+3 and Pcl+3. Chem. Phys. Lett.*, **1993**, *203*, 477.

[21] Binning, R. C.; Curtiss, L. A. Theoretical study of GeHn, AsHn, and SeHn: Ionization energies. *J. Chem. Phys.*, **1990**, *92*, 3688.

[22] Berkowitz, J. Photoionization mass spectrometric studies of AsHn (n=1–3). *J. Chem. Phys.*, **1988**, *89*, 7065.

[23] Balasubramanian, K.; Tao, J. X. Electronic states, ionization potentials, and bond energies of TlHn, InHn, TlH+n, and InH+n (n=1–3). *J. Chem. Phys.*, **1991**, *94*, 3000. The bond lengths of InH^{3+} and TlH^{3+} are the results after average (C_{2v}).

[24] Dai, D.; Balasubramanian, K. Geometries and energies of electronic states of AsH3, SbH3, and BiH3 and their positive ions. *J. Chem. Phys.*, **1990**, *93*, 1837, 1990.

[25] Pople, J. A.; Lake, B. T.; Frisch, M. J.; Binkley, J. S. Theoretical thermochemistry. 1. Heats of formation of neutral AHn molecules (A = Li to Cl). *J. Phys. Chem.*, **1985**, *89*, 2198.

[26] Ruscic, B.; Schwarz, M.; Berkowitz, J. Photoionization studies of GeHn (n=2–4). *J. Chem. Phys.*, **1990**, *92*, 1865.

[27] Reed, K. J.; Brauman, J. I. Photodetachment of electrons from Group IVa binary hydride anions: The electron affinities of the SiH3 and GeH3 radicals. *J. Chem. Phys.*, **1974**, *61*, 4830.

[28] Binning, R. C.; Curtiss, L. A. Theoretical study of GeHn, AsHn, and SeHn@B: Bond dissociation energies. *J. Chem. Phys.*, **1990**, *92*, 1860.

[29] Crutiss, L. A.; Pople, J. A. Theoretical thermochemistry. 4. Ionization energies and proton affinities of AHn species (A = lithium to boron and sodium to aluminum): geometries and enthalpies of formation of their cations. *J. Phys. Chem.*, **1988**, *92*, 894.

[30] Pople, J. A.; Luke, B. T.; Frisch, M. J.; Binkley, J. S. Theoretical thermochemistry. 1. Heats of formation of neutral AHn molecules (A = Li to Cl). *J. Phys. Chem.*, **1985**, *89*, 2198.

[31] Zahedi, M.; Harrison, J. A.; Nibler, J. W. 266 nm CH3I photodissociation: CH3 spectra and population distributions by coherent Raman spectroscopy. *J. Chem. Phys.*, **1994**, *100*, 4043.

[32] Shimanouchi, T., *Tables of Molecular Vibrational Frequencies Consolidated Volume I*, National Bureau of Standards, **1972**, 1-160.

[33] Kawaguchi, K. Fourier transform infrared spectroscopy of the BH3 ν3 band. *J. Chem. Phys.*, **1992**, *96*, 3411.

[34] Davis, S.; Anderson, D. T.; Duxbury, G.; Nesbitt, D. J. Jet-cooled molecular radicals in slit supersonic discharges: Sub-Doppler infrared studies of methyl radical. *J. Chem. Phys.*, **1997**, *107*, 5661.

[35] Pullumbi, P.; Bouteiller, Y.; Manceron, L.; Mijoule, C. Aluminium, gallium and indium trihydrides. an IR matrix isolation and ab initio study. *Chem. Phys.*, **1994**, *185*, 25.

[36] Sumiyoshi, Y.; Tanaka, K.; Tanaka, T. Diode laser spectroscopy of SiH3 in the 4.5 μm region. *Appl. Surf. Sci.*, **1994**, *79/80*, 471.

[37] Kawaguchi, K.; Butler, J. E.; Yamada, C.; Bauer, S. H.; Minowa, T.; Kanamori, H.; Hirota, E. Observation of the gas-phase infrared spectrum of BH3. *J. Chem. Phys.*, **1987**, *87*, 2438.

[38] Wormhoudt, J.; McCurdy, K. E. A measurement of the strength of the ν2 band of CH3. *Chem. Phys. Lett.*, **1989**, *156*, 47.

[39] Johnson, R. D.; Tsai, B. P.; Hudgens, J. W. Multiphoton ionization of SiH3 and SiD3 radicals: Electronic spectra, vibrational analyses of the ground and Rydberg states, and ionization potentials. *J. Chem. Phys.*, **1989**, *91*, 3340.

[40] Johnson, R. D.; Tsai, B. P.; Hudgens, J. W. The electronic spectrum of the GeH3 radical. *J. Chem. Phys.*, **1988**, *89*, 4558.

[41] Kawaguchi, K. Fourier transform infrared spectroscopy of the BH3 ν2 band. *Can. J. Phys.*, **1994**, *72*, 925.

[42] Tam, S.; Macler, M.; Fajardo, M. E. Matrix isolation spectroscopy of laser ablated carbon species in Ne, D2, and H2 matrices *J. Chem. Phys.*, **1997**, *106*, 8955.

[43] Wickham-Jones, C. T.; Moran, S.; Ellison, G. B. *J. Chem. Phys.*, **1989**, *90*, 795.

[44] Yamada, C.; Hirota, E.; Kawaguchi, K. Photoelectron spectroscopy of BH−3. Diode laser study of the ν2 band of the methyl radical. *J. Chem. Phys.*, **1981**, *75*, 5256.

CHAPTER 5

The GRANCH Techniques for Analysis of DNA, RNA and Protein Sequences

Ashesh Nandy*

Centre for Interdisciplinary Research and Education, 404B Jodhpur Park, Kolkata 700068, India

Abstract: The very rapid growth in molecular sequence data from the daily accretion of large gene and protein sequencing projects have led to issues regarding viewing and analyzing the massive amounts of data. Graphical representation and numerical characterization of DNA, RNA and protein sequences have exhibited great potential to address these concerns. We review here in brief several different formulations of these representations and examples of applications to diverse problems based on what this author had presented at the Second Mathematical Chemistry Workshop of the Americas in Bogota, Colombia in 2010. In particular, we note several insights that were gained from such representations, and the applications to the bio-medicinal field.

Keywords: Alignment free sequence comparison, bio-medical applications, chaos generator representation, DNA graphical representation, DNA numerical characterization, DNA sequences global characteristics, DNA visualization, GRANCH applications, GRANCH techniques, identifying potential drug target regions in viruses, lattice graphs in medical applications, long range correlations, long range palindromic sequences, phylogenetic trees, protein graphical representation, protein numerical characterization, purine pyrimidine relationship, QSAR.

INTRODUCTION

My first brush with a DNA sequence, in around 1990, left me totally puzzled: I could not "see" nor get a "feel" of anything noteworthy in the apparent jumble of characters that symbolized a DNA, not the least because I had never studied biology myself. My background was physics, and I began a search for, to me, a more meaningful exposition of the sequence of characters that represented the

***Corresponding author Ashesh Nandy:** Centre for Interdisciplinary Research and Education, 404B Jodhpur Park, Kolkata 700068, India; Tel: +91 33 2473 0577; E-mail: anandy43@yahoo.com

Subhash C. Basak, Guillermo Restrepo & José L. Villaveces (Eds)
Copyright © 2015 Bentham Science Publishers Ltd. Published by Elsevier Inc. All rights reserved.
10.1016/B978-1-68108-053-6.50005-3

DNA sequence. My studies led me to appreciate and anticipate the immense potential opening up with the sequencing of genome length sequences and the concomitant need for rapidly scanning and analyzing DNA sequences for matters of interest [1], and to get excited at the new insights bcing gained from a global perspective of the DNA sequences: Jeffreys [2] had shown through his Chaos Generator Representation that such sequences had an inherent fractal nature; Peng *et al.* [3] speculated that DNA sequences had long-range correlations, an observation that raised a storm of papers in very short order; and Voss [4] showed that long range fractal correlations existed in DNA sequences with the degree of correlation varying with evolutionary divergence.

But a close up look at a DNA sequence and how the bases were distributed along it still lacked an appealing representation. Experimenting with various formats I determined that a 2D graphical representation, as explained later in this chapter, was what I could relate to on a purely personal basis. After many graphs of various sequences and discussions with some eminent persons to ensure that such a simple stratagem was not already familiar to cutting edge biologists, I published a paper on it in Current Science (Bangalore) in 1994 [5]. Imagine my consternation when I was informed soon after that Gates had already anticipated such a device, albeit with different axes assignments, way back in 1986 [6], but which seemed to have been in limbo since! A short note had to be published soon after informing of this oversight and explaining the differences although both used Cartesian co-ordinate system to plot the graphs [7].

However, a physics background demands some quantitative appraisal of whatever nature has to offer. I had observed certain similarities and changes in plots of conserved gene sequences of various species, but coming up with some way to measure the changes posed difficulties with these plots of discrete numbers. I had done some number crunching with individual gene segments like introns and exons [8, 9], but now the need was for whole sequences for which we came up with a geometrical interpretation to describe in general a macro-molecular sequence and measure sequence differences. We presented our scheme at the First Indo-US Workshop on Mathematical Chemistry in Shantiniketan, West Bengal, India in 1998 [10] where we reported, as stated in the abstract, that *"Geometrisation of macromolecular sequences in the form of a graphical representation provides one ... technique where the nucleotides in a gene sequence can be viewed as objects in a 4-dimensional space; the method can be extended, in principle, to include, say proteins, in a 20-dimensional space. We have found a reduced 2-dimensional representation of DNA sequences very useful in studies of nucleotide distribution and composition. .... We here propose a new measure of the dispersion of DNA graphs that can be used to quantify the differences between two or more graphs of genes of various organisms .... It also appears that once standardized the proposed scheme may help study molecular phylogeny in evolutionary time scale."*

Although the participants in the Shantiniketan Workshop included stalwarts in the field like Prof. Milan Randić, Prof. Haruo Hosoya, Prof. Paul Mezey and others, our scheme did not seem to evoke any response, not surprising since they apparently did not know about DNA issues. But Prof. Subhash Basak of the University of Minnesota, USA and co-organiser of the Workshop was intrigued enough by our work and its potential to describe DNA sequences through graph invariants to meet me in Kolkata after the workshop to discuss the possibility of using invariants for DNA sequences as descriptors. Subsequently Prof. Basak invited me the following summer to Duluth to carry out further research on DNA mathematical descriptors in his group funded by the Natural Resources Research Institute (NRRI). Prof. Milan Randić and some other distinguished scientists were also invited there to begin to work on DNA descriptors in a project funded by NRRI.

It began with a talk I gave at the University of Minnesota, Duluth about my work on mathematical descriptors of DNAs arising from my graphical representation method. Among the attendees was Prof. Milan Randić who, with Prof. Basak, immediately saw the potential for converting a DNA sequence graph to a matrix and thereby extract numerical invariants which could be a more meaningful way to characterize DNA sequences. We collaborated then on a proposal for a 3D graphical representation and a matrix method for extracting graph invariants for the first exons of beta globin sequences of several species. This was published in 2000 in the Journal of Chemical Information and Computer Science [12] and led very soon to a whole host of papers on the DNA graphical representations and numerical characterisations and applications of them that continues still as more and more areas keep opening up and a new field of research seems to have begun.

This review is a brief introduction to the readers of this new and exciting field of research on graphical representation and numerical characterization (GRANCH) of bio-molecular sequences, based on the talk I presented at the Second Mathematical Chemistry Workshop of the Americas in Bogota, Colombia, in July 2010 [13]. Some of the various applications made to date using these techniques are also covered briefly, with special emphasis on our recent work that provides a possible approach to anti-viral vaccine design that could be expected to be less susceptible to invalidation through mutational changes in the viral proteins. More details can be found in the several reviews [14-18] and book chapters [19-21] that have appeared on the subject, and of course there are always the original papers. (*Note added in proof: See also bibliography in Ref [82].*)

GLOBAL CHARACTERISTICS OF DNA SEQUENCES

As sequence data on long stretches of DNA began to become available in the late 1980's, there arose a problem on how to view them and a curiosity to know whether any systematics lay hidden in the apparently random arrangement of characters representing the bases in the sequence. H J Jeffrey [2] came up with the idea of plotting them in a square grid where the four corners were identified with the four bases A, C, G, T. The algorithm was to start from the center of the square and for the first base plot a point midway between the origin and the home corner of the base. For the second base he started from the point representing the first base and plotted a point midway between it and the home corner of the second base. Continuing in this way filled up the square with a series of points until the entire sequence was plotted. This diagram he called the Chaos Generator Representation (CGR) of the DNA sequence. He noticed that different animal kingdoms showed different patterns – double scoop depletion regions for vertebrates, striped patterns for plant sequences, an apparently random distribution for bacterial genomes. Overall, each sub-square section of the CGR pattern seemed a replica of the whole, *i.e.* DNA sequences had properties of self-similarity or fractal nature.

The CGR diagrams of various sequences were investigated by several researchers to find different properties of biological interest. Burma *et al.* [22] showed the structures observed in CGR diagrams arises from skews in base composition and presence of repetitive sequences or specific motifs. Dutta and Das [23] reported that a CGR plot can be reproduced by suitable algorithms by manipulating different combinations of strings of bases with appropriate frequencies. Thus, the double scoop depletion patterns seen in vertebrate CGRs arises from scarcity of CG dinucleotides, and so on. Baranidharan *et al.* [24] developed quantitative methods to generate similarity/dissimilarity maps of genomic sequences and showed that for certain mitochondrial genomes species wise characteristic features could be seen when nucleotide stretches of 7 or more bases at a time were analysed.

In a slightly different vein, Peng *et al.* [3] considered the structure of the bases in a DNA sequence and analysed them on the basis of their being pyrimidines (C,T) or purines (A,G) only. On an X-Y graph where the x-axis counted the nucleotide number, they plotted the appearance of the bases in a sequence by taking a step diagonally upwards if it was a purine or downwards if it was a pyrimidine to the next nucleotide number. Plotting the whole sequence step by step in this manner they generated a graph with an irregular up-down structure which they called a "DNA landscape". By taking subsections of the graph they found that the subsections also looked similar to the up-down structure of the whole, and the same was true of sub-sub-sections and so on showing that the purine-pyrimidine structure of a DNA sequence had self-similarity, which was what Jeffrey had remarked two years ago. Peng *et al.* then searched for possible correlations by estimating frequencies of different lengths of nucleotide stretches and found that all gene sequences with the mosaic structure of introns and exons had long-range correlations whereas intronless genes did not show this feature. The implications of such an observation, on the face of it, are huge: the beginning of the DNA sequence should, theoretically, be knowing what the end would be like! Such an observation quite naturally led to a storm of papers on the subject until it quitened down after the observation that since DNA sequences are known to elongate by duplicating long stretches of subsequences, it was possible that such sequences showed apparent long range correlations.

Around the same time Voss [4] conducted a rigorous analysis of 25000 DNA sequences with over 50 million bases covering organisms of all classes to search for long range correlations. Using a spectral density function analysis he concluded that "(a) long range fractal correlations exist in DNA sequences, (b) the degree of correlation as measured by a spectral exponent varies systematically with evolutionary category and (c) short range periodicities of period 3 are prominent while other periods, *e.g.* 9, are also present. The fractal correlations have been seen to extend over long ranges of nucleotide positions, with the smallest for phage and bacteria and extending to over 100,000 bases for the higher classes" [14].

VISUALIZING THE DNA

To get a feel for the actual distribution of bases along a DNA sequences you need a more direct graphical depiction than what the abstract representations of Jeffrey [2] or Peng *et al.*, [3] can offer. This problem was addressed many years ago by Hamori and Ruskin [25] with their proposal for a 3-dimensional graphical representation of a DNA sequence. They proposed a hypothetical square on the x-y plane with four corners (NW, NE, SE, SW) identified with the four bases A, C, G, T and the nucleotide number to be counted along the z-axis. Thus for a DNA sequence like ACGGT, one would plot a point on the A-corner at z-coordinate 1, then draw a line to the next base, C in this case, in its corner with z now equal to 2, and so on. For a sequence like ACGTACGTACGT this would generate a spiral around the z-axis; in case there was a preponderance of one base or the other the curve would flow along those corners. These curves the authors called H-curves. Visualizing such a 3D image on the 2D plane of the paper is admittedly difficult. However, the authors suggested that drawing two such curves at slightly different angles would allow stereoscopic vision so that the DNA could be seen in 3D. Taking the bacteriophage M13 as an example they showed that in their representation they could easily identify regions of sharp changes in base composition through visualization that would be difficult to determine from the normal character representation.

This author's search for a meaningful display of DNA sequence information led him to propose a 2-dimensional graphical representation where the four cardinal directions are associated with the four bases [5]. The method is to take a walk in the negative x-direction if there is an adenosine in the sequence, in the positive y-direction for a cytosine, positive x-direction for a guanine and the negative y-direction for thymine. Proceeding to walk in succession in the appropriate direction in the order of the bases making up the particular DNA sequence generates a path that visually depicts the arrangement of bases in the sequence. These DNA plots were found to be characteristic of the types of gene sequences and that the same genes from different species showed almost the same pattern. Since we know that specific genes from different species have significant homology, and in fact that is how often new genes are recognized, it is not surprising that their graphical plots will show basically the same shape.

It was found later that Gates [6] had already proposed a similar scheme to depicting gene sequences, although his assignment of bases were different from Nandy's scheme. A year later Leong and Morgenthaler [26] independently proposed another 2D scheme, where the base assignments were again different from the two just mentioned. On a 2D Cartesian co-ordinate system the assignments of the bases with the cardinal directions in the three schemes are, starting from the negative x-direction and going clockwise, GTCA (Gates), ACGT (Nandy) and CTAG (Leong and Morgenthaler). It is interesting to note that these three axes representations exhaust all possible 2D schemes of this type, and these can be seen to be like 2D projections of the Hamori-Ruskin H-curves.

The 2D plots can be scaled to accommodate from the largest to the smallest DNA sequences depending on the level of detail one wishes to observe. In reference 27 (the illustration (3)) depicts the 73326 thousand base long human beta globin sequence that contains the beta, eta, delta and the gamma globins of less than 2000 bases each, which can individually be plotted on a smaller scale. Plots such as these provide a quick estimation of base composition and distribution along a DNA sequence. An inspection of the human beta globin sequence graph shows that it has two sections that are mainly A-T rich with one part in between that is T-dominated. A plot of a sequence like the chicken myosin heavy chain gene is represented in illustration 2 (*loc. cit.,*) shows that it also is *AT*-rich; from the angle it makes with the axes, it is evident that the sequence is dominated by larger percentage of T's than A's, and likewise one can determine preponderance of structures like A_mT_n from inspections of such plots. Further applications of these graphs are taken up later in this chapter.

The 2D representations, however, suffer from degeneracy in that nucleotide pairs, like AG or CT in the Nandy scheme, will result in only one step instead of two. Bielińska-Wąż *et al.* [28] have shown that this can be accounted for by a mathematical method of using a weight parameter for each visit to the same location, but a number of researchers has been to propose different ways to represent DNA sequences graphically that reduces or removes this degeneracy. An extensive coverage of these methods can be found in Nandy, Harle and Basak [16], but we may mention here that one of the first proposals to reduce the degeneracy was the scheme of Guo, Randić and Basak [29] where the unit vectors for the four bases were aligned at a small angle to the cardinal directions. Yau *et al.* [30] used a two-quadrant representation in 2D space where A,G were inclined to the x-axis in the 4th quadrant and T,C were inclined to the x-axis but in the first quadrant, and the nucleotide count was recorded along the x-axis; this generated a DNA graph extending in the positive x-direction and had no degeneracy. He *et al.,* [31] proposed to characterize a DNA sequence by their chemical (amino, keto), structural (purine, pyrimidine) and bond strengths (weak, strong) and plotted this set of three reduced sequences as characteristic curves that extended along the x-axis with nucleotide number thus avoiding degeneracy altogether. Randić proposed several constructs, among them a 4 horizontal line scheme [32] where the four bases were plotted in order of the sequence along four lines parallel to the x-axis and placed unit distance apart wile the nucleotide number was again counted along the x-axis, a compact "worm curve" representation [33], four-color maps [34], "spectrum-like" curves [35] among others which reduced or eliminated the degeneracy inherent in the classical 2D approach.

3D and higher dimensional representations have been proposed to more faithfully reproduce the features of a DNA sequence or enable more accurate calculations. Hamori and Ruskin [25] had originally proposed a 5D model where the four bases were plotted in four dimensions and the fifth was for nucleotide count, but since this was difficult to visualize he had moved to the 3D H-curve representation. 3D representations and variations were also proposed by Randić, Vracko, Nandy and Basak [36], and Li and Wang [37] to name a few. A 4D method was proposed by

Chi and Ding [38], a 6D method by Liao and Wang [39] and an 8D method by Liu and Wang [40]. The interested reader can refer to the reviews [*e.g.* Ref 16] and the literature for details of these interesting developments.

NUMERICAL CHARACTERIZATION OF DNA SEQUENCES

Thus, the study of DNA sequences is facilitated in many ways by graphical representation, but making intra- and inter-sequence comparison becomes meaningful when the similarities and differences can be quantified in some manner. The difficulty is that since the graphical plots are composed of a set of discrete points, one has to apply either novel geometrical methods or use graph theory where the points are considered as nodes and the connections between the nodes as edges. We describe below first the geometrical methods and then the graph theoretic methods.

Geometrical Methods for Numerical Estimations

Two techniques were devised, one for intra-sequence comparison and another for inter-sequence comparison. For the variations within a sequence arising from the base distribution, we had observed [27] that coding regions of mammalian gene sequences appeared as a dense cluster of points in the 2D graphical representations implying high degree of mixing of the four bases in almost equal proportions, whereas the non-coding regions that were A-T or G-C rich usually appeared as long filaments. We therefore devised a cluster density measurement by enclosing such regions in a square grid and dividing the number of points in the grid by the area of the square. This was complemented by an inverse displacement method and a fractal coefficient method to numerically assess the differences between these two types of regions. Analysis of 386 introns (non-coding regions of a gene) and exons (coding regions) of 35 genes from various species by these measures showed [27] that (a) cluster density of non-coding regions are very small and fall off exponentially rapidly, (b) the cluster density of coding regions grows to about 0.8 per unit area and falls off gradually, (c) exons of evolutionarily later genes have higher cluster densities, (d) cluster densities of intronless genes like the phage M13 genome or the bacteriophage lambda are very low, closely paralleling intron densities and (e) more recent genes show greater fragmentation and smaller lengths of the exons. The cluster density measure also enabled us to propose a way of predicting protein coding regions in new DNA sequences [41] and was used to analyse the human chromosome 3 contig 7 and predict existence of several genes [42].

Gates [6] had proposed a Manhattan distance computation to compare two or more sequences, but this method is suitable for equal length sequences, whereas gene sequences are not generally of equal lengths. To study similarities and dissimilarities of genes from various species we devised a new and different methodology, which was reported for the first time in the First Indo-US Workshop 1998 [10] and published the following year [11], as mentioned earlier. Since we have in the 2D graphical representation a set of discrete points comprising each gene sequence, we defined a function to describe the sequence as

$$C_{map} \sim C(S_0, S_1, S_2, \ldots)$$

where S_0 is the zeroth-order term representing the coordinates x_f, y_f of the end points, S_1 is the linear term representing the first-order moments about the two axes, S_2 the second-order term representing the variance about the mean, S_3 the third order term representing the skewness, *etc.*, all of which taken together became a descriptor for the sequence. For the initial presentation we computed the first order moments as weighted center of mass only and defined a graph radius, g_R, the distance of the weighted center of mass from the origin, for each sequence and a Δg_R to estimate the difference between two sequences plotted on the same scale; this scheme gave a reasonable fit to the dispersion of the beta globin genes from various species [11]. Because of the cumulative nature of the sequence plots, differences in base distributions will lead to progressively increasing differences in the plots. Closely related sequences with less mutational changes between them will have smaller Δg_R while unrelated sequences can be expected to lead to larger values of the Δg_R. As remarked by the authors, this method could clearly be generalized to apply to the case of protein and other sequences where one may represent the sequences in a multidimensional hyperspace with a view to eventually develop phylogenetic trees. These techniques have been used by several authors (*e.g.*, [43, 48]).

Bielińska-Wąż *et al.* [28] have computed the moments to various higher orders in a 2D dynamic graph with statistical moments of mass–density distributions as new descriptors. Computing the moments for a set of histone genes, they showed that the larger number of descriptors improved the characterization of the object and different aspects of the DNA could be compared separately while retaining the simplicity of the 2D graphs. Nandy and Nandy [44] showed that the g_Rs were quite sensitive measures where base composition or base arrangement differences caused the g_R to change and that two or more sequences will not have the same g_R value except in some pathological cases.

Graph Theoretic Method for Numerical Characterization of DNA Sequences

The graph theoretic method arose out of deliberations after the first presentation of the 2D graphical representations in Duluth in 1999. The method described in the paper by Randić, Vracko Nandy and Basak [36] was to first represent a DNA sequence graphically in a 3D Cartesian grid and then convert the points to elements of a matrix by computing the ratio of Euclidean distance to graph theoretic distance between all possible pairs of points taken systematically. Matrix methods are well studied and have well recognized properties. The D/D matrix generated by the distance measures was analysed to yield a set of eigenvalues with the leading eigenvalue being taken as invariant of the matrix and therefore of the sequence. Differences between the leading eigenvalues of various gene sequences could then be taken as indicative of their evolutionary distances, although this seminal paper limited itself to computation on the basis of the first exons of 11 beta globin genes only. The interesting point to note is that this paper led to generation of intense interest among researchers and many different ways of representing DNA sequences and computation of evolutionary distances subsequently ensued (see review Nandy *et al.* [16]). Authors such as Randić *et al.* [33], Randić [35], He and Wang [45], Song and Tan [46] and many others proposed different ways to graphically represent DNA sequences and convert the plots to mathematical objects, and derive leading eigenvalues as invariants of the sequences. For example, He and Wang [45] reduced the DNA sequences to a set of three sequences based on their structural, chemical and bonding nature and devised a vector of the three leading eigenvalues of the matrices associated with each of the reduced sequences which they proposed as being characteristic sequences of the original DNA sequence. Distances between two sequences then were computed by determining the distances between the end points of the two vectors. Song and Tan [46] similarly devised a 24-component vector characterizing a sequence, others came up with other ways of computing the inter-sequence distances based on vectors devised out of the matrix eigenvalues. Such matrix invariants from their own representations were used by Liao *et al.* [47], Wang *et al.* [43], Liao *et al.* [48] and others to draw phylogenetic trees for mitochondrial genes, SARS coronavirus genomes, *etc.* The graph theoretic method, however, does not seem to have been applied so far to determine specific features within a sequence.

GRAPHICAL REPRESENTATION AND NUMERICAL CHARACTERIZATION OF PROTEIN SEQUENCES

Graphical Representation of Protein Sequences

Developments in the graphical representation and numerical characterization of DNA sequences raised the possibilities of using similar analysis of protein sequences, albeit with difficulty arising from the fact that now we have to contend with 20 amino acids making up a protein chain whereas DNA sequences were made up of only four nucleotides. Although Meeta Rani [49] had shown as early as 1998 the presence of statistical self-affinity, a kind of self-similarity, in protein sequences that implies a fractal nature, graphical representation methods for proteins drew attention with the paper of Randić [50]. The basic idea here was to start with the CGR method of Jeffrey to plot a RNA sequence drawing triangles for every triplets of bases, *i.e.*, the genetic codes, and taking the centers of each such triangle as corresponding to the residue the triplet would code for. Thus starting with the mRNA, this method generates a CGR-equivalent 2D graphical representation for the protein sequence. Randić *et al.* [51] carried the method further to construct a zigzag curve for the A-chain of human insulin which allows a direct conversion of a protein sequence into a numerical sequence of (x,y) coordinates that can be used subsequently for construction of the graph-theoretic matrices and sequence invariants. The technique was refined to remove some arbitrariness that were inherent in the 2D scheme by converting the 2D graph to a 3D graphical representation where the triplets were assigned to the corners of a tetrahedron structure; although visual inspection of the graphical patterns had to be discarded in this scheme, the authors claimed that construction of graph invariants in this manner was more accurate and unique.

Randić *et al.* [52] proposed a Magic Circle representation where the protein sequence graph starts from the centre following the sequence by moving half way towards the corresponding amino acids which are positioned equally spaced on the circumference of a unit circle. The result of the complete execution of the protein sequence within the circle produces a typical graph for a particular protein, except for large protein sequences which are often found to have lesser visual benefits.

Bai and Wang [53] considered the triplet codon concept and using a complex co-ordinate scheme constructed a purine–pyrimidine graph on the left half of the complex plane, with purines (A and G) in the first quadrant and pyrimidines (T and C) in the fourth quadrant. A protein sequence can then be drawn from the triplet codons extending along the x-axis allowing visual inspection of the trends and also from the co-ordinates generate graph-theoretic matrices and their leading eigenvalues as descriptors of the sequences.

Bai and Wang [54] next proposed a 3D graphical representation for protein sequences where the 20 amino acids are represented as end points in a dodecahedron embedded in the 3D space, *i.e.* each amino acid is represented at one of the vertices of the dodecahedron. This allows construction of a sequence graph following the amino acids in the sequence where each point in the plot can be considered as a node of the graph, from which one can again generate matrices and sequence invariants.

Liao *et al.* [48] used a 2D graphical representation method to compare 24 coronavirus sequences where the four cardinal directions were associated with particular properties of the amino acids. They classified the 20 amino acids of a protein sequence into four separate groups according to the chemistry of their R groups: amino acids A,V,F,P,M,I,L to the hydrophobic chemical group; amino acids D,E,K,R to charged chemical group; amino acids S,T,Y,H,C,N,Q,W to polar chemical group; and the G amino acid to glycine chemical group. Starting with the nucleotide sequence, this enabled them to construct three 2D graphs (one for each reading frame) for each gene sequence and compute a distance matrix. In a similar construction, Aguero-Chapin *et al.* [55] grouped the 20 amino-acids into four categories: acidic, basic, polar and non-polar and assigned the four groups to the four cardinal directions of a Cartesian frame to compute numcric descriptors of 108 sequences of polygalacturonases.

In recent years the field has progressed rapidly to numerically characterize protein sequences for application to different issues. González-Díaz and collaborators have extended these representations to the study of protein sequences [56] and applied to mass spectral data of proteins and protein serum profiles in parasites [57]. Gonzalez-Diaz has found that using different type of numerical indices derived from the protein 2D molecular graphics to perform QSAR studies is simpler than having to work with the protein 3D structures [58]. Integrated QSARs [59] developed using chemodescriptors for ligands and biodescriptors of a molecular entity connect structural information of drug molecules, DNA and RNA sequences or RNA secondary and protein tertiary structures. Basak *et al.* [60] using a new differential QSAR approach for study of dihydrofolate reductases (DHFR) from multiple strains of *Plasmodium falciparum* showed that DHFR from the wild strain is substantially different from four mutant strains of their study; this indicated that the protocols indicated in the paper can be used for the development of drugs to combat drug-resistant pathogens arising continuously in nature due to mutations. Nandy *et al.* [61] showed that their 20D graphical representation of protein sequences (explained later) was useful in generating phylogenetic relationships between sequences without necessity of multiple alignments and for determining conserved surface exposed stretches on viral proteins that could be useful in drug and vaccine designs [62].

We mention in passing that Randić [63], Basak and Gute [64] had developed mathematical techniques for analysis of proteomics data drawing parallels with DNA GRANCH techniques, but we do not go into any details about this topic in this review.

A detailed review of graphical representations of proteins including of proteomics has been made by Randić *et al.* [65].

APPLICATIONS OF GRANCH TECHNIQUES

Any new technique needs to be tested through applications to real problems and these methods of graphical representation and numerical characterization of bio-molecular sequences are no exception. The intense interest which these GRANCH techniques have evoked amongst researchers have led to many and varied applications which shows the wide applicability and great potential of the methods. We cover some of these applications in brief here, with a novel application to anti-virus drug targeting in slightly more detail.

As a natural application of the graphical representation of DNA sequences, consider the visualization of patterns in base arrangements that are otherwise difficult to see in the normal character representation. As already mentioned, Gates [6] had noticed large scale repeats that were revealed by his 2D graphical plots and Nandy [5] showed that conserved genes have shapes on the 2D maps that are similar across species. From a detailed analysis of the graphical plots of families of conserved gene sequences that these altered with evolution such that the constituent bases appear to tend to greater homogeneity in base composition and higher complexity in base composition in the protein coding sequences [66].

Also, visual inspection of the graphical plots can enable new insights into similarities of different stretches of DNA sequences. Larionov *et al.* [67] had thus found long range palindromes in the mouse and human chromosomes. Nandy, Gute and Basak [68] reported on a stretch of the H5N1 avian flu neuraminidase gene that appeared to be well conserved among the various strains of the avian flu and reported on the possibility of using this site as drug or vaccine target so that these can be effective over many mutational changes (see below). Further observations and numerical computations on over 600 H5N1 neuraminidase sequences showed the wide dispersion and mutations of the gene sequences and especially the possible exchange of structural parts of the genes, which was a new observation for this type of virus [69].

Based on the observations of the plots of several conserved gene sequences, Nandy [70] showed that the base arrangements of these sequences could be conceived as bound by a characteristic function of the instantaneous population of the four bases as one moves along the sequence. Based on spot mutations, Nandy proposed an equation connecting the instantaneous values of the purine and pyrimidine population asymmetries. It was hypothesized that this may have important consequences for genetic engineering since it implied that stability of engineered gene sequences required these constraints to be followed.

An important issue in molecular biology is identification of protein coding regions in DNA sequences. Nandy showed from the 2D graphical representations that exon and intron regions of mammalian genes showed distinctly different patterns and how these could be used to discriminate between the exons and introns [41]. This method was used by Ghosh *et al.* [42] to analyse a newly sequenced human chromosome III contig 7 DNA to identify coding regions and predict, using web-based tools, possible genes in the sequence. He, Li and Wang [31] used the numerical characterization of characteristic sequence representation of He and Wang [45] to suggest a protein coding gene finding algorithm specific for the yeast genome and found that the total number of protein coding genes in the yeast genome was 5897, which matches very well with estimates from other methods of 5800–6000. Discrimination between protein coding and non-coding regions was also proposed on an entropy-based approach [71] differentiating the DNA sequence into three subsequences and using Shannon's formula.

Wiesner and Wiesnerova [72] did an interesting application of GRANCH techniques to study plant germplasm identificators. For their study of multiallelic marker loci from 18 *Begonia* × *tuberhybridas,* they used a 2D random walk digitization of the DNA sequences by three transform classes according to the prescription of Bai *et al.* [73] and derived invariants from the respective matrices to compute sequence similarities and dissimilarities. Principal component analysis done to compare the 18 marker loci to the DNA invariants found statistical correlations between the genetic diversity of the marker loci and the random walk invariants. Based on their results, the authors concluded that "DNA walk representation may function as an efficient pre-scanning procedure, which can predict allele-rich genomic loci as highly informative DNA markers solely using the information from their primary sequence."

One of the early observations was that these graphical and numerical techniques allowed comparison of DNA and protein sequences without having to do multiple sequence alignment since here we are dealing with numbers derived from the method rather than having to compare base by base or residue by residue. Almost all proposals of schemes for graphical representations have computed distances between DNA sequences to determine similarities and dissimilarities without multiple sequence alignment and obtained fairly good, though not uniform, results. For example, Liao *et al.* [47] used a 2D graphical representation proposed by Liao [74] to derive a phylogenetic tree from the elements of a similarity matrix for eleven mitochondrial gene sequences without having to go through any multiple alignment procedure. They constructed a 2x2 covariance matrix of the weighted centers of masses from the co-ordinates of each base of a sequence and computed the Euclidean distance between pairs of sequences to obtain their similarity/dissimilarity matrix.

Liao *et al.* [48] also investigated the phylogeny of 24 SARS coronavirus genomes by their 2D graphical representations for protein sequences where they could draw three plots for each sequence by considering the three reading frames. These generate three eigenvalues for each sequence which are then used to compute a distance matrix from which they could diagrammatically show the relationships of various strains of the virus. In another exercise, Bai and Wang [75] compared nine different neurocan nerve protein sequences in their 3D dodecahedron representation scheme. A direct comparison of these protein sequences through alignments is difficult since these protein sequences have different lengths. Using 10- and 35-component vectors from their model, they compared the distances between end-points of the vectors corresponding to each of the nine genes and built phylogenetic trees.

Nandy *et al.* [61] used their 20D representation of protein sequences to compute distances between sequences of the families of globin, the rat and human voltage gated sodium channel alpha subunit and their phylogenetic relationships. It is to be noted that deriving phylogenetic trees from protein sequences is usually a difficult matter when the sequences are of different lengths; but with the GRANCH techniques where D/D and other matrices can be computed for any length sequence and only the eigenvalues compared, the sequence length differences become irrelevant.

Jayalakshmi *et al.* [76] generalized these methods to compute alignment free sequence comparison using n-dimensional similarity space.

H Gonzalez-Diaz and his group have used 2D graphical methods for extensive work in the bio-medicine field. Based on pseudo-folding Lattice Network (LN) and Star-Graphs (SG) topological indices they proposed two DNA promoter QSAR models to predict promoter sequences in the function regulation of several mycobacterial pathogens [57]. Aguero-Chapin *et al.* [55] using their reduced four groups of amino acids on a 2D Cartesian co-ordinate framework computed numerical descriptors for 108 polygalacturonases through a Markov model and were able to discriminate between these and other proteins and predict polygalacturonase activity of a new protein.

Comparison of RNA secondary structures are important to understand their catalytic properties. Bai *et al.* [77] considered a 3D graphical representation of RNA characteristic sequences taken 2 bases at a time to compare similarities and dissimilarities in viral RNAs of nine species. They computed three modular lengths and three phases for each sequence from which they constructed a 6-component vector characteristic for each viral sequence. Two sequences were considered to be similar if their vectors pointed in the same direction and difference between sequences could be quantified by computing the Euclidean distance between the end points of the two vectors: the bigger the distance the less similar the sequence. The resultant difference table showed how methods such as these could be used to do cluster analysis without having to use alignment tools which are time consuming and requires several assumptions. In another instance, Gonzalez-Diaz *et al.* [78] has computed 2D-RNA coupling numbers by adapting the 2D graphical representation method for DNA sequences.

Application of GRANCH Technique to Viral Proteins

A novel application of GRANCH techniques was proposed by Ghosh *et al.* [62] to determine targets on viral proteins for drug and vaccine design. Viruses are known to mutate very fast and therefore become resistant to drugs and vaccine sin short time scales; the virulence of the avian flu led to an apprehension that it might mutate to a form that would enable human to human transmission of the disease and thus cause widespread infection and possible death as had happened in the case of the Spanish Flu outbreak in 1918 when millions died. New drugs and vaccines, especially ones that could be readily moved from table to dispensaries were badly needed.

We had already noticed in early 2006 that certain parts of the neuraminidase gene appeared to be fairly well conserved [68]. The neuraminidase, along with hemagglutinin, are surface proteins that enable the viral particles to enter and leave the human cells where they proliferate, and of these the neruaminidase is the preferred target of the currently available drug, Tamiflu. We therefore determined to search the neuraminidase protein for surface residues that were well conserved.

Our procedure was to scan a small stretch of the neuraminidase protein sequences of 600+strains of the H5N1 virus and then slide the window by one base and scan again to calculate the protein graph radius in our 20D representation system. We know that these radii are very sensitive to any changes in the sequence, so equal values of the radii in one stretch over all the strains implied that this stretch was conserved. By covering the entire sequence for all strains we could get a good profile of regions of least variability.

The next step was to determine which parts of the sequences were surface exposed. There are several on-line engines available to scan a sequence and assign parameters to predict the degree of probability that certain portions were surface exposed. Matching these predictions with the hard facts we had on low variability we were able to identify six regions in the neuraminidase protein that were surface exposed and largely stable to mutational changes. These included the peptide we had identified earlier as being exceptionally stable. However, in a recent report on influenza virus RNA structure [79], it has been noted that the structures seen in the crystalline form may be one of several structural forms *in vivo* and confirmation will need to be experimentally determined.

The results of the analysis on the H5N1 neuraminidase protein sequence was published in 2010 [62]. Subsequently we have done a similar study on the VP7 protein of the rotavirus, a mainly tropical disease responsible for causing deaths to over half a million children every year. We identified four regions on the VP7 which appeared to be surface situated and quite stable. Our findings were reported at the 2nd Mathematical Chemistry Workshop at Bogota, Colombia in 2010 [13] and the Indian Biophysics Conference, Delhi 2011 [80].

PROBLEMS AND PROSPECTS OF THE GRANCH TECHNIQUE

While a number of applications have shown the usefulness of the GRANCH approach to analyzing DNA, RNA and Protein sequences, this remains as yet a nascent field where many issues need to be looked into and problems resolved for the potential to be well realized. An early indication of some of these areas was outlined some years ago [81], but they are worth recapitulating along with some more issues that may bear scrutiny.

The intense interest in this field of graphical representation and numerical characterization of bio-molecular sequences have led to proposals for a vast array of models for depicting the sequences, some real and some virtual, more for DNA sequences, less for protein and RNA sequences. This has almost become an intellectual sport, with new ideas being propounded on regular basis, generally without a proper rationale for yet another method or critical comparison with earlier proposals. What appears to be lost in the process is the target: How useful are these representations to the practicing biologist? Critical to this issue is the problem of determining the domains of applicability of the various representations if different, *i.e.*, which model is best suited to address which classes of problems. As of now, the vast majority of proposals have addressed themselves to comparisons of similarity and dissimilarity, but as we have seen in the previous section, the issues that we can address and which biologists need answers for are more varied.

From the applications made to date, the 2D graphical representations where the sequence data are easily viewed have generated the most interest. Even aside for the global characteristics revealed by the investigations of Jeffreys [3] and Peng *et al.* [3], the particular patterns of intron and exon segments [9] or characteristic curves of He *et al.* [31] have led to models to predict protein coding regions, determination of long-range palindromes [67], identification of target segments for vaccine development for viral proteins [62] and determination of allele-rich genomic loci for plants [72] among other applications have been based on 2D representation schemes. Hamori had identified regions of sharp changes in base compositions from his 3D H-curves [25], but for almost all other 3D, 4D and higher dimensionality representations applications have been restricted to sequence similarities and generation of phylogenetic trees.

The mathematical technique involved in generating the descriptors and characterizers for DNA sequences are still at a preliminary level. While the first moments in the geometrical method for generating descriptors have generally yielded reasonable results in comparing intra- and inter-genic sequences, attempts to calculate higher moments to increase the accuracy and effectiveness of these descriptors have only lately begun [28, 82]. The leading eigenvalues from the Euclidean and graph theoretic distance ratios matrix have so far been used mainly to compute inter-sequence distances; given the rigorous mathematics of matrix mechanics, it may be worthwhile to try and extend the applications to other areas. For the benefit of users of these methods, it would be useful to have a comparison of the geometrical and graph theoretic models to determine at what level the two could give comparable results.

In the case of 2D graphical representations using Cartesian co-ordinates, we had seen that gene sequences take characteristic shapes [5]. This raises the possibility that some day we could create an Atlas of Gene Sequences where samples of each gene would be depicted and the descriptor parameters listed for easy reference and rapid visual identification. We have described quantitatively the gross features of the graphical plots in the 2D representations by using the first moments in a geometrical method [10, 11]; better descriptors can be determined through higher order moments [28] to quantify the curvature, skewness and other properties. These, and the leading eigenvalues from the graph theoretic approach, could be considered as a list of parameters describing the sequence, akin to the quantum numbers that are used to describe elementary particles. Such a scheme then provides a method to electronically store, retrieve and compare data between various sequences more efficiently, especially with a view to quickly scan newly sequenced DNA, RNA and proteins to determine the genes and functions.

We have considered the moments calculated from the geometrical approach to 2D graphical representations as numerical "descriptors" of the DNA sequence and taken tentative steps to enhance the number of descriptors of a sequence by computing higher order moments to more completely describe the sequences. In the matrix method applied to different varieties of graphical representation, leading eigenvalues arising from the matrices have been taken as "invariants" of the sequence in the strict mathematical sense. However, the concept of invariants derived from these matrix methods of numerically characterizing DNA sequences may require some modification to account for the fact that DNA sequences constantly change due to mutations in the bases. The vast majority of these changes do not affect the functioning of the protein or the enzyme coded by the gene due to synonymous mutations in the coding segments or in the non-coding part; *e.g.* in the case of intronless gene like the neuraminidase of the avian flu H5N1 we had found [69] 447 out of total 682 sequences prevalent over the period 1997 to 2008 had undergone mutations in one or more bases in the gene, but even then, all of these variants coded for a functioning flu neuraminidase protein. For a beta globin gene, the common standard example of most graphical representation schemes, a sample from one person may differ by a base or more from the next person due to mutational changes. Determining an "invariant" from one sample sequence of these genes, while being mathematically precise, may not adequately express or characterize a gene sequence from a practical point of view. Perhaps a biologically more relevant measure would be a sampling of several such sequences and from them to compute an average eigenvalue with a standard deviation and derive a numerical to characterize the gene. In fact, in the absence of a sensitivity analysis or a standard deviation, it would be difficult to accept that the computations through leading eigenvalues of distances between several sequences that are only a few percentage points apart could be statistically meaningful. The descriptors are no exception either.

Once these basic issues are attended to, the GRANCH techniques can become a useful tool in the medicare field. Since the computations of the numerical descriptors/characteristics are quite simply done, they can be incorporated into the DNA sequencing schemes so that there will be automatic computations of, *e.g.*, g_R and p_R values which would enable the physician to immediately ascertain the presence of any harmful genetic disorders; Huntington's potential to degenerate into a disease for the patient, or some similar genetic problem areas could be easily read out as the genome is sequenced provided we know the characteristic locus and have a standard genome, for example the readout for a normal person from the family, available for comparison. The viral application already discussed in detail in the previous section could be automated and extended to other viruses and bacterial genomes to promote new generation of drugs and vaccines. The researches of Gonzalez-Diaz [55, 58, 78] and Basak [59, 60] are already pointers to new directions.

Many potential application areas remain to be explored. Since the numerical descriptors mentioned previously are seen to be quite sensitive to changes in base composition and distribution, the potential exists to devise schemes to index various aspects related to the bio-molecular sequences. Initial attempts have been made to index toxic chemicals that have damaging effects on DNA sequences [83], and to index SNP gene sequences measured against some standard sequences [84]. However, these need to be refined and made more useful for the confidence to be generated for their use in laboratory situations.

One area that requires in depth study is how to address non-contiguous sequence segments. For example, in the case of epitopes, it is found that there could be continuous epitopes and discontinuous epitopes; in the latter case the folded protein brings residues from different parts of the amino acid sequence close together, which then become sites for the antibodies to act upon. The methods delineated so far for g_R and p_R or leading eigenvalue evaluation require contiguous span of the bases or residues for the numbers to be calculated. One way to circumvent this difficulty is to work on small segments of the sequence at a time as had been done in Ref. [62]. However, this is time consuming and inefficient, and more improved methods to be able to focus on regions of interest and calculate a minimum number of the parameters could offer better rewards.

CONCLUSION

In summary it is apparent that graphical representation and numerical characterization of molecular sequences hold far-reaching potential of rapidly analyzing the sequences to extract numerous information. It opens up new ways to look at these sequences, and to gain new insights such as long range palindromes, fractal properties and intra-purine intra-pyrimidine relationships not seen by any other means. It allows one to compute many aspects of biological and medicinal interest and provide novel methods of tackling old problems; we have seen examples of gene identification, analysis of evolutionary trends and generation of phylogenetic trees, identification of conserved sites on viral proteins for drug and vaccine targeting, predict promoter sequences and new properties of polygalacturose proteins, among others and many possibilities remain unexplored, or barely scratched. Still, from plants to viruses, from mammalian genes to mitochondrial genomes, a varied series of applications have been formulated. Although many issues doubtless remain yet such as handling non-contiguous stretches of bases and residues like discontinuous epitopes, it is apparent that the GRANCH techniques hold a lot of promise for a new direction in molecular analysis.

ACKNOWLEDGEMENTS

I would like to express my gratitude to the Department of Science and Technology, Government of India for funding my research in its initial stages and to the Natural Resources Research Institute (NRRI), University of Minnesota, Duluth, USA for sponsoring my first trip to the NRRI in 1999. We are also thankful to the Consortium for Bioinformatics and Computational Biology, University of Minnesota, for funding the project awarded to Subhash Basak entitled "*Numerical Characterization of DNA Sequences and Investigations into Structural aspects of Gene Additions and Mutations*" that supported my second trip to NRRI in 2005 to do further research on the mathematical characterization of DNA sequences.

CONFLICT OF INTEREST

The author confirms that this chapter contents have no conflict of interest.

ABBREVIATION

GRANCH = Graphical Representation and Numerical Characterization

REFERENCES

[1] Nandy, A. Recent investigations into characteristics of long DNA sequences. *Ind. J. Biochem. Biophys.*, 1994, *31*, 149-155.

[2] Jeffrey, H. J. Chaos game representation of gene structure. *Nucleic Acids Res.*, **1990**, *18*, 2163–2170.

[3] Peng, C.-K.; Buldyrev, S.V.; Goldberger, A.L.; Havlin, S.; Sciortino, F.; Simons, M.; Stanley, H.E. Long range correlation in nucleotide sequences. *Nature (London)*, **1992**, *356*, 168-170

[4] Voss, R. Evolution of long-range fractal correlations and 1/f noise in DNA base sequences. *Phys. Rev. Lett.*, **1992**, *68*, 3805-3808.

[5] Nandy, A. A new graphical representation and analysis of DNA sequence structure: I. Methodology and application to globin genes. *Curr. Sci.*, **1994**, *66*, 309–314.

[6] Gates, M.A. A Simple way to look a DNA. *J. Theor. Biol.*, **1986**, *119*, 319–328.

[7] Nandy, A. Graphical representation of long DNA sequences. *Curr. Sci.*, **1994**, *66*, 821.

[8] Nandy, A. Graphical analysis of DNA sequence structure: III. Indications of evolutionary distinctions and characteristics of introns and exons. *Curr. Sci.* **1996**, *70*, 661–668.

[9] Nandy, A. Two dimensional graphical representation of DNA sequences and intronL-exon discrimination in intron-rich sequences. *Comput. Appl. Biosci.*, **1996**, *12*, 55-62.

[10] Raychaudhury, C.; Nandy, A. Indexation Schemes and Similarity Measures for Macromolecular Sequences. Paper presented at the Indo-US Workshop on Mathematical Chemistry, Shantiniketan, West Bengal, India, December 1998.

[11] Raychaudhury, C.; Nandy, A. Indexing scheme and similarity measures for macromolecular sequences, *J. Chem. Inf. Comput. Sci.* **1999**, *39*, 243–247.

[12] Randić, M.; Vracko, M.; Nandy, A.; Basak, S.C. On 3–D representation of DNA primary sequences,.*J. Chem. Inf. Comput. Sci.* **2000**, *40*, 1235–1244.

[13] Nandy, A. A decade and more of GRANCH: Novel analysis of DNA and Protein sequences through Graphical Representation and Numerical Characterization techniques. Second Mathematical Chemistry Workshop of the Americas, Universidad de los Andes, Bogota, Colombia, July 19 - 24, 2010.

[14] Roy, A.; Raychaudhury, C.; Nandy, A. Novel Techniques of Graphical Representation and Analysis of DNA Sequences - ***A Review***. *J. Biosc.*, **1998**, *23*(1), 55-71.

[15] Berger, J.A.; Mitra, S.K.; Carli, M.; Neri, A. Visualization and analysis of DNA sequences using DNA walks. *J. Franklin Instt.*, **2004**, *341*, 37–53.

[16] Nandy, A.; Harle, M.; Basak, S.C. Mathematical descriptors of DNA sequences: development and applications. *Arkivoc,* **2006**, *9*, 211-238.

[17] Nandy, A.; Basak, S.C. New Approaches to Drug-DNA Interactions Based on Graphical Representation and Numerical Characterization of DNA Sequences, *Current Computer-Aided Drug Design*, **2010**, *6*, 283-289.

[18] Ghosh, A.; Nandy, A. Graphical representation and mathematical characterization of protein sequences and applications to viral proteins. *Adv. in Prot. Chem. and Struc. Biol.*, **2011**, *83*, 1-43.

[19] Chang, H.T. DNA Sequence Visualization. In *Advanced data mining technologies in Bioinformatics*, Hui-Huang Hsu Ed.; Idea Group Publishing: Hershey, PA, USA, **2006**; pp 63-84.

[20] Chun Li; Zhi-fu Wang, Z-f. Charcaterizations of DNA Primary Sequences. In *Trends in DNA Research*, Woods, C.R. Ed.; Nova Science Publishers, Inc.: New York, USA, **2006**; pp 123-144

[21] Todeschini, R.; Consonni, V. *Molecular Descriptors for Chemoinformatics*, Methods and Principles in Medicinal Chemistry, Mannhold, R.; Kubinyi, H.; Folkers, G. Series Eds.; Wiley-VCH,Weinheim, Germany, **2009**, Vol 41.

[22] Burma, P.K.; Raj, A.; Deb, J. K; Brahmachari, S.K. Genome analysis: A new approach for visualisation of sequence organisation in genomes. *J. Biosci.*, **1992**, *17*, 395-411.

[23] Dutta, C.; Das, J. Mathematicalc haracterisationo f chaos, game representation: New algorithms for nucleotide sequence analysis. *J. Mol. Biol.*, **1992**, *228*, 715-719.

[24] Baranidharan, S.; Sankaranarayanan,B.; Brahmachari, S.K. Chaos game representation of similarities and differences between genomic sequences. *Int. J. Genome Res.*, **1994**, *1*, 309-319.

[25] Hamori, E.; Ruskin, J. H curves, a novel method of representation of nucleotide series especially suited for long DNA sequences. *J. Biol. Chem.*, **1983**, *258*, 1318–1327.

[26] Leong, P.M.; Morgenthaler, S. Random walk and gap plots of DNA sequences. *Comput. Applic. Biosc.* **1995**, *11*, 503–507.

[27] Nandy, A. Graphical analysis of DNA sequence structure: III. Indications of evolutionary distinctions and characteristics of introns and exons. *Curr. Sci.* **1996**, *70*, 661–668.

[28] Bielińska-Wąż, D.; Nowak, W.; Wąż, P.; Nandy, A.; Clark, T. Distribution moments of 2D-graphs as descriptors of DNA sequences. *Chem. Phys. Lett.*, **2007**, *443*, 408–413.

[29] Guo, X.; Randić, M.; Basak, S.C. A novel 2-D graphical representation of DNA sequences of low degeneracy. *Chem. Phys. Lett.,* **2001**, *350,* 106-112.

[30] Yau, S. S. T.; Wang, J.; Niknejad, A.; Lu, C.; Jin, N.; Ho, Y. DNA sequence representation without degeneracy. *Nucleic Acids Res.,* **2003**, *31*, 3078.

[31] He, P.; Li, C.; Wang, J. Finding Protein Coding Genes in the Yeast Genome Based on the Characteristic Sequences, *Internet Electron. J. Mol. Des.* **2005**, *4*, 613–624,

[32] Randić, M.; Vracko, M.; Lers, N.; Plavsic, D. Analysis of similarity/dissimilarity of DNA sequences based on novel 2-D graphical representation. *Chem. Phys. Lett.,* **2003**, *371*, 202–207.

[33] Randić, M.; Vracko, M.; Zupan, J.; Novic, M. Compact 2-D graphical representation of DNA. *Chem. Phys. Lett.,* **2003**, *373*, 558.

[34] Randić, M.; Lers, N.; Plavsic, D.; Basak, S. C.; Balaban, A. T. Four-color map representation of DNA or RNA sequences and their numerical characterization. *Chem. Phys. Lett.*, **2005**, *407*, 205–208.

[35] Randić, M. Spectrum-like graphical representation of DNA based on codons. Acta Chim. Slov. **2006**, *53*, 477–485.

[36] Randić, M.; Vracko, M.; Nandy, A.; Basak, S.C. On 3–D representation of DNA primary sequences,. *J. Chem. Inf. Comput. Sci.* **2000**, *40*, 1235–1244.

[37] Li, C.; Wang, J. On a 3-D Representation of DNA Primary Sequences. *Combinatorial Chem. & High Throughput Screening* **2004**, *7*, 23.

[38] Chi, R.; Ding, K. Novel 4D numerical representation of DNA sequences. *Chem. Phys. Lett.,* **2005,** *407*, 63.

[39] Liao, B.; Wang, T. Analysis of Similarity/Dissimilarity of DNA Sequences Based on Nonoverlapping Triplets of Nucleotide Bases. *J. Chem. Inf. Comput. Sci.***,** **2004**, *44*, 1666.

[40] Liu, Y. Z., Wang, T. M. Vector representations and related matrices of DNA primary sequence based on L-tuple. *Math. Biosci.*, **2010**, *227*, 147–152.

[41] Nandy, A. Two-dimensional graphical representation of DNA sequences and intron-exon discrimination in intron-rich sequences, *Comput. Appl. Biosc.*, **1996**, *12*(1), 55-62.

[42] Ghosh, S.; Roy, A.; Adhya, S.; Nandy, A. Identification of New Genes in Human Chromosome 3 Contig 7 by Graphical Representation Technique. *Curr. Sci.* **2003**, *84* (12), 1534-1543.

[43] Wang, W.; Liao, B.; Wang, T.; Zhu, W. A Graphical Method to Construct a Phylogenetic Tree. *International J, of Quant. Chem.*, **2006**, *106*, 1998–2005.

[44] Nandy, A.; Nandy, P. On the uniqueness of quantitative DNA difference descriptors in 2D graphical representation models. *Chem. Phys. Lett.*, **2003**, *368*, 102–107.

[45] He, P.A.; Wang, J. Characteristic Sequences for DNA Primary Sequence. *J. Chem. Inf. Comput. Sci.* **2002**, *42*, 1080–1085.

[46] Song, J.; Tang, H. A new 2-D graphical representation of DNA sequences and their numerical characterization. *J. Biochem. Biophys. Methods* **2005**, *63*, 228.

[47] Liao, B.; Tan, M.; Ding, K. Application of 2-D graphical representation of DNA sequence *Chem. Phys. Lett.,* **2005**, *414*, 296-300.

[48] Liao, B.; Liu, Y.; Li, R.; Zhu, W. Coronavirus phylogeny based on triplets of nucleic acids bases. *Chem. Phys. Lett.*, **2006**, *421*, 313–318.

[49] Rani, M. Dynamics of protein evolution. *J Biosc.* **1998**, *23*(1), 47-54.

[50] Randić, M. 2-D Graphical representation of proteins based on virtual genetic code. *SAR & QSAR* **2004**, *15*, 147-157.

[51] Randić, M.; Zupan, J.; Balaban, A.T. Unique graphical representation of protein sequences based on nucleotide triplet codons. *Chem. Phys. Lett.*, **2004**, *397*, 247–252.

[52] Randić, M.; Butina, D.; Zupan, J. Novel 2-D graphical representation of proteins. *Chem. Phys. Lett.*, **2006**, *419*, 528–532.

[53] Bai, F.; Wang, T. A 2-D graphical representation of protein sequences based on nucleotide triplet codons. *Chem. Phys. Lett.*, **2005**, *413*, 458–462.

[54] Bai, F.; Wang, T. On graphical and numerical representation of protein sequences. *J. Biomol. Struct. Dyn.* **2006**, *23*, 537–546.

[55] Aguero-Chapin, G.; Gonzalez-Diaz, H.; Molina, R.; Varona-Santos, J.; Uriarte, E.; Gonzalez-Diaz, Y. Novel 2D maps and coupling numbers for protein sequences. The first QSAR study of polygalacturonases; isolation and prediction of a novel sequence from Psidium guajava *L., FEBS Letters* **2006***, 580***,** 723-730.

[56] Aguero-Chapin, G.; Varona-Santos, J.; de la Riva, G. A.; Antunes, A.; Gonzalez-Vlla, T.; Uriarte, E.; Gonzalez-Diaz, H. Alignment-free prediction of polygalacturonases with pseudofolding topological indices: experimental isolation from Coffea arabica and prediction of a new sequence. *J. Proteome Res.* **2009**, *8*, 2122–2128.

[57] Gonzalez-Diaz, H.; Prado-Prado, F.; Ubeira, F. M. Predicting antimicrobial drugs and targets with the MARCH-INSIDE approach. *Curr. Top. Med. Chem.* **2008**, *8*, 1676–1690.

[58] Gonzalez-Diaz, H.; Perez-Montoto, L. G.; Duardo-Sanchez, A.; Paniagua, E.; Vazquez-Prieto, S.; Vilas, R.; Dea-Ayuela, M.A.; Bolas-Fernández, F.; Munteanu, C.R.; Dorado, J.; Costas, J.; Ubeira F.M. Generalized lattice graphs for 2D-visualization of biological information. *J. Theor. Biol.* **2009**, *261*, 136–147.

[59] Basak, S.C.; Mills, D.; Gute, B. D.; Natarajan, R. Predicting pharmacological and toxicological activity of heterocyclic compounds using QSAR and molecular modeling. In *QSAR and Molecular Modeling Studies of Heterocyclic Drugs I*, Gupta, S. P. Ed.: Springer-Verlag, New York, **2006**, pp 39-80.

[60] Basak, S.C.; Mills, D.; Hawkins, D. M. Characterization of dihydrofolate reductases from multiple strains of Plasmodium falciparum using mathematical descriptors of their inhibitors. Chem. Biodivers., **2011**, *8*, 440–453.

[61] Nandy, A.; Ghosh, A.; Nandy, P. Numerical Characterization of Protein Sequences and Application to Voltage-Gated Sodium Channel Alpha Subunit Phylogeny. *In Silico Biol.* **2009**, *9*, 77-87.

[62] Ghosh, A.; Nandy, A.; Nandy, P. Computational analysis and determination of a highly conserved surface exposed segment in H5N1 avian flu and H1N1 swine flu neuraminidase. *BMC Structural Biology* 2010, **10**:6 doi:10.1186/1472-6807-10-6.

[63] Randić, M. On Graphical and Numerical Characterization of Proteomics Maps. *J. Chem. Inf. Comput. Sci.* **2001**, *41,* 1330-1338

[64] Basak, S. C.; Gute, B. D. Mathematical biodescriptors of proteomics maps: background and applications. *Curr. Opin. Drug Discov. Devel.*, **2008**, *11*, 320–326.

[65] Randić, M.; Zupan, J.; Balaban, A.T.; Vikic-Topic, D.; Plavsic, D. Graphical Representation of Proteins. *Chem. Rev.* **2011**, *111,* 790–862.

[66] Nandy, A. Investigations on Evolutionary Changes in Base Distributions in Gene Sequences, *Internet Electron. J. Mol. Des.* **2002**, *1*, 545–558, http://www.biochempress.com.

[67] Larionov, S.; Loskutov, A.; Ryadchenko, E. Chromosome evolution with naked eye: palindromic context of the life origin. Chaos **2008**, *18*, 013105.

[68] Nandy, A.; Basak, S.C.; Gute, B.D. Graphical representation and numerical characterization of H5N1 avian flu neuraminidase gene sequence. *J. Chem. Inform. Model.* **2007**, *47,* 945-951.

[69] Ghosh, A.; Nandy, A.; Nandy, P.; Gute, B.D.; Basak, S.C. Computational study of dispersion and extent of mutated and duplicated sequences of the H5N1 influenza neuraminidase over the period 1997–2008. *J. Chem. Inf. Model.*, **2009**, *49*, 2627–2638.

[70] Nandy, A. Empirical relationship between intra-purine and intra-pyrimidine differences in conserved gene sequences. PLoS ONE **2009**, *4*, e6829.

[71] Li, C.; Wang, J. Relative entropy of DNA and its application. *Physica A* **2005**, *347*, 465-471,

[72] Wiesner, I.; Wiesnerová, D. 2D random walk representation of *Begonia × tuberhybrida* multiallelic loci used for germplasm identification. Biologia Plantarum **2010**, *54*(2): 353-356.

[73] Bai, F-L.; Liu, Y-Z.; Wang, T-M. A representation of DNA primary sequences by random walk. *Math. Biosci.*, **2007**, *209,* 282-291.

[74] Liao, B. A 2D graphical representation of DNA sequence. *Chem. Phys. Lett.*, **2005**, *401*, 196–199.

[75] Bai, F.; Wang, T. On graphical and numerical representation of protein sequences. *J. Biomol. Struct. Dynamics*, **2006**, *23*(5), 537-545.

[76] Jayalakshmi, R.; Natarajan, R.; Vivekanandan, M.; Natarajan, G.S. Alignment-Free Sequence Comparison Using N-Dimensional Similarity Space, *Current Computer-Aided Drug Design*, **2010**, *6*, 290-296.

[77] Bai, F.; Zhu, W.; Wang, T. Analysis of similarity between RNA secondary structures. *Chem. Phys. Lett.* **2005**, *408*, 258–263.

[78] Gonzalez-Diaz, H.; Aguero-Chapin, G.; Varona, J.; Molina, R.; Delogu, G.; Santana, L.; Uriarte, E.; Podda, G. 2D-RNA-Coupling Numbers: A new computational chemistry approach to link secondary structure topology with biological function. *J. Comput. Chem.*, **2007**, 28, 1049-1056.

[79] Gultyaev,A.P.; Fouchier, R.A.M.; Osthoorn, R.C.L. Influenza Virus RNA Structure: Unique and Common Features, *International Rev. Immunol.*, **2010**, *29*, 533–556.

[80] Ghosh, A.; Chattopadhyay, S.; Chawla-Sarkar, M.; Nandy, P.; Nandy, A. Characterization of Conserved Regions in Rotaviral VP7 Proteins: A Graphical Representation Approach towards Epitope Prediction. In: *7th Asian Biophysics Association (ABA) Symposium & Annual Meeting of the Indian Biophysical Society (IBS).* Indian Habitat Center, New Delhi, India: ABAIBS-2011; 2011: 88.

[81] Nandy, A. Mathematical Characterization of DNA Sequences - Towards New Directions, CP963, Vol 1, *Computational Methods in Science and Engineering, Theory and Computation: Old Problems and New Challenges,* G. Maroulis and T. Simos, Ed. **2007**, pp. 596-602.

[82] Bielińska-Wąż, D. Graphical and numerical representations of DNA sequences: statistical aspects of similarity. *J. Math. Chem.,* **2011**, *49*, 2345-2407.

[83] Nandy, A.; Basak, S.C. Simple numerical descriptor for quantifying effect of toxic substances on DNA sequences, *J. Chem. Infor and Comput Sc* **2000**, *40*, 915-919.

[84] Nandy, A.; Nandy, P.; Basak, S.C. Quantitative Descriptor for SNP Related Gene Sequences. *Internet Electron. J. Mol. Des*., **2002**, *1*, 367–373

CHAPTER 6

Linear Regression, Model Averaging, and Bayesian Techniques for Predicting Chemical Activities from Structure

Jarad B. Niemi[1,*] and Gerald J. Niemi[2]

[1]Department of Statistics, Iowa State University, Ames, IA 50011, -1210, USA and [2]Natural Resources Research Institute and Department of Biology, University of Minnesota, 5013 Miller Trunk Highway, Duluth, MN 55811, USA

Abstract: A primary goal of quantitative structure-activity relationships (QSARs) and quantitative structure-property relationships (QSPRs) is to predict chemical activities from chemical structure. Chemical structure can be quantified in many ways resulting in hundreds, if not thousands, of measurements for every chemical. Chemical activities measures how the chemical interacts with other chemicals, *e.g.* toxicity, biodegradability, boiling point, and vapor pressure. Typically there are more chemical structure measurements than chemicals being measured, the so-called large-*p*, small-*n* problem. Here we review some of the statistical procedures that have been commonly used to explore these problems in the past and provide several examples of their use. Finally, we peek into the future to discuss two areas that we believe will see dramatically increased attention in the near future: model averaging and Bayesian techniques.

Keywords: AIC, Bayesian analysis, BIC, cross-validation, elastic net, *k*-means clustering, LASSO, model averaging, model selection, modeling, partial least squares, prediction, principal component analysis, principal component regression, regression, ridge regression.

INTRODUCTION

The science of quantitative structure-activity relationships (QSAR) has a varied but relatively recent history [1]. The central tenet of structure-activity relationships (SAR) is that *form follows function* and this idea has probably been in existence for ages. Yet, the quantitative aspect of QSAR from a computer-age perspective is relatively young; perhaps only in existence for about 30 to 40 years.

***Corresponding author Jarad B. Niemi:** Department of Statistics, Iowa State University, Ames, IA 50011-1210, USA; Tel: (515) 294-8679; Fax: (515) 294-4040; E-mail: niemi@iastate.edu

Subhash C. Basak, Guillermo Restrepo & José L. Villaveces (Eds)
Copyright © 2015 Bentham Science Publishers Ltd. Published by Elsevier Inc. All rights reserved.
10.1016/B978-1-68108-053-6.50006-5

Hence, this is a very young area of science and ripe for opportunities and advancement. The age of computers and our ability to compile, quantify, and analyze information is unprecedented.

The basic theory of QSAR is that the structure of a chemical determines its activity [2-4]. The mystery of chemicals and of chemistry is how structure or substructures are related with activity. Any change in chemical structure (*e.g.*, the addition of a methyl group or element) results in different chemical behavior. It is of great societal interest to predict how chemical activity changes with chemical structure. If we could do so, then more effective drugs can be developed as well as the development of more effective, but safer chemicals for societal use.

The emergence of computers has dramatically increased the use statistics to problems in chemistry. Before the modern computing age, the calculations for many of the statistical procedures were too time-consuming to perform by hand especially for large datasets. In addition, hand calculations are subject to considerable error.

Large datasets in QSAR and in the field of computational chemistry have emerged for both chemical structure and activity. For instance, many software programs are now available such as Molconn-Z [5], Polly [6], DRAGON [7], and CODESSA [8] that calculate measurements of chemical structure. A past limitation existed when chemical activity data were available, but there were few structure measurements available to allow predictions of those activities. Similarly, in the past the exploration of activity-to-activity correlation approaches used in predictive pharmacology and toxicology failed because experimental activity data were unavailable. As the computer and information age has emerged, there are many additional databases available that are based on standardized endpoints such as toxicology [9, 10], mutagenicity [11], and chemical activities [12].

Today, statistical applications are common in chemistry and they have a variety of names such as chemometrics and pattern recognition. Here our primary goal is to 1) summarize several statistical techniques that have been used extensively in the past 30 years, 2) explore the recent use and potential for Bayesian statistical analysis in QSARs, and 3) provide examples of these statistical techniques in past

QSAR studies. This chapter is not intended to be an exhaustive review of all statistical procedures used in QSARs, QSPRs, or other associated analyses on the relationships between chemical structure and their properties or activities.

A STATISTICAL GOAL

The databases that house chemical structure and property measurements are ever increasing; a fundamental issue is that there are typically more structure measurements than chemicals being measured. Even as more chemicals are added to the database, scientists will increase the number of ways we can measure them and the issue will remain. If we use p to refer to the number of different measurements taken and n to refer to the number of different chemicals in the database, then this situation is referred to as the large-p, small-n problem.

The QSAR goal discussed here is to use the structure and activities measurements on a set of chemicals to predict the unknown activities of a different set of chemicals for which structure measurements are available. We concentrate on a single chemical activity at a time, although the methods below could be used individually for each activity that requires prediction. Further, we restrict our attention to predicting activities that are continuous, *e.g.* boiling point, as opposed to properties that are categorical, *e.g.* mutagenicity. The latter maybe analyzed by methods such as logistic regression and discriminant analysis.

Throughout the following we will use the following notation:

- Y: $a\ n \times 1$ vector of chemical activity measurements; and
- X: $a\ n \times p$ matrix of chemical structure measurements.

where, typically, the first column of X is a vector of ones. The ith chemical has property measurement y_i and structure measurements x_i, the ith row of X. We are then typically interested in predicting the chemical property measurements of a new chemical, X^*, based on its chemical structure measurements, Y^*.

MODELING

To predict continuous chemical activities from measurements of chemical structure, we focus on the linear regression model. Although other methods such as generalized additive models and recursive partitioning allow more flexibility, they require a larger sample size, *n*, which is often not available.

Multiple Linear Regression

Multiple linear regression defines a model that has the form $Y = X\beta + \epsilon$, where β is a set of unknown regression parameters and the random deviation has $E(\epsilon) = 0$ and $V ar(\epsilon) = \sigma^2 I$ where I is an identity matrix of order *n*. The model says that the activity for chemical *i* is a linear combination of the structure measurements and a random error, *i.e.* $Y_i = \beta_1 X_{i1} + \beta_2 X_{i2} +::: + \beta_p X_{ip} + \epsilon_i$ where, again, $X_{i1}$is often set equal to 1 to provide a model intercept. From this relationship, it should be clear that if $\beta_j = 0$, then X_{ij} does not affect the activity and therefore the jth structure measurement is not important for determining the activity in this model. The ordinary least squares (OLS) estimate for the vector parameter β is $\hat{\beta}_{OLS} = (X'X)^{-1}X'Y$ which is found by minimizing the quantity $\| Y - X\beta \|_2 = (Y - X\beta)'(Y - X\beta)$. To then predict the property for a new chemical, we use $\hat{Y}^* = X^* \hat{\beta}_{OLS}$

Niemi *et al.,* [13] used multiple regression to develop a prediction model for octanol/water partition coefficient. The independent variables used were 70 variables algorithmically-derived from information content and from molecular connectivity indices [3]. The analysis used a best-subsets (see Section 4.2) regression model for a dataset of over 4,000 chemicals with measured values of octanol/water partition coefficients. Explained variation ranged from 63 to 90% among 14 different groups of chemicals; the groups were formed on the basis of the degree of hydrogen bonding. Both information content and molecular connectivity indices were equally as effective in the prediction equations.

This example of regression uses a combination of simple grouping of a large dataset of over 4,000 chemicals using a theoretical basis that degree of hydrogen bonding is related to octanol/water partitioning. There are a multitude of variations that have been applied to regression analysis in QSAR studies [14] and variations of multivariate techniques combined with regression for making predictions about chemical properties such as partial least squares regression [15, 16]. All of these approaches likely have merit because the chemical universe is diverse and simple changes in chemical structure can have profound changes in chemical properties. Most of these statistical approaches, however, should be viewed as exploratory techniques subject to extensive scrutiny, further experimental testing, and ultimately the development of mechanistic understanding for the relationship between structure and activity.

One area of prediction using regression in QSAR that has received some scrutiny is the statistics of validation. Several authors [17-19] suggest that many publications present a naive q^2 and provide an improved means to present validation for a predictive relationship. Furthermore, in a comparative study of principal components regression, partial least squares, and ridge regression, ridge regression out-performed the other two [20].

Two important assumptions exist for multiple linear regression that typically make its direct use in QSAR studies dubious. The first is that the number of observations must be larger than the number of structure measurements (small-p, large-n). The second is that the structure measurements are uncorrelated which is questionable when many structure measurements are made. We now introduce two other statistical approaches: principal component regression and ridge regression that are useful for regression analysis when there exists a large- p, small-n problem and when the structure measurements are correlated.

Principal Component Regression

Principal component regression (PCR) is a two-step procedure that initially utilizes principal component analysis (PCA) to select principal components and then performs multiple regression using the selected principal components. Principal component analysis (PCA) is a multivariate statistical technique that uses correlations between and among variables to identify new components that are linear combinations of the original variables [21, 22]. PCA is part of a family of statistical procedures (*e.g.*, factor analysis) that are used when there are a large number of variables, many of which are highly correlated. This is often the case with the algorithmically-derived variables used in QSAR such as regression when collinearity among independent variables violates statistical assumptions. Furthermore, in datasets where the number of independent variables is large relative to the number of chemicals (*n*) available in the dataset, then spurious correlations can be a problem. A relevant solution is to use a dimension reduction procedure like PCA to reduce the number of independent variables by eliminating pairs of variables that are highly correlated or using the principal components as new uncorrelated, independent variables in the analysis. If the principal components are used, it is often difficult to interpret the results so calculations of the correlations between the original variables and the principal components are useful.

For instance, Basak *et al.,* [23] used PCA for 151 topological indices for a training set of 220 compounds. About 60% of the variation in the 151 indices could be explained by the first principal component and more than 95% of the variation could be explained by the first 12 principal components. This indicated substantial redundancy among the topological indices. PCA allowed the number of independent variables to be reduced to 60 and subsequently used in further analysis of the dataset. In these cases where there are a large number of potential explanatory variables there is no option except to reduce the complexity of the problem by using a dimension reduction procedure like PCA or in combination with regression approaches [24].

Numerous additional examples of this type of procedure exist in the QSAR literature [1, 13, 25]. There are a wide variety of additional dimension reduction procedures available. As the name implies, their purpose is to reduce the dimensionality of the data to the essential and important dimensions. PCA is one of the most common forms and seeks to identify orthogonal factors that are very useful in analysis such as multiple regression that assumes orthogonality among the independent variables. More complex dimension reduction procedures use various mathematical variations of factoring the independent variables or rotations of the factor axes to increase the interpretation of the resulting variables, *e.g.*, varimax rotation.

This approach to principal components regression where PCA is run first on the independent variables alone followed by regression using top principal components is typically driven by a desire to eliminate multicolinearity in the independent variables. Since the PCA is run without regard to the dependent variable this leads to shortcomings of the PCR methodology. First, there is no reason to believe the top principal components are related to the dependent variable and thus elimination of lower components may eliminate the important relationships. Second, use of principal components as independent variables leads to an analysis that is hard to interpret. Third, PCA is useless in designed experiments since the principal components are determined entirely by the experimental design. To alleviate some of these shortcomings [26], provides an approach to dimension reduction of the independent variables that generates a *sufficient reduction* of the these variables which depends on the observed dependent variable values.

Penalized Regression

Ridge Regression

Ridge regression (RR) is an alternative option to PCR that does not require eliminating highly collinear structure measurements. The basic idea behind RR is to shrink the OLS regression coefficient estimates toward zero by adding a penalty for large coefficients. Rather than minimizing the quantity $\| Y - X\beta \|_2$ which results in the OLS estimates, ridge regression minimizes the quantity $\| Y - X\beta \|_2 + k \| \beta \|_2$ for a chosen $k \geq 0$. If $k = 0$, the OLS estimate is the RR estimate and no shrinkage is observed. In contrast, for $k > 0$ the RR estimate is $\hat{\beta}_{RR} = (X'X + kI)^{-1}X'Y$ and as k increases the estimates for β get closer and closer to zero [24]. For predicting the activity of a new chemical, the OLS estimate is replaced with the RR estimate, *i.e.* $\hat{Y}^* = X^*\hat{\beta}_{RR}$. The choice of k is left to Section 4.

Many articles have utilized ridge regression for dealing with the large-*p*, small-*n* problem in the QSAR literature [17, 24, 27-29]. In particular [30], used ridge regression to determine whether biodescriptors provide additional information over chemodescriptors in predicting eight toxicity measures in 14 halocarbons. The biodescriptors, which were obtained by exposing the halocarbons to hepatocytes and producing a two-dimensional electrophoresis gel, were found to provide additional information over the use of chemodescriptors alone.

Although ridge regression is gaining popularity much is still unknown about its theoretical properties in the $p >> n$ situation. For example, are the ridge regression estimators consistent, *i.e.* do they recover the truth as the number of observations increases? The difficulty here is that to ensure $p >> n$ when the number of observations increases, the number of independent variables must also increase.

LASSO

Ridge regression is one specific type of *regularized regression* which also includes LASSO (least absolute shrinkage and selection operator) [31, 32] and the elastic net [33]. LASSO minimizes the quantity $\| Y - X\beta \|_2 + k \| \beta \|_1$ where $\| \beta \|_1 = \sum_{j=1}^{p} |\beta_j|$. So whereas ridge regression penalizes the square of the deviation of _ from zero, LASSO penalizes the absolute value of the deviation from zero. The adaptive LASSO improves on the original by allowing adaptively determined weights for penalizing individual coefficients [34]. Group LASSO is an extension to LASSO for predefined groups of independent variables that are included or removed as a whole [35].

Elastic Net

Unlike ridge regression, LASSO cannot select more structure measurements than observations and therefore [18] suggest it may not be appropriate for use in QSAR studies. The elastic net penalizes both the square and the absolute deviation from zero and therefore is somewhere between ridge regression and LASSO. The elastic net is a promising approach for selecting important structure measurements while still retaining predictive ability [18].

Additional Penalized Regression Approaches

Other statistical techniques are available for dealing with the large-*p*, small-*n* problem, particularly PLS (partial least squares/projection to latent spaces) [36, 37]. The QSAR literature appears to be favoring the use of RR [24, 38], although at least one has suggested that using PCR and RR together is preferable [39]. More recent work has generalized LASSO for use in the $p >> n$ situation by combining a Bayesian regression approach with a loss function to set some coefficients to zero [40]. Another option that is closely related to LASSO and RR, is the horseshoe [41].

Clustering Techniques

In the modeling discussed above, we have implicitly assumed that all chemicals being analyzed are equally described by the one model that is chosen. Given the heterogeneity in chemical structures and activities, it is intuitive that certain chemical groups would follow one model while another would follow a quite different model. Therefore it seems reasonable to cluster chemicals into groups with similar structures. Statistical cluster analysis encompasses many different algorithms and methods for grouping objects, *e.g.*, chemicals, of similar type into respective groups. In QSAR applications there are situations where the chemical database may be relatively large and contain compounds of many different types, *e.g.*, halogens, phenols, alkanes, *etc.* It may be difficult to find a statistical model that will produce satisfactory results when a database contains chemicals of many different types [25] or different modes of action [23]. Cluster analysis can be useful to *a priori* group chemicals into similar groups based on chemical structures or activities. Individual prediction models within a cluster can then subsequently be developed.

A common clustering technique is k-means clustering in which the user can determine *a priori* the number of clusters or one can iterate the analysis to determine an optimal number of clusters in an exploration of a dataset (see Section 4). Niemi *et al.,* [42] used k-means clustering to explore the persistence or degradation of 287 chemicals tested with the standard biochemical oxygen demand (BOD) procedure. The 287 chemicals were derived from an extensive literature search of available BOD values, plus scrutiny of the quality of the BOD procedure used. The dataset was diverse and consisted of a wide variety of chemical groups, *e.g.*, halogens, aldehydes, hydrocarbons, acids, and sulfonates. Fifty-four molecular connectivity indices were calculated and five chemical properties were either available or estimated. To reduce the dimensionality, PCA was used and resulted in eight principal components that explained more than 94% of the variation in the original data. The eight principal components were calculated in a k-means clustering algorithm that was iterated many times to identify an optimum number of clusters that provided the best discrimination of biodegradable and persistent chemicals. Once the analyses were completed, a series of structural features were identified that were associated with degradable and persistent chemicals. The overall model correctly classified 85% of the degradable chemicals and 94% of the persistent chemicals. In addition, several chemicals that were misclassified as degradable or persistent were retested. In many cases, retesting of the chemicals indicated that the biodegradability model was correct and the original biodegradability test values were erroneous.

MODEL SELECTION

It is often the goal of an analysis to choose one final model based on the data at hand. The choices to arrive at this final model are extensive including which structure measurements to include in the model, how many principal components to include, what the ridge regression penalty should be, and how to cluster chemicals. Here we discuss a number of statistical tools used to compare models for the ultimate goal of selecting one model for prediction purposes.

F-Test

In some cases, the models under consideration *are nested.* Model A is nested in model B if the parameters in model B can be set to particular values to recover model A. Consider the two regression models:

$$A: Y_i = X_{i1}\beta_1 + \epsilon_i$$

$$B: Y_i = X_{i1}\beta_1 + X_{i2}\beta_2 + \epsilon_i:$$

Model A is nested in model B since setting $\beta_2 = 0$ recovers model A. Often we are interested in determining whether model A or model B is preferable. Model B will always fit the data better than model A since it has an additional parameter. Unfortunately adding this additional parameter may simply fit noise and therefore harm our predictions. Therefore we need a way to distinguish when the model is fitting noise and when it is fitting signal.

A formal approach to compare two nested models is an *F*-test. This test determines whether the larger model is a statistically significant improvement over the smaller model. If it is, this suggests the additional parameters involved in creating the larger model are likely modeling signal rather than noise.

This is the approach used in [24] where the structures are grouped into categories: topostructural (TSI), topochemical (TCI), geometrical (3D), and semiempirical quantum chemical (QC) variables. Models were tested in hierarchical lists where TSI was added first followed by TCI, then 3D, and finally QC. The analysis showed that in most of the datasets incorporating all four categories provided the best model, each category provided a statistically significant improvement over the smaller model.

Akaike/Bayesian Information Criterion

Often we are not interested simply in nested models. Consider a simple example where there are two predictor variables and we consider the four models consisting of every combination of variable inclusions. Then the model that has only the first variable is not nested in the model that has only the second variable rendering the F-test ineffective. The most common approach to determining which variables to include is to use either Akaike Information Criterion (AIC) [43] or Bayesian Information Criterion (BIC) [44].

Both of these criteria put penalties on the number of parameters in a model and thereby encourage model parsimony. If the number of models is small enough, then the criterion can be computed for all models and the model with the best criterion, called the *best subsets* model, can be chosen [13]. Typically p is too large to enumerate all models in a reasonable amount of time and then the criterion is combined with a stepwise selection procedure to find a reasonable model [45, 46], but no guarantee is made that this procedure finds the best subset.

Cross-Validation

The F-test, AIC, and BIC are useful tools when we are interested in which structure measurements to include, but these tools are not useful in determining the number of principal components to use, the ridge regression parameter, or how many clusters to use. A good approach for these choices is cross-validation [17].

Although many variants of cross-validation exists, we only describe leave-one-out cross-validation. This approach calculates the prediction sum of squares

(PRESS) for each candidate model, *e.g.* each number of principal components. PRESS is calculated according to the following procedure:

1. For $i = 1; \ldots; n$.
   (a) Fit the model while leaving out chemical *i*,
   (b) Predict the property of chemical *i*, $\hat{Y}_i$,
   (c) Calculate the squared prediction error for chemical *i*, $(Y_i - \hat{Y}_i)^2$,
2. Sum all the squared prediction errors (PRESS).

The candidate model with lowest PRESS is chosen.

An alternative to this cross-validation approach is to separate the dataset into two groups: the training and hold-out testing data. All candidate models are fit using the training data and then a model is chosen based on performance among the testing data. Although computationally faster than cross-validation, this hold-out testing approach is only reliable when both the training and testing data are numerous [47]. Due to the small sample sizes typically available in QSAR studies, the cross-validation approach described here will be more reliable and less wasteful than a hold-out approach [17, 47].

MODEL AVERAGING

The previous section outlined methodology for selecting **one** best model and assuming it is the true model to make predictions. But, as George Box once wrote [48]:

All models are wrong, but some are useful.

It is often useful to interpret the best model to suggest a mechanism that describes the property being analyzed, but we should not pretend that this model is the truth. For the purposes of making a prediction, we should instead acknowledge our uncertainty about model truth and account for that uncertainty. This is exactly what model averaging does.

Suppose we consider a total of J models, *e.g.* if we have 10 structure measurements then we could consider the set of J $=2^{10}$ = 1024 models that includes all combinations of those measurements. Now suppose that our prediction for Y^* based on X^* from each model j is $\hat{Y}_j^*$, then the *model averaged prediction* is $\sum_{j=1}^{J} w_j Y_j$ where w_j are model weights such that $\sum_{j=1}^{J} w_j = 1$.

One approach to determining these weights is to use the AIC values for each model [49]. Let AIC_j be the AIC value for model j, AIC_{min} be the minimum AIC among the J models, and $\Delta AICj = AIC_j - AIC_{min}$. Then the *Akaike* weight for each model is

$$w_j = \frac{e^{-\Delta AIC_j/2}}{\sum_{i=1}^{J} e^{-\Delta AIC_j/2}}$$

A difficulty with the use of model averaging in practice is the number of possible models. If p is in the hundreds and we consider the models consisting of all combinations of predictors being in the model, then we have as many models as atoms in the universe, $10^{80} \approx 2^{266}$. It is infeasible to estimate the parameters, predict new values, and calculate the weight for all models. Fortunately, we can approximate the model averaged prediction if we can find the models with large weight, w_j. Methods, such as shotgun stochastic search [50], are currently being developed to efficiently find these large weight models.

In [51], AIC model averaging was used in conjunction with PLS, PCR, and cross-validation to determine the key biological predictors responsible for generating a specific cytokine response. A specific difficulty for their study was the use of time-course measurements which provide a profile of ligand-induced changes in protein phosphorylation state and cytokine output response in macrophage like RAW264.7 cells. These time-course measurements are highly correlated and therefore when used as predictors can severely violate independence assumptions. Through the use of model averaging and variable selection techniques, the authors were able to relax this assumption and provide both a predictive and, possibly, mechanistic understanding of the cytokine reponse.

BAYESIAN STATISTICS

The use of Bayesian statistics is increasing all fields of science including QSAR studies. An appealing aspect of Bayesian statistics is the coherence of all methodologies through the use of conditional probability and Bayes' rule [52] as in equation (1),

$$P(A|B) = \frac{P(B|A)P(A)}{P(B)} \tag{1}$$

In all Bayesian analyses, A represents anything we don't know whereas B represents everything we know or assume. For example, B includes the data, *e.g.* measured chemical activities. In contrast, A represents model parameters, *e.g.* regression coefficients, or predictions, *e.g.* unmeasured chemical activities. The goal of a Bayesian analysis is to obtained the *posterior*, $P(A|B)$, based on the information provided in the *prior*, $P(A)$, the statistical model, $P(B|A)$, and the *marginal likelihood*, $P(B)$. In this way, we can view the Bayesian approach as a formal mathematical tool to move from the information we have before an experiment is observed, *i.e.* the prior, to the information we have after an experiment is concluded, *i.e.* the posterior.

The interpretation of a Bayesian analysis is much different from the interpretation of a frequentist analysis. For example, in the model selection context a frequentist produces a p-value where a Bayesian produces a posterior model probability. The interpretation of the p-value *is the probability of observing a test statistic as or more extreme than that observed, if the null hypothesis is true* while a posterior model probability (for the null hypothesis) has the interpretation *as the probability that the null hypothesis is true given the data we observed.* Similarly for parameter uncertainty a frequentist produces a confidence interval where a Bayesian produces a credible interval. The interpretation of a 100(1-α)% confidence interval *is over repeated realizations of the data, the constructed confidence intervals will contain the true parameter* 100(1-α)% *of time time* while a 100(1-α)% credible interval has the interpretation *the probability the true parameter value is in the interval is* 100(1-α)%. In both cases, the latter is a more natural interpretation (at least to us), but comes at the cost of requiring a prior distribution for parameters and, for model probabilities, a prior probability for models.

In the rest of this section, we will show the natural connection between previously mentioned techniques, *e.g.* regression, ridge regression, and model averaging, and Bayesian methods. For a more thorough review of Bayesian background and approaches please see [53, 54].

Bayesian Regression

In the regression problem described in Section 3.1, we are typically interested in estimating the unknown parameters β and σ^2 based on the available data. Equation (2) provides a rewriting of Bayes' rule to utilize the notation previously introduced where lower case ps are now used since we are talking about continuous distributions.

$$p(\beta,\sigma^2|y) = \frac{p(y|\beta,\sigma^2)p(\beta,\sigma^2)}{p(y)} \tag{2}$$

In this statement, $\mathrm{p}(\mathrm{y}|\beta;\, \sigma)^2$, represents the regression model $\mathrm{y} = \mathrm{X}\beta + \epsilon$, $\mathrm{p}(\beta, \sigma^2)$ represents prior information available concerning the model parameters, and $\mathrm{p}(\beta, \sigma^2|\mathrm{y})$ represents the information available after analyzing the new data. It is common, albeit confusing, to eliminate *X* and the model itself from the conditional probability statements in equation (2) since neither of these are included in A and B of equation (1).

A convenient computational choice for the prior, $p(\beta;\, \sigma^2)$, is to choose a normal-inverse gamma prior for β and σ^2, $p(\beta,\sigma^2) = p(\beta|\sigma^2)p(\sigma^2) = N(\beta;\beta_0,\sigma^2\Sigma_o\ )Ga(\sigma^{-2};\alpha_0,\beta_0)$ where $\mathrm{N(a;\, b, c)}$ represents a normal distribution for a with mean b and variance matrix c and $\mathrm{Ga(d;\, e, f)}$ represents a gamma distribution with shape e and rate f. If we simultaneously let b, c, and Σ_0^{-1} approach zero, then we obtain the prior $p(\beta,\sigma^2)\alpha\,\sigma^{-2}$ where the proportionality symbol is used to indicate that this is not a proper distribution since it does not integrate to one. Nonetheless, the posterior is a proper distribution and is

$$p(\beta;\, \sigma^2|y) = N(\beta;\, \hat{\beta}_{OLS},\sigma^2\,(X'X)^{-1})Ga(\sigma^{-2};\, n/2,b_n) \tag{3}$$

where $b_n = (y - X\hat{\beta}_{OLS})'(y - X\hat{\beta}_{OLS})/2$. Therefore the posterior expectation of β, $E[\beta|y]$, is exactly the same as the ordinary least squares estimate.

Informative Priors

For simplicity, assume now that σ^2 is known and we are interested in providing an informative prior for β. A computationally convenient choice will be a normal distribution with mean 0 and variance Σ_0. If we further assume that $\Sigma_0 = r^2 I$, then the posterior expectation for β is $\hat{\beta}_{RR}$, the ridge regression estimate where $k = \sigma^2/r^2$ [55].

A computationally less convenient choice is the Laplace [56], also called the double exponential, prior distribution. If this prior, centered at zero, is used, then the posterior expectation for β is a LASSO estimate. If the prior is a mixture of a normal and Laplace prior both centered at zero, then the resulting posterior expectation for β is an elastic net estimate [57].

Rather than strictly providing better parameter estimates, informative priors can also be used to formally incorporate scientific knowledge. This was used in [58] to combine information across multiple experiments to build a predictive model of ligand-receptor binding affinities. It has been suggested that Bayesian regression be further explored for its benefit in decision making [59].

Bayesian Prediction

To predict a new chemical activity from its structure, we use Y* as unknown while Y is known. Utilizing the rules of probability, we arrive at the following prediction equation:

$$p(Y^*|Y) = \int p(Y^*|\beta,\sigma^2)p(\beta,\sigma^2|Y)d\beta d\sigma^2$$

This equation describes the entire distribution for our prediction for Y^* which can be helpful in understanding how much uncertainty we have in the predicted point estimate. The point estimate is found by taking the expectation and using the law of iterated expectations:

$$E[Y^*|Y] = E[E[Y^*|\beta,\sigma^2,Y]] = E[X^*\beta|Y] = X^*\hat{\beta}$$

where $\hat{\beta}$ will be the point estimate for β for the model under consideration, *e.g.* for ridge regression, it is $\hat{\beta}_{RR}$. Therefore, to obtain a point estimate under the Bayesian approach we have exactly the same two-step process: 1) estimate the parameters in the model and 2) predict the new data point based on those estimates.

Bayesian Model Averaging

As discussed in Section 5, there is no reason to presume that the one model we have selected is actually the true model and predictions can be improved if, rather than selecting a single model, all models are entertained as possibilities and our prediction is based on a weighted average over all these models. The Bayesian approach provides a formal derivation of this approach called *Bayesian model averaging* [60, 61] which we outline here.

Using the laws of probability, we have,

$$p(Y^*|Y) = \sum_{j=1}^{J} p(Y^*|M_j)P(M_j|Y)$$

where the upper case P is used since this is an actual probability. To find a point estimate for Y*, we calculate its expectation

$$E[Y^*|Y] = \sum_{j=1}^{J} E[Y^*|M_j]P(M_j|Y)$$

The expectation for each model is calculated according to the previous section, *i.e.* estimate the parameters and then predict Y* based on those estimates. Therefore this approach is exactly consistent with the model averaging approach in Section 5 if we set $w_j = P(M_j|Y)$.

To find the posterior model probability $P(M_j|Y)$, we use Bayes' rule,

$$P(M_j|Y) = \frac{p(Y|M_j)P(M_j)}{p(Y)} \tag{4}$$

where $P(M_j)$ is our prior probability for model j, $p(Y\,|M_j)$ is the distribution for our data under model M_j, and $p(Y) = \sum_{i=1}^{p} p(Y\,|M_j)P(M_j)$ assures that the posterior probability over all models sums to unity. Bayesian model averaging in regression models can be accomplished using the **BMA** package [62] in the statistical software R [63].

SUMMARY

In this article, we covered the use of linear regression techniques for continuous valued activities in QSAR and suggested model averaging and Bayesian approaches as possible future directions to extend the use of these techniques. We would be remiss not to mention that there are several other approaches to dealing with the large-*p*, small-*n* problem including PLS [36, 37] and Bayesian neural networks [64-66]. Bayesian neural networks can provide extremely good predictive power under cross-validation scrutiny, but we prefer the interpretability afforded regression models which can lead to mechanistic understanding of how structure affects activity. A number of authors have tried to compare these different methods [67-70]. We also did not cover the vast field of binaryor categorical-valued activities [71-73], but even there the idea of Bayesian model averaging has improved predictive power [72].

Statistical analysis and particularly multivariate statistics provide the mathematical chemist with a powerful arsenal of tools to improve our understanding of SARs. Here we have provided several examples of their applications to problems in QSAR. With the recent emergence of the Internet and outstanding search engines, there is an extensive amount of information describing and illustrating the use of these statistical techniques. However, it is important to recognize that much of the information has not been peer-reviewed and we urge the reader to seek standard textbooks and the vast peer-reviewed literature that has developed and been accepted by the scientific community. In addition to the Internet, there are many excellent statistical packages that are now available for most of the standard, classical statistical tests such as regression, PCA, and clustering techniques as well as their many variations. Many of the manuals that come with these statistical packages are also well-documented. Exceptions for available software still apply to many of the Bayesian approaches, but this is likely to improve in the future.

In this brief review of some older statistical techniques and some new approaches, we have tried to provide a flavor for how many of the complex problems in SAR can be simplified with the use of multivariate statistics. However, statistics is in itself a vast field of science and we certainly cannot do it justice in a brief review. It is incumbent upon the scientist to clearly articulate the question(s) he/she seeks to address and fully understand the potential statistical techniques that could address the question(s). We strongly encourage the scientist to also seek professional advice from a statistician and include a statistician in team approaches to solving these complex problems in QSAR. Moreover, it is wise to include or consult a statistician in the start of a project rather than expecting one to fix a problem or analyze data after it has been gathered.

ACKNOWLEDGEMENTS

We dedicate this chapter in memory of Dr. Gilman Veith – a superb mentor, colleague, and friend. Gil provided the wisdom and inspiration for our involvement and contributions to the study of quantitative structure-activity relationships (QSAR). We also thank Dr. Subhash Basak for the invitation to contribute to this book. His unrelenting perseverance to theoretical and mathematical advancements in QSAR has been astonishing.

CONFLICT OF INTEREST

The authors confirm that this chapter contents have no conflict of interest.

REFERENCES

[1] Basak, S.C. Chemo-bioinformatics based mathematical descriptors and their applications in computational drug design. *Curr Comput Aided Drug Des*., **2010**, *6*(4), 223-224.

[2] Hansch, C. Quantitative structure-activity relationships in drug design. *Drug Des*, **1971**, 1, 271-342.

[3] Basak, S.C.; Grunwald, G.D.; Niemi, G.J., Use of graph-theoretic and geometrical molecular descriptors in structure-activity relationships, In From *Chemical Topology to Three-dimensional Geometry*. Balabah, Plenum Press, New York **1997**.

[4] Cronin, M.T.D., Quantitative structure-activity relationships (QSARs) applications and methodology, In Puzyn, Tomasz; Leszczynski, Jerzy; Cronin, Mark T., editors, *Recent*

Advances in QSAR Studies volume 8 of Challenges and Advances in Computational Chemistry and Physics, pp. 3-11. Springer Netherlands **2010**.

[5] eduSoft LC, VAAshland, Molconn-Z 4.10 Manual, http://www.edusoftlc.com/molconn/manuals/400/**2011**.

[6] Basak, S.C.; Gieschen, D.P.; Harriss, D.K.; Magnuson, V.R. Physicochemical and topological correlates of the enzymatic acetyltransfer reaction. *J Pharm Sci-Us*, **1983**, 72(8), 934-937.

[7] Todeschini, R.; Consonni, V. DRAGON-software for the calculation of molecular descriptors, version 1.0 for Windows, Milano chemometrics and QSAR research group. *Freely available at http://www.vcclab.org/lab/edragon/,* **2000**.

[8] Katritzky, A.R.; Lobanov, V.S.; Karelson, M. CODESSA software. *University of Florida, SemiChem, Shawnee*, KS, **1994**, p. 211.

[9] Veith, G.D. On the nature, evolution and future of quantitative structureactivity relationships (QSAR) in toxicology. *SAR QSAR Environ. Res.*, 15, **2004**, 5(6), 323-330.

[10] Schultz, T.W.; Carlson, R.E.; Cronin, M.T.; Hermens, J.L.; Johnson, R.; O'Brien, P.J.; Roberts, D.W.; Siraki, A.; Wallace, K.D.; Veith, G.D. A conceptual framework for predicting the toxicity of reactive chemicals: modeling soft electrophilicity. *SAR QSAR Environ. Res.*, **2006**, 17(4), 413-428.

[11] Basak, S.C.; Mills, D. Prediction of mutagenicity utilizing a hierarchical QSAR approach. *SAR QSAR Environ. Res.*, **2001,** 12(6), 481.

[12] Mackay, D., *Handbook of physical-chemical properties and environmental fate for organic chemicals: Introduction and hydrocarbons* volume 1, CRC Press/Taylor & Francis **2006**.

[13] Niemi, G.J.; Basak, S.C.; Grunwald, G.; Veith, G.D. Prediction of octanol/water partition coe_cient (KOW) with algorithmically derived variables. *Environ. Toxicol. Chem*., **1992**, 11(7), 893-900.

[14] Yap, C.W.; Li, H.; Ji, Z.L.; Chen, Y.Z. Regression methods for developing QSAR and QSPR models to predict compounds of speci_c pharmacodynamic, pharmacokinetic and toxicological properties. *Mini Rev. Med. Chem.*, **2007**, 7(11), 1097-1107.

[15] Wold, S.; Sjöström, M.; Eriksson, L., *Partial least squares projections to latent structures (PLS) in chemistry*, Wiley Online Library **2006**.

[16] Wold, S.; Eriksson, L.; Kettaneh, N., PLS *in Data Mining and Data Integration,* Springer **2010**.

[17] Hawkins, D.M.; Basak, S.C.; Mills, D. Assessing model fit by crossvalidation. *J Chem Inf Comp Sci*, **2003**, 43(2), 579-586.

[18] Kraker, J.J.; Hawkins, D.M.; Basak, S.C.; Natarajan, R.; Mills, D. Quantitative structure-activity relationship (qsar) modeling of juvenile hormone activity: Comparison of validation procedures. *Chemometr. Intell. Lab*, **2007,** *87*(1), 33-42.

[19] Basak, S.C.; Mills, D.; Hawkins, D.M.; Kraker, J.J., Proper statistical modeling and validation in QSAR: A case study in the prediction of rat fat-air partitioning, In *AIP Conference Proceedings* volume 963, p. 548 **2007**.

[20] Basak, S.C.; Mills, D.; Hawkins, D.M. Characterization of dihydrofolate reductases from multiple strains of plasmodium falciparum using mathematical descriptors of their inhibitors. *Chem Biodivers*, **2011**, 8(3), 440-453.

[21] Pearson, K. Liii. on lines and planes of closest _t to systems of points inspace. *PHILOS Magseries* 6, **1901**, 2(11), 559-572.

[22] Jolli_e, I. Principal component analysis. *Encyclopedia of Statistics in Be-havioral Science*, **2002**.

[23] Basak, S.C.; Grunwald, G.D.; Host, G.E.; Niemi, G.J.; Bradbury, S.P. Acomparative study of molecular similarity, statistical, and neural methodsfor predicting toxic modes of action. *Environ. Toxicol. Chem.*, **1998**, 17(6), 1056-1064.

[24] Hawkins, D.M.; Basak, S.C.; Shi, X. QSAR with few compounds and many features. *J Chem Inf Comp Sci*, **2001**, 41(3), 663-670.

[25] Basak, S.C.; Magnuson, V.R.; Niemi, G.J.; Regal, R.R. Determining structural similarity of chemicals using graph-theoretic indices. *Discrete Appl Math*, **1988**, 19(1-3), 17-44.

[26] Cook, R.D. Fisher lecture: Dimension reduction in regression. *Stat Sci*, **2007**, 22(1), 1-26.

[27] Basak, SC; Mills, D.; Hawkins, DM; El-Masri, HA Prediction of tissue-airpartition coefficients: A comparison of structure-based and property-based methods. *SAR QSAR Environ. Res.*, 13, **2002**, 7(8), 649-665.

[28] Hawkins, D.M.; Basak, S.C.; Mills, D. QSARs for chemical mutagens from structure: ridge regression _tting and diagnostics. *Environ. Toxicol. Pharmacol.*, **2004**, 16(1-2), 37-44.

[29] Basak, S.C.; Mills, D.; Hawkins, D.M.; Kraker, J.J. Quantitative structure activity relationship (QSAR) modeling of human blood: Air partitioning with proper statistical methods and validation. *Chem. Biodivers*, **2009**, 6(4), 487-502.

[30] Hawkins, D.M.; Basak, S.C.; Kraker, J.; Geiss, K.T.; Witzmann, F.A. Combining chemodescriptors and biodescriptors in quantitative structureactivity relationship modeling. *J Chem Inf Model*, **2006**, 46(1), 9-16.

[31] Tibshirani, R. Regression shrinkage and selection *via* the lasso. *J Roy Stat Soc B Met*, **1996**, 58, 267-288.

[32] Hastie, T.; Tibshirani, R.; Friedman, J., *The elements of statistical learn- ing: data mining, inference and prediction*, Springer 2 edition **2009**.

[33] Zou, H.; Hastie, T. Regularization and variable selection *via* the elastic net. *J Roy Stat Soc B*, **2005**, 67(2), 301-320.

[34] Zou, H. The adaptive lasso and its oracle properties. *J Am Stat Assoc*, **2006**, 101(476), 1418-1429.

[35] Meier, L.; Van De Geer, S.; Buhlmann, P. The group lasso for logistic regression. *J Roy Stat Soc B*, **2008**, 70(1), 53-71.

[36] Dunn, W. J.; Wold, S.; Edlund, U.; Hellberg, S.; Gasteiger, J. Multivariate structure-activity relationships between data from a battery of biological tests and an ensemble of structure descriptors: The PLS method. *Quant. Struct.-Act. Relat*, **1984**, 3(4), 131-137.

[37] Roy, P.P.; Roy, K. On some aspects of variable selection for partial least squares regression models. QSAR & *Comb Sci*, **2008**, 27(3), 302-313.

[38] Al-Hassan, Y.M.; Al-Kassab, M.M. A Monte Carlo comparison between ridge and principal components regression methods. *Appl. Math. Sci.*, **2009**, 3(42), 2085 - 2098.

[39] Vigneau, E.; Devaux, MF; Qannari, EM; Robert, P. Principal component regression, ridge regression and ridge principal component regression in spectroscopy calibration. *J Chemometr*, **1997**, 11(3), 239-249.

[40] Bondell, Howard; Reich, Brian, Consistent high-dimensional bayesian vari- able selection *via* penalized credible regions, Accepted for publication in the *J Am Stat Assoc.*

[41] Carvalho, C.M.; Polson, N.G.; Scott, J.G. The horseshoe estimator for sparse signals. *Biometrika*, **2010**, 97(2), 465-480.

[42] Niemi, G.J.; Veith, G.D.; Regal, R.R.; Vaishnav, D.D. Structural features associated with degradable and persistent chemicals. *Environ. Toxicol. Chem.*, **1987**, 6(7), 515-527.

[43] Akaike, H. A new look at the statistical model identification. *IEEE Trans. Autom. Control*, **1974**, 19(6), 716-723.

[44] Schwarz, G. Estimating the dimension of a model. *Ann. Statist*, **1978**, 6(2), 461-464.

[45] Rose, R.M.; St. J. Warne, M.; Lim, R.P. Quantitative structure{activity relationships and volume fraction analysis for nonpolar narcotic chemicals to the australian cladoceran. *Ceriodaphnia cf. dubia. Arch. Environ. Contam. Toxicol*, **1998**, 34(3), 248-252.

[46] Vighi, M.; Migliorati, S.; Monti, G.S. Toxicity on the luminescent bacterium Vibrio _scheri (beijerinck). i: QSAR equation for narcotics and polar narcotics. *Ecotoxicol. Environ. Saf.*, **2009**, 72(1), 154-161.

[47] Hawkins, D.M. The problem of over_tting. *J Chem Inf Comp Sci*, **2004**, 44(1), 1-12.

[48] Box, G.E.P.; Draper, N.R., Empirical *model-building and response surfaces*, John Wiley & Sons **1987**.

[49] Burnham, K.P.; Anderson, D.R., *Model selection and multimodel inference: a practical information-theoretic approach*, Springer Verlag **2002**.

[50] Hans, C.; Dobra, A.; West, M. Shotgun stochastic search for \large p" regression. *J Am Stat Assoc*, **2007**, 102(478), 507-516.

[51] Wu, Y.; Johnson, G.; Gomez, S. Data-driven modeling of cellular stimulation, signaling, and output response in RAW 264.7 cells. *J. Mol. Signaling*, **2008**, 3, 1-14, 10.1186/1750-2187-3-11.

[52] Bayes, M.; Price, M. An Essay towards solving a Problem in the Doctrine of Chances. By the late Rev. Mr. Bayes, FRS communicated by Mr. Price, in a letter to John Canton, AMFRS. *Philos. Trans*, **1763**, 53, 370.

[53] Armstrong, N.; Hibbert, D.B. An introduction to bayesian methods for analyzing chemistry data:: Part 1: An introduction to bayesian theory and methods. *Chemom. Intell. Lab. Syst*, **2009**, 97(2), 194-210.

[54] Hibbert, D.B.; Armstrong, N. An introduction to Bayesian methods for analyzing chemistry data: Part II: A review of applications of Bayesian methods in chemistry. *Chemom. Intell. Lab. Syst*, **2009**, 97(2), 211-220.

[55] Lindley, D.V.; Smith, A.F.M. Bayes estimates for the linear model. *J Roy Stat Soc B Met*. Series B (Methodological), **1972**, 34(1), 1-41.

[56] Norton, R.M. The double exponential distribution: Using calculus to find a maximum likelihood estimator. *Amer. Statist*, **1984**, 38(2), 135-136.

[57] Li, Q.; Lin, N. The Bayesian elastic net. *Bayesian Anal*, **2010**, 5(1), 151-170.

[58] Murray, C.W.; Auton, T.R.; Eldridge, M.D. Empirical scoring functions. II. the testing of an empirical scoring function for the prediction of ligandreceptor binding affinities and the use of bayesian regression to improve the quality of the model. *J Comput Aid Mol Des*, **1998**, 12, 503-519, 10.1023/A:1008040323669.

[59] Sahlin, U.; Filipsson, M.; Oberg, T. A risk assessment perspective of current practice in characterizing uncertainties in qsar regression predictions. *Mol. Inf,* **2011**, 30(6-7), 551-564.

[60] Raftery, A.E.; Madigan, D.; Hoeting, J.A. Bayesian model averaging for linear regression models. *J Am Stat Assoc*, **1997**, 92(437), 179-191.

[61] Hoeting, J.A.; Madigan, D.; Raftery, A.E.; Volinsky, C.T. Bayesian model averaging: A tutorial. *Stat. Sci*, **1999**, 14(4), 382-401.

[62] Raftery, A.; J.Hoeting; Volinsky, C.; I.Painter; Yeung, K.Y., BMA: *Bayesian Model Averaging* **2011**, R package version 3.14.1.

[63] R Development Core Team, R: A Language *and Environment for Statistical Computing*, R Foundation for Statistical Computing Vienna, Austria **2011**, ISBN 3-900051-07-0.

[64] Frank, R.; Winkler, D.A. A quantitative structure-activity relationships model for the acute toxicity of substituted benzenes to tetrahymena pyriformis using Bayesian-regularized neural networks. *Chem. Res. Toxicol*, **2000**, 13(6), 436-440.

[65] Qin, Y.; Deng, H.; Yan, H.; Zhong, R. An accurate nonlinear QSAR model for the antitumor activities of chloroethylnitrosoureas using neural networks. *J. Mol. Graphics Modell*, **2011**, 29(6), 826- 833.

[66] Jalali-Heravi, M.; Mani-Varnosfaderani, A. QSAR modelling of integrin antagonists using enhanced bayesian regularised genetic neural networks. *SAR QSAR Environ. Res.*, **2011**, 22(3-4), 293-314.

[67] Eriksson, L.; Jaworska, J.; Worth, A.P.; Cronin, M.T.D.; McDowell, R.M.; Gramatica, P. Methods for reliability and uncertainty assessment and for applicability evaluations of classification-and regression-based QSARs. *Environ. Health Perspect*, **2003**, 111(10), 1361.

[68] Yao, X.J.; Panaye, A.; Doucet, J.P.; Zhang, R.S.; Chen, H.F.; Liu, M.C.; Hu, Z.D.; Fan, B.T. Comparative study of QSAR/QSPR correlations using support vector machines, radial basis function neural networks, and multiple linear regression. *J Chem Inf Comp Sci*, **2004**, 44(4), 1257-1266.

[69] Nandi, Sisir; Vracko, Marjan; Bagchi, Manish C. Anticancer activity of selected phenolic compounds: QSAR studies using ridge regression and neural networks. *Chem. Biol & Drug Des*, **2007**, 70(5), 424{436.

[70] Basak, S.C.; Mills, D. Quantitative structure-activity relationships for cycloguanil analogs as PfDHFR inhibitors using mathematical molecular descriptors. SAR QSAR Environ. Res., 21, **2010**, 3(4), 215-229.

[71] McDowell, R.M.; Jaworska, J.S. Bayesian analysis and inference from QSAR predictive model results. *SAR QSAR Environ. Res.*, **2002**, 13(1), 111-125.

[72] Angelopoulos, N.; Hadjiprocopis, A.; Walkinshaw, M.D. Bayesian model averaging for ligand discovery. *J Chem Inf Model*, **2009**, 49(6), 1547-1557.

[73] Bender, A. Bayesian methods in virtual screening and chemical biology. *Methods Mol. Biol.*, **2011**, 672, 175.

CHAPTER 7

Marine Algal Toxicity Models with *Dunaliella tertiolecta: In Vivo* and *In Silico*

Melek T. Saçan[1], Marjana Novic[2,*], M. Doğa Ertürk[1] and Nikola Minovski[2]

[1]Boğaziçi University, Institute of Environmental Sciences, 34342, Hisar Campus, Bebek, Istanbul, Turkey and [2]National Institute of Chemistry, Hajdrihova 19, 1000 Ljubljana, Slovenia

Abstract: The importance of marine alga, namely *Dunaliella tertiolecta*, in toxicity determination of organics, inorganics, and mixtures, as well as for raw and treated industrial effluents is emphasized. Ultrastructural changes for metals below the toxic concentration in *D. tertiolecta* are also highlighted. Examples for synergistic, antagonistic and hormetic effects in case of exposure of *D. tertiolecta* to chemicals are given. In a case study, we focus on modeling the toxicity of selected phenols to *D. tertiolecta*. Quantitative structure-activity relationship (QSAR) methodology is employed to model the toxicities of phenolic compounds to *D. tertiolecta* using counter-propagation artificial neural networks (CP ANN). The endpoint for the toxicity determination is growth inhibition of algae exposed to chemicals in a batch system containing natural sea water enriched by the modified f/2 medium. Results reveal that QSAR methodology can be successfully applied to fill the data gap present in marine algal ecotoxicity data.

Keywords: Antagonistic, artificial neural network, counter-propagation, *Dunaliella tertiolecta*, hormetic exposure effects, *in silico* modelling, *in vivo* toxicity, industrial effluents, marine alga, mechanistic interpretation, molecular structure descriptors, non-linear model, phenols, QSAR, synergistic.

INTRODUCTION

Industrial, agricultural and municipal effluent discharges into aquatic environment continue to be a significant environmental problem. The impact of human activities on the environment affects non-target organisms (*e.g.* use of pesticides) and a chain of interconnecting ecosystems. To evaluate the risks of the vast majority of chemicals that may pose to the environment, information on their

***Corresponding author Marjana Novič:** National Institute of Chemistry, Hajdrihova 19, POB 660, 1001 Ljubljana, Slovenia; Tel: +386 1 4760 253; Fax: +386 1 4760 300; E-mail: marjana.novic@ki.si

Subhash C. Basak, Guillermo Restrepo & José L. Villaveces (Eds)
Copyright © 2015 Bentham Science Publishers Ltd. Published by Elsevier Inc. All rights reserved.
10.1016/B978-1-68108-053-6.50007-7

potential to adversely affect ecosystems is required for various trophic levels. Among the organisms used in ecotoxicity testing, algae represent the primary producers in aquatic ecosystems. They play a major role in the sustainability of biodiversity in aquatic environments by providing food for higher trophic levels and producing oxygen. The importance of algae is better understood considering the fact that approximately half of the planetary primary production is generated by marine phytoplankton, which also affects the abundance and diversity of marine organisms. Furthermore, algae strongly influence biochemical cycles, most notably the carbon cycle [1]. Apart from the crucial role they play for the sustainability of ecosystems, their ubiquitous distribution throughout the globe, ease of culturing and rapid growth rate makes them ideal test species for toxicological studies. Their large surface area to volume ratio is an important attribute for the uptake and interactions with the chemicals [2]. Due to their ecological importance, studying the toxic effects of chemicals on algae has attracted considerable scientific interest [3-6].

Although many studies have been conducted on the impacts of chemicals to fresh water green algae [7-12] and some regulations recommend marine microalgae as test species to be used in toxicity assessments [13, 14], little is known about the toxicity of chemicals to marine green algae. For this reason, there is a need to provide more toxicity data on marine organisms as recommended by expert groups [15] and regulatory schemes [16]. The use of marine algal species such as *Dunaliella tertiolecta* for ecotoxicity testing may be extremely important for the assessment of environmental impact on aquatic systems for geographical areas where the freshwater bodies are not as abundant as the marine ecosystems. Thus, information on the toxic potential of chemicals on marine organisms and the comparative sensitivity to the freshwater organisms is needed for the reassurance of the same degree of protection for the marine as for the freshwater ecosystems.

Ben-Amotz *et al.* [17] provided extensive information for the alga *D. tertiolecta.* It is not only used for the measurement of toxicity but also used to determine treatment efficiencies of industrial effluents. Conventional water quality parameters such as chemical oxygen demand and suspended solids may not detect toxic compounds present in a variety of industrial wastewaters and treated wastes, and chemical procedure alone cannot provide sufficient information on the potential harmful effects of chemicals on the aquatic environment. In some cases, the treated effluent does not exceed the discharge limits; the results of toxicity tests show potential toxicity [18]. In fact, some chemicals are not totally eliminated because the conventional technology of treatment used in wastewater treatment plants appears to be insufficient for completely removing these specific compounds [19].

Therefore, in many developed countries, toxicity tests on industrial effluents are required to ensure that such discharges will not have adverse effects on the aquatic organisms in receiving water. The toxic effects of unknown substances in complex mixtures or with possible synergistic effects among compounds on effluents can be detected only by toxicity testing. The growth responses of *D. tertiolecta* to the effluents of an aluminum plating plant and to the wastewater from an analgesic/anti-inflammatory drug-producing pharmaceutical plant were investigated by measuring two response parameters, optical density (OD_{640}) and *in vitro* chlorophyll fluorescence, for a period of 14 days [4]. All wastes affected algal growth either by inhibition only or by stimulation at low concentrations and inhibition at high concentrations. Although the water quality parameters of treated effluent of both plants were in the permitted range reported by the Local Control Act, they inhibited growth at higher concentrations, implying that the two treatment plants were inefficient. Therefore, the importance of toxicity tests in wastewater discharge regulations is emphasized in this study.

Toxicity testing have also been used to evaluate whether the method proposed and applied for the treatment of effluent is suitable for monitoring the efficiency of the treatment with respect to the removal of micro pollutants. Two commercial reactive dyes, the azo dye Reactive Black 5 and the copper phthalocyanine dye Reactive Blue 21, have been treated with advanced oxidation processes [20]. The relative growth inhibition of the azo dye towards the marine algae *D. tertiolecta* relative to control that was initially 70%, did not exhibit an increase during the studied advanced oxidation reactions and complete detoxification at the end of the treatment period could be achieved for all investigated treatment processes. In another study, toxicity tests with *D. tertiolecta* were performed both on raw and treated effluents from pulp and paper mills [21] which have been evaluated as highly toxic [22, 23].

To improve the treatment performance of pulp and paper mill effluents, and the quality of treated wastewater, the ozonation process was integrated with biological treatment performed by algae. The pre-treated bleaching wastewater was combined with the other streams of the Kraft pulp mill. In the toxicity tests carried out with *D. tertiolecta* on all of these raw and treated effluents, a stimulatory effect was observed for both the raw and Granulated active carbon (GAC)-equilibrated effluents, and this effect was significantly higher for the GAC-equilibrated sample. Although both ozonation processes enabled the elimination of toxicity, treatment of bleaching effluents with O_3/GAC reduced the treatment time necessary for producing nontoxic effluents. Extending the ozonation time led to the formation of toxic products. The reason for these responses was discussed in the reported study [21].

Chemicals rarely exist individually in the environment. Toxic concentrations of organics in the environment can be greatly influenced by the presence of other chemicals [24]. The combined effect of chemicals is an important issue in ecotoxicology. Saçan and Özkovalak [25] examined the interaction between selenium (Se) and azo dyes. The authors determined the effect of synthetic dye bath, containing six azo dyes, on the growth of *D. tertiolecta* in the presence and absence of Se in batch culture under two light intensities (38.4 μE m^{-2} s^{-1} and 76.8 μE m^{-2} s^{-1}). Considering the average 8-d IC_{25} and SC_{20} values expressed as per cent dilution (v/v) the order of toxicity was found as Dye Bath= Remazol Black 5>Reactive Yellow 37>Reactive Orange 69=Assisting chemicals. Se increased the stimulatory effect of all the tested chemicals at low concentration range, whereas it decreased the inhibitory effect at high concentration range, except assisting chemicals and Reactive Orange 69. Similar results were obtained for lead (Pb) and aluminum (Al) by Saçan *et al.* [26]. Both metals had a stimulatory effect at low concentration and an inhibitory effect at high concentration (hormesis). The IC_{25} values of Pb are 8.43, 7.29, and 6.74 mg L^{-1} for 24h, 48h, and 72h, respectively. The corresponding values for Al are 30.54, 22.42, and 18.16 mg L^{-1}.

Although environmental factors such as pH and suspended sediments may enhance/decrease the acute or chronic toxicity of chemicals, at low concentrations, the release of chemicals to the environment separately or in combination can lead to uncontrolled algal growth and, consequently, can exacerbate the eutrophication problem [4, 25, 26]. Both stimulation and inhibition of algal growth influence the entire aquatic ecosystem by changing algal populations. Moreover, if algal growth is affected, the biomass at higher tropic levels is impacted as well. On the other hand, long-term effects of continuous low-level exposure to chemicals and their metabolites are not well understood. Biological effects of low-level (below toxic dose) exposure of Pb and Al to *D. tertiolecta* were determined by Saçan *et al.* [26]. They reported that low concentrations of both metals, alone and in combination, affected the ultrastructure of *D. tertiolecta.* The most obvious effects of Pb were disrupted thylakoidal membranes, accumulated polyphosphate bodies and vacuoles, and Pb precipitates on the cell surface. These ultra-structural alterations were partially present in Al-treated and Pb–Al-treated cells. In joint exposure, the most important change was the lysis of the cell membrane. Al and Pb seem to act synergistically on the cell membrane leading to cell membrane lysis.

Heavy metals pose a great threat to the health of the environment because of their non-biodegradability and hazardous characteristics. The accumulation of heavy metals by phytoplankton in laboratory cultures has been the focus of numerous studies, because some microalgal species can be used for the removal of metals from industrial effluents and they act as an indicator of heavy metal levels in both estuarine and coastal waters throughout the world. Saçan *et al.* [27] determined the bioaccumulation of three heavy metals; namely, copper (Cu), Pb and Se, in the unicellular green alga *D. tertiolecta* in natural sea water enriched with f/2 medium in the laboratory conditions. The pH effect on the bioaccumulation of each metal has also been investigated. The bioaccumulation of metals occurred in a dose-dependent manner. When pH decreased, the bioaccumulation of Cu decreased significantly, but the bioaccumulation of Pb and Se was almost constant. Cu interacted antagonistically with Pb in various bimetallic combinations. Pb showed no effect on the bioaccumulation of copper. Coexistence of Se with Pb and Cu showed antagonistic behavior on the growth of alga microscopically. Saçan and Balcıoğlu [28] also investigated the bioaccumulation of different concentrations of Al in *D. tertiolecta* at two pH levels in 24h. Bioaccumulation of Al was reported as pH dependent. They compared the bioaccumulation of aluminum to that of two essential (copper and selenium) and one nonessential (lead) element. They reported antagonistic effect of Al on Pb bioaccumulation in *D. tertiolecta.*

As the brief literature survey points out, *in vivo* toxicity studies utilizing *D. tertiolecta* as test species is well established and provides invaluable information regarding the toxicity of a wide range of chemicals. However, considering the fact that more than eighty thousand compounds (organic, inorganic, organometalic *etc.*) are in commercial use, more toxicity data is required for hazard identification and risk assessment purposes. Currently, there are many databases, which store toxicity information of chemicals to various aquatic organisms [29, 30]. However, more chemicals are waiting to be evaluated than current databases list and also *D. tertiolecta* are not available in these databases. Therefore, the available knowledge gained through *in vivo* toxicity studies should be linked to or exploited through *in silico* studies.

Quantitative Structure-Activity Relationships (QSARs)

QSAR as a methodology is based on the general principle, by which the activity, in our study explained as biological toxicity, of a given compound is related to its molecular structure. Consequently, structurally similar molecules may have similar biological toxicity [31]. Since the structural properties of compounds are usually encoded in a form of molecular descriptors, QSAR can also be defined as a mathematical relationship between the molecular descriptors and the biological toxicity values of *in vitro* tested known compounds. As a consequence, the result obtained will be in a form of predictive mathematical model, which can successfully be used for prediction of the biological toxicity values for structurally similar unknown compounds.

Considering the increasing number of chemicals, it is almost impossible to supply the necessary toxicological information to regulators using only the data generated through exhaustive bioassays. As such, several regulatory schemes have already recommended the use of theoretical mathematical models [32, 33] to be able to fill the data gaps present in various domains of ecotoxicity to be able to develop fast and reliable environmental monitoring strategies. Such methodologies are frequently being utilized to cut the cost and time involved in toxicity testing.

To speed up the progression of evaluation of a large number of different toxicants, many new computational techniques have been developed. The methodology of establishing a quantitative relationship between the chemical structures for a given compound's class and their experimentally determined properties is commonly known as quantitative structure-property relationship (QSPR) [34]. In toxicity research, the term property usually refers to experimentally determined activity/toxicity values (*e.g.* $\log(IC_{50}^{-1})$), and therefore the acronym QSPR could be renamed as QSAR (quantitative structure-activity relationship) or QSTR (quantitative structure-toxicity relationship), respectively [35, 36]. In the late 1964, when QSAR as a concept was introduced for the first time by Free/Wilson and Hansch/Fujita (the well known Free-Wilson and Hansch-Fujita models) [37, 38], a wide variety of QSAR methodologies have been invented. Beside the popular Free-Wilson and Hansch-Fujita models, which belong to the class of two-dimensional QSAR methods (2D-QSAR), to date the development of QSAR approaches incredibly increased and today several *n*-dimensional QSAR congeners (such as 3D-, 4D-, 5D-, and even 6D-QSAR) have been introduced [39, 40]. Although 2D-QSAR methods are quite old, today they are still very useful, particularly for numerical predictions of the biological toxicity values for compounds, which have not yet been synthesized [41].

The general procedure of QSAR modeling consists of several consecutive steps: dataset preparation, dataset division on the so-called training and test set, calculation of molecular descriptors for the compounds from both sets, modeling on the training set and selection of statistically-significant molecular descriptors, and finally testing the practical predictive performances of the established QSAR model for prediction of the biological toxicity values for the previously excluded test set compounds, Fig. (**1**).

As presented in Fig. (**1**), QSAR modeling as a mathematical procedure of establishing a correlation between the calculated molecular descriptors and biological toxicity values, for each compound separately, is performed utilizing only the training set. To date, several QSAR modeling algorithms are introduced, which can be divided in two main categories: linear modeling algorithms (*e.g.* multiple linear regression (MLR)) and non-linear modeling algorithms (*e.g.* artificial neural networks (ANN)). No matter what QSAR modeling strategy is utilized (linear, non-linear, or their combination), the result is usually expressed in a form of mathematical equation which describe the correlation between the calculated molecular descriptors selected for establishing the model and the biological toxicity in the series (*Eq. 1*):

$$pT = c_0 + c_1 d_1 + c_2 d_2 + c_3 d_3 + ... + c_i d_i \tag{1}$$

where c_i are coefficients (the fitted parameters), while d_i are molecular descriptors.

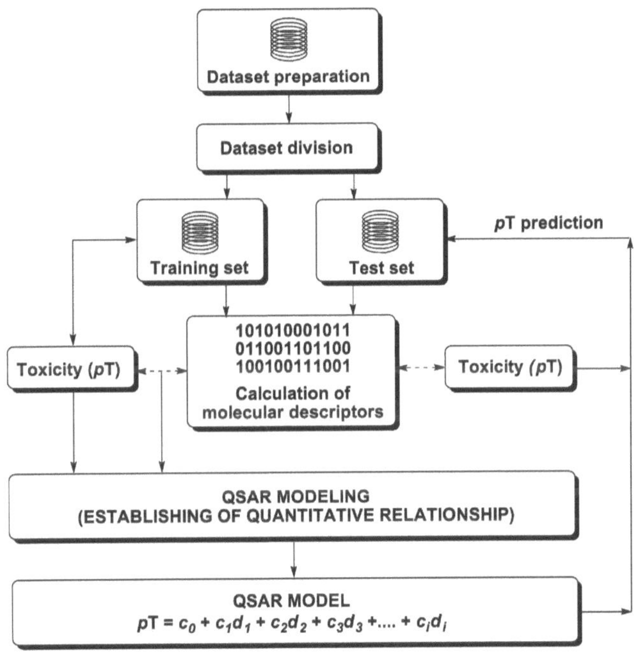

Figure 1: The complete QSAR methodology workflow.

As a measure for the quality of the established QSAR model, the squared correlation coefficient is used and it can be calculated using the following equation (*Eq. 2*) [42]:

$$R^2 = 1 - \frac{\sum_{i=1}^{n} (\hat{y}_i - y_i)^2}{\sum_{i=1}^{n} (y_i - \bar{y})^2} \tag{2}$$

where $\hat{y}_i$ and y_i designate the predicted and observed values for the *i*-th compound in the dataset, respectively, whereas $\bar{y}$ indicates the average of the observed values. Moreover, during the modeling, internal model validation procedures such as cross-validation leave-one-out (CV LOO) are often performed. The validation of the model is important for the statistical stability of the QSAR model and the estimation parameter can be calculated by employing the same equation (*Eq. 2*) if the predicted value $\hat{y}_i$ is obtained from the *i-th* model in the LOO procedure, in which the *i-th* object was removed from the training data set.

The above mentioned measures of quality of the models have to be in accordance with the Organization for Economic Co-operation and Development (OECD) guidance document on the validation of (Q)SAR models used for regulatory purposes. The appropriate validation procedure should be chosen regarding the specificity of the data set and modeling method used. The basic modeling method in this study is counter-propagation artificial neural network (CP ANN).

Counter-Propagation Artificial Neural Networks (CP ANNs)

Comparing to the linear QSAR modeling strategies (*e.g.* MLR, PLS), where the relationship between the calculated molecular descriptors for a set of compounds and their corresponding experimentally determined biological toxicity values is established in a linear manner (linear relationship), the non-linear QSAR methods such as artificial neural networks (ANNs) transform the input variables (*e.g.* calculated molecular descriptors for investigated compound's set) into corresponding output values (responses) utilizing non-linear transforming functions. Similar to the linear methods, ANN methods have also found a wide range of applications [43, 44]. Since ANNs are of different types and it is difficult to describe them with only a single definition, they can most easily be presented as a statistical tool, Fig. (**2**), consisting of multivariate input $X = (x_1, x_2, x_3, \ldots, x_m)$ and multi-response output $Y = (y_1, y_2, y_3, \ldots, y_n)$.

The primary building-block of any ANN is the so-called computer-simulated or artificial neuron, which in its most common form can be defined as a mathematical/statistical tool constructed to mimic the function of the biological neurons in the brain [45]. Therefore, similarly to the biological neurons which are capable to accept a huge number of information in a form of electrical impulses and to process them in an appropriate way, the artificial neuron must be capable to input a huge number of input variables of different nature (x_i) from the neighboring neurons and to process them in a non-linear way in order to give a result (y_j).

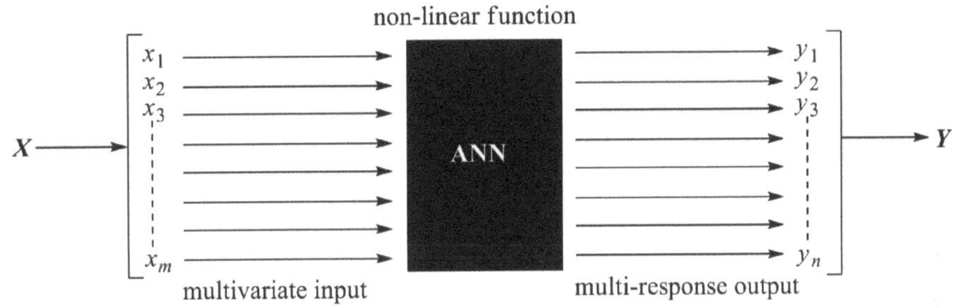

Figure 2: Schematic representation of ANNs.

ANNs can be constructed of different number of neurons, which form the so-called network architecture. In QSAR modeling, the number of neurons in the ANNs mainly depends on the number of training set objects, which will be used to train the network. Moreover, the neurons in ANNs can be organized in different number of parallel layers. One of the most important issues when using ANNs as a modeling tool is to know the nature of the problem for which we want to establish a model. Therefore, two main types of learning approaches could be distinguished: a supervised learning approach (*e.g.* building a mathematical model that can be used for prediction of biological toxicity values for a series of toxicants), and an unsupervised learning approach (*e.g.* some experimentally determined data are examined for the first time, so they can be represented in the most informative way) [46].

To date, several different ANN methodologies are devised, of which for the purposes of ecotoxicological studies, precisely for building a robust and predictive QSAR model for prediction of biological toxicity values for toxicants, two ANNs deserve a special attention: Kohonen Artificial Neural Networks (KANNs) as well as Counter-Propagation Artificial Neural Networks (CP ANNs). Both methodologies are described in the literature in detail and therefore we give here only a short summary [44, 45, 47].

The Kohonen Artificial Neural Networks (KANNs) alias Self Organizing Maps (SOM) represents a basic type of ANNs. Here, the input variables (*e.g.* calculated molecular descriptors for a series of compounds) that form the multi-dimensional descriptor space are mapped onto a two-dimensional network of neurons [48]. This mapping procedure is progressing through a non-linear algorithm commonly known as training (or learning of the network). During the learning phase, KANN is automatically suiting itself to the input objects, so the similar objects are associated with topologically close neurons on the network (*i.e.*, the similar objects will excite the neurons that are physically located close to each other on the network). The Kohonen learning can be literally described as "*winner takes all strategy*", *i.e.*, for each input only the most excited (or so-called "*winning neuron*") is selected and therefore the correction is performed around this "*winning neuron*". The selection of the most excited neuron ("*winning*" or central neuron) is based on the Euclidean distances between the neurons (between the vectors of weights and objects), *i.e.*, the minimal determined distance gives the "*winning neuron*" (*Eq. 3*) [45]:

$$d_{j,s}^{Eucl} = \sqrt{\sum_{i=1}^{m}\left(W_{j,i} - X_{s,i}\right)^2}, \min\left\{d_{j,s}^{Eucl}, j = 1 \cdots N_{net}\right\} \rightarrow W_c \tag{3}$$

where W_j are the vectors of weights ($w_{j1}, w_{j2}, w_{j3}, \ldots, w_{ji}, \ldots, w_{jm}$), X_s are the objects ($x_{s1}, x_{s2}, x_{s3}, \ldots, x_{si}, \ldots, x_{sm}$), N_{net} is the total number of neurons in the network, whereas W_c designates the selected "*winning*" or central neuron. Once the "*winning neuron*" is found and selected, the correction of its weights as well as the weights in the neighboring neurons is performed (*Eq. 4*) [45]:

$$\Delta w_{j,i} = \eta\left(1 - \frac{d}{p+1}\right)\left(x_{s,i} - w_{j,i}^{old}\right), d = 0,1,2,\ldots, p \tag{4}$$

where η designates the learning rate, d is topological distance counting rings of neurons between the central neuron (W_c) and the j-th neuron for which the correction should be performed, while p designates the maximum topological distance to which the correction is applied (*Eq. 5*) [45]:

$$p = \left(1 - \frac{n_{epochs}}{n_{tot}}\right) N_{net} \tag{5}$$

At the beginning of the training procedure the correction is applied to the whole network, $n_{epochs} = 0$, and therefore $p = N_{net}$, whereas at the end of the training, $n_{epochs} - n_{tot}$, so $p = 0$, what mcans that in the last epoch of the learning process, only the weights of the central neuron are corrected. The result of the training represents a two-dimensional array of neurons (also known as *top-map*), which are occupied with objects in such a way that similar objects are positioned close to each other or on the same neuron. However, here we would like to stress that the trained network is not only a look up table of objects, but also a model, because the so-called empty neurons that were not hit by any object at the end of the training, contain interpolated information from neighboring neurons.

The Counter-Propagation Artificial Neural Networks (CP ANN) can be defined as an extension of KANN [45, 49]. In comparison with KANN, here the toxicity or target values (*e.g.* pT) are additionally included in the modeling procedure. Pictorially, they can be represented exactly as KANNs (which contain only one layer of neurons), but extended by another layer of neurons positioned exactly below the Kohonen layer. This additional layer is functioning exactly as the Kohonen layer, *i.e.*, it accepts the target variables (t_{si}) of the target vector T_s in exactly the same way as the Kohonen layer accepts the input variables (x_{si}) of the input vector X_s, Fig. (**3**), only that it does not influence to the positioning of objects (does not contribute to the Euclidean distance, *Eq. 3*).

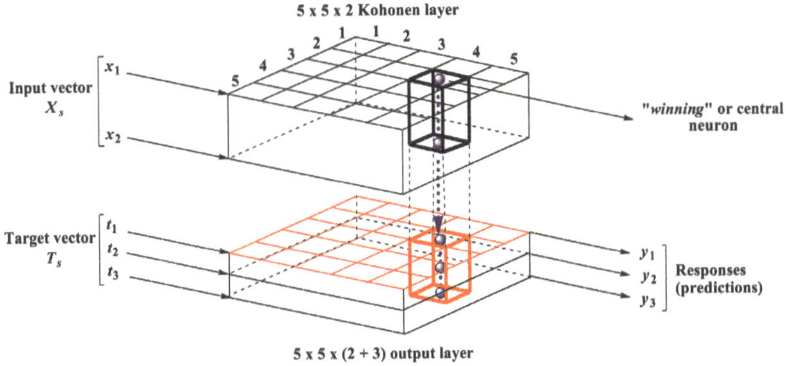

Figure 3: Counter-propagation artificial neural networks (CP ANNs) architecture.

Compared to the KANN, where the learning procedure is performed in one-step (*e.g.* solving unsupervised learning problems) the learning algorithm in CP ANN runs in two steps. The first step (or unsupervised learning procedure) that corresponds to the mapping of objects onto the input layer (Kohonen layer) is exactly the same as the KANN learning procedure described previously, whereas the second step corresponds to a supervised learning procedure, *i.e.*, for each input in the learning procedure, the target or response value is required. Consequently, the training of the network is possible with a set of input-target pairs $\{X_s, T_s\}$, where T_s designates the input vector (target vector) that carries the dependent variables (*e.g.* toxicity values). The output layer is consisting of different number of planes of weights as the target vector T_s has responses. In CP ANN, the correction of the weights (u_{ji}) in the output layer is performed according to the following equation (*Eq. 6*) [48, 49]:

$$\Delta u_{j,i} = \eta\left(1-\frac{d}{p+1}\right)\left(x_{s,i}-u_{j,i}^{old}\right), d = 0,1,2,\ldots,p \qquad (6)$$

which is the same as (*Eq. 4*), except that in this case the weights (u_{ji}) are adapted to the target values (t_{si}), and not to the input variables (x_{si}) like in the Kohonen learning. The output layer is carrying the predictive capability of the CP ANN and therefore it can be successfully employed for prediction of the biological toxicity values not only for existing known compounds, but also for newly synthesized compounds.

Case Study: Predicting the Marine Algal Toxicity of Phenols Using QSAR Approach

Industrial production of organic compounds carrying the structure of phenol dates back to 1860's [50]. Since then they have been used in the production of synthetic resins, dyes, pharmaceuticals, pesticides, synthetic tanning agents, perfumes, lubricating oils and solvents [51]. Owing to their widespread use, they have been detected in terrestrial and aquatic food chains [52] and in environmental samples, particularly in those obtained from aquatic ecosystems [53]. Due to the environmental concern regarding the persistence and adverse effects of this class of compounds, numerous studies have focused on determining their toxicity to various aquatic organisms [29].

Toxicity data has been generated throughout decades and this resulted in the accumulation of valuable information regarding the toxic potency of chemicals, which in turn have been exploited by regulatory schemes to counter the possible threats of toxicants to biodiversity and ecosystems. However, due to the wider availability of test methods and the early assumption that freshwater ecosystems were at greater risk [15], such information is more abundant for freshwater species as opposed to organisms living in marine environments.

The data gap in marine ecotoxicity has been highlighted in literature [54, 55] with emphasis on the lack of sufficient algal data [56, 57]. Recently, the European Center for Ecotoxicology and Toxicology of Chemicals (ECETOC) workshop on the probabilistic approaches for marine hazard assessment [15] once again stressed the need for marine algal toxicity data to be able to fill the data gap present in marine ecotoxicity. However, the data requirement in marine algal toxicity is almost impossible to be supplied through exhaustive laboratory testing considering the time and money needed to fulfill such an epic undertaking. As the experimental determination of toxicity is costly and time consuming, approaches such as QSARs can be exploited in efforts to predict the toxicity of untested chemicals.

We present a novel dataset comprising the toxicity of phenol, 19 chlorophenols and 10 poly-phenols to marine alga *Dunaliella tertiolecta* [58]. The data will be explored from a modeling perspective in search for statistically sound and mechanistically interpretable QSARs through the use of powerful CP ANN methodology. Using the toxicity data generated for phenols in algal growth inhibition bioassays employing *D. tertiolecta* as test species, we attempt to demonstrate how QSAR methodology can be applied successfully to marine algal data.

MATERIALS AND METHODS

Test Chemicals

The test compounds were purchased from Sigma-Aldrich, Co., with the exception of phenol, which was obtained from Merck & Co., Inc. The majority of the chemicals had purity ≥98%. No further purification was undertaken.

Stock solutions of the compounds were prepared in artificial seawater prepared according to standard methods [13]. However, the stock solutions of tetrachlorophenols, pentachlorophenol, chlorohydroquinone and tetrachlorohydroquinone were prepared in dimethyl sulfoxide (DMSO) because of their low water solubility. For the tests using these compounds, an additional solvent control containing the maximum DMSO concentration (0.1% v/v) was employed. Statistical analysis using *t*-test revealed no significant difference (p>0.05) between the growth in controls with and without DMSO. The inhibitory concentration of these chemicals was calculated taking the growth in solvent controls into account.

Algal Growth Inhibition Assay Using *D. tertiolecta*

Algal growth inhibition tests were performed in batch cultures according to the standard procedures [13, 59] using marine alga *D. tertiolecta*. The inoculum was prepared with algae harvested from four to five-day old cultures. Each milliliter of inoculum contained approximately 10^4 cells. Experiments were carried out in a temperature controlled growth chamber (18 ± 0.5 °C) under continuous illumination (2000 lux ± 10% at the level of test solutions). Experiments were conducted in three replicates using five concentrations of the test chemical. 100 mL test medium with algae was dispensed into sterile 500 mL borosilicate Erlenmayer flasks.

Bioassays were carried out in filtered (GF/C Glass microfiber Whatman filters, England) seawater enriched with modified f/2 medium [60]. Seawater samples were collected from the coast of Samatya in Istanbul and stored in a freezer at –24 °C in a plastic container after filtration. Natural seawater characterization was made based on the standard procedures [13]. Additionally, the concentration of environmentally significant heavy metals in seawater was measured using Inductive Coupled Plasma (Perkin Elmer, Optima 2100 DV) and were found to be in trace levels to produce an adverse effect on algae (results not shown). Following the recommendations of OECD [59] that the increase in *p*H between the beginning and end of the test should not exceed 1.5 units and the coefficient of variation (CV) should be less than ≤10% among the controls, any replicate that did not concur with these criteria was removed from analysis.

Toxicity Determination

The growth response of *D. tertiolecta* exposed to each of the studied chemicals was determined by daily measurements of optical density at 680 nm (OD_{680}) with a spectrophotometer (Schimadzu, UV-1208) over 96 hours.

The response variable, yield, for control and treatments, was calculated as recommended in OECD Guidelines [59]. We verified that there is a linear relationship between algal cell counts and optical density (R^2=0.97). Therefore, optical density was used as a surrogate measure for the calculation of response variable for *D. tertiolecta* to express biomass increase during the tests.

Percent inhibition relative to the controls against the test substance concentration was fitted using polynomial regression to calculate the concentration that inhibited the algal growth by 50% (IC_{50}). Curve fitting was carried out in SPSS 18.0 [61] and the roots of the polynomials were calculated using the Scientific Workplace software [62].

Descriptor Calculation

The three dimensional structures of the test compounds were sketched and geometrically optimized using semi-empirical PM3 method in Spartan 06 [63]. Using the appropriate representations of the molecules, 1664 and 818 molecular descriptors were originally retrieved from the Dragon 5.4. [64] and the Codessa Pro [65], respectively. Constant descriptors were excluded to minimize the redundant information. The reduced set of 1762 molecular descriptors was then imported to the Codessa 2.2. [66], where the significant descriptor pairs were selected using the heuristic search method for the subsequent model construction process. The heuristic search was restricted to two variables per model to be able to develop a simple QSAR that would allow a straightforward mechanistic interpretation of the relationship.

Model Development, Data Splitting and Validation

CP ANN models were developed using the modules developed at the Slovenian National Institute of Chemistry [45]. Model quality was assessed for fit based on the coefficient of determination (R^2), and the leave-one-out cross-validation (R^2_{cv}).

To be able to test the predictive ability of the CP ANN model, the marine algal toxicity data was split into training/test sets prior to model development as follows: The compounds were ordered according to their descending toxicity values. Out of 30 compounds, 6 clusters were formed which consist of 5 compounds. The most toxic and the least toxic compounds were assigned to the training set. The 3rd compound in each cluster was assigned to the test set. As a result, training and test sets were formed with 24 and 6 compounds, corresponding to 80% and 20% of the dataset, respectively. The test set was used after model development for the external validation of the proposed model.

The predictive capacity of the model was estimated on the predicted values in test sets by Q^2_{F3} (*Eq. 7*) [42].

$$Q^2_{F3} = 1 - \frac{\left[\sum_{i=1}^{n_{test}} (\hat{y}_i - y_i)^2\right] / n_{test}}{\left[\sum_{i=1}^{n_{tr}} (y_i - \bar{y}_{tr})^2\right] / n_{tr}} \quad (7)$$

where y_i and $\hat{y}_i$ are, the measured and predicted (over the test set) values of the dependent variable, respectively; $\bar{y}_{tr}$ is the mean value of the dependent variable for the training set. The summations in the nominator and denominator cover all the compounds in the test and training set, respectively.

RESULTS AND DISCUSSION

Toxicity data [$\log(IC_{50}^{-1})$, in mmol/L, denoted as pT] of 30 phenols determined at the end of 48h (pT_{48}), 72h (pT_{72}) and 96h (pT_{96}) exposures to *D. tertiolecta* are presented in Table **1** together with the values of the descriptors used in the models. In all exposure durations, resorcinol was found to be the least toxic and pentachlorophenol was the most toxic compound tested. As expected, the correlation of the toxicity data determined at the end of 48h, 72h or 96h was in excellent agreement among themselves, Fig. (**4**).

The results revealed that the toxicity of phenols decreased with increasing exposure duration. The decrease in the toxicity of phenols was attributed to an increase in pH of the test medium due to photosynthetic activity of algae, which in turn led to reduced bioavailability of phenols. Also, it is likely that algae acclimated to the test compounds during the bioassays which rendered phenols less toxic as exposure duration increased.

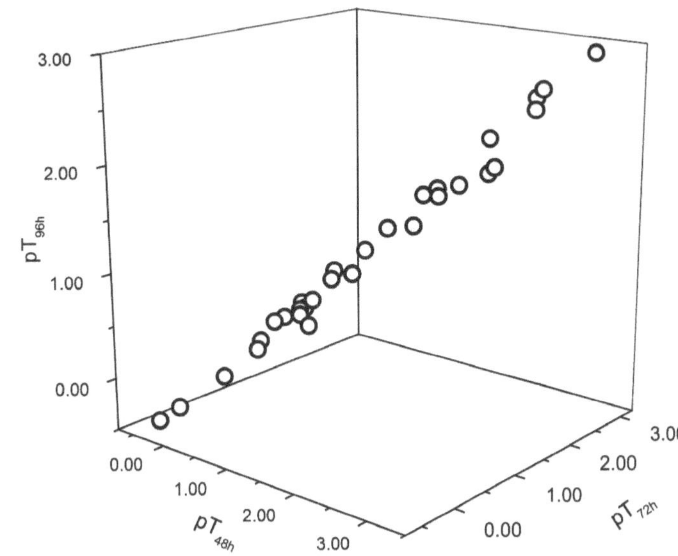

Figure 4: 3D-plot of the toxicity values determined at the end of 48h, 72h and 96h exposures.

The results revealed that the toxicity of phenols decreased with increasing exposure duration. The decrease in the toxicity of phenols was attributed to an increase in *p*H of the test medium due to photosynthetic activity of algae, which in turn led to reduced bioavailability of phenols. Also, it is likely that algae acclimated to the test compounds during the bioassays which rendered phenols less toxic as exposure duration increased.

Marine Algal QSAR

The aim of this case study was to develop QSARs with predictive ability using the toxicity of 30 phenolic compounds to the marine algae *D. tertiolecta.* In order to find a relationship between the marine algal toxicity of phenols and structural features of the chemicals, a diverse set of molecular descriptors was used, which

Table 1: Marine algal toxicity of phenols and descriptor values

ID	Compound	pT_{48h}	pT_{72h}	pT_{96h}	Mor24m	C_{ortho}
1	Phenol	-0.24	-0.28	-0.30	0.068	-0.1444
2	2-chlorophenol	0.44	0.34	0.33	0.031	-0.1314
3	3-chlorophenol[P]	0.91	0.62	0.49	0.124	-0.1471
4	4-chlorophenol	0.81	0.67	0.63	0.107	-0.1656
5	2,3-dichlorophenol	0.98	0.92	0.88	0.170	-0.1350
6	2,4-dichlorophenol	1.23	1.20	1.14	0.103	-0.1551
7	2,5-dichlorophenol	0.98	0.97	0.95	0.042	-0.1331
8	2,6-dichlorophenol	0.43	0.30	0.25	0.004	-0.1207
9	3,4-dichlorophenol[P]	1.43	1.35	1.35	0.253	-0.1777
10	3,5-dichlorophenol	1.71	1.46	1.39	0.155	-0.1508
11	2,3,4-trichlorophenol	1.88	1.68	1.65	0.329	-0.1671
12	2,3,5-trichlorophenol	1.78	1.78	1.69	0.209	-0.1388
13	2,3,6-trichlorophenol	0.80	0.60	0.58	0.144	-0.1248
14	2,4,5-trichlorophenol[P]	1.99	1.90	1.73	0.199	-0.1659
15	2,4,6-trichlorophenol	0.75	0.68	0.67	0.099	-0.1472
16	3,4,5-trichlorophenol	2.22	2.15	2.15	0.373	-0.1895
17	2,3,4,5-tetrachlorophenol	2.59	2.46	2.41	0.448	-0.1786
18	2,3,4,6-tetrachlorophenol	2.29	2.04	1.85	0.318	-0.1593
19	2,3,5,6-tetrachlorophenol	2.32	2.11	1.90	0.254	-0.1300
20	Pentachlorophenol	3.03	2.96	2.91	0.515	-0.1714
21	Pyrogallol (1,2,3-trihydroxybenzene)	0.59	0.57	0.53	0.141	-0.1969
22	Hydroquinone	2.56	2.52	2.51	0.097	0.0595
23	Chlorohydroquinone[P]	2.59	2.59	2.58	0.151	0.0706
24	Tetrachlorohydroquinone	1.68	1.66	1.64	0.104	0.0910
25	Catechol[P]	0.73	0.67	0.60	0.107	-0.1042
26	4-chlorocatechol	0.86	0.74	0.70	0.123	-0.1262
27	3,5-dichlorocatechol	1.19	1.03	0.94	0.182	-0.1077
28	Resorcinol	-0.42	-0.42	-0.43	0.099	-0.2387
29	4-chlororesorcinol[P]	0.16	0.04	0.00	0.133	-0.2641
30	4,6-dichlororesorcinol	0.51	0.49	0.49	0.146	-0.2538

[P] Compound in the test set.

was finally reduced to two descriptors, Mor24m and C_{ortho}. The former one, Mor24m, depicts the 3D molecular structure representation based on electron diffraction weighted by molecular mass. It is obtained from 3D MoRSE descriptors [67, 68] that are derived from infrared spectra simulation using a generalized scattering function [69]. The second selected descriptor, C_{ortho}, reflects the partial charge on the carbon atom located at the ortho position relative to the OH moiety. A more detailed description is given below, under "Mechanistic interpretation of the CP ANN model".

The entire dataset was previously divided into training set and test set and good agreement was determined between the screened independent variables (selected molecular descriptors, *i.e.*, Mor24m and C_{ortho}) and toxicity values (pT_{48h}, pT_{72h}, and pT_{96h}). The modeling was performed simultaneously for all three exposure durations (pT_{48h}, pT_{72h}, and pT_{96h}) on the training set (24 compounds), utilizing the internal cross-validation leave-one-out (CV LOO) procedure for optimization of the technical parameters. For these purposes, different network architectures (determined by the dimensionality of the two-dimensional network, *i.e.*, the number of neurons in X and Y direction) as well as different number of epochs (training iterations) were employed (Table **2**). In order to cover all investigated objects (compounds from the training set (24)), as starting optimal network dimensionality, a network of $n_x = n_y = 5$ (n_x x n_y = 5 x 5 = 25) was selected, where n_x is the total number of neurons in X direction, while n_y is total number of neurons in Y direction. A total of 36 CP ANN models were built, by increasing the dimensionality of the network by one more neuron ((n_x + 1) x (n_y + 1)) as well as the number of training epochs. For our purposes, three different network architectures (5x5, 6x6, and 7x7) were trained with different number of training epochs (Table **2**). Since the experimentally determined toxicity values between each case (pT_{48h}, pT_{72h}, and pT_{96h}) are highly correlated, Fig. (**4**), the selection of the best CP ANN model was performed according to the best averaged $R^2_{cv\_avg}$ value (Table **3**) obtained using the following equation (*Eq. 8*):

$$R^2_{cv_{avg}} = \frac{R^2_{cv_{48}h} + R^2_{cv_{72}h} + R^2_{cv_{96}h}}{3} \tag{8}$$

Where $R^2_{cv\_48h}$, $R^2_{cv\_72h}$, and $R^2_{cv\_96h}$, are the coefficients of correlation for cross-validation leave-one-out (CV LOO). Such a procedure resulted in obtaining of a

Table 2: Training set learning strategy details using CV LOO[1]

Training Set		NN-Architecture																	
24 Compounds		5 x 5						6 x 6						7 x 7					
		48h		72h		96h		48h		72h		96h		48h		72h		96h	
		R^2	R^2_{cv}	R^2	R^2_{cv}	R^2	R^2_{cv}	R^2	R^2_{cv}	R^2	R^2_{cv}	R^2	R^2_{cv}	R^2	R^2_{cv}	R^2	R^2_{cv}	R^2	R^2_{cv}
Learning Epochs	10	0.8827	0.7040	0.8714	0.6839	0.8709	0.6818	0.8881	0.6390	0.8759	0.6209	0.8765	0.6269	0.8855	0.6918	0.8818	0.6778	0.8822	0.6788
	15	0.8927	0.6868	0.8853	0.6710	0.8829	0.6666	0.8939	0.6765	0.8875	0.6657	0.8887	0.6656	0.9011	0.6763	0.8971	0.6551	0.8975	0.6562
	20	**0.9105**	**0.7131**	**0.9034**	**0.7076**	**0.9022**	**0.7088**	0.9043	0.6853	0.8997	0.6821	0.9007	0.6847	0.9178	0.6780	0.9119	0.6638	0.9111	0.6637
	25	0.9136	0.6682	0.9059	0.6602	0.9057	0.6621	0.9172	0.6929	0.9108	0.6830	0.9085	0.6820	0.9255	0.6623	0.9208	0.6523	0.9181	0.6566
	50	0.9340	0.6536	0.9238	0.6555	0.9198	0.6567	0.9450	0.6633	0.9379	0.6449	0.9384	0.6454	0.9401	0.6468	0.9374	0.6385	0.9369	0.6403
	100	0.9370	0.6431	0.9264	0.6391	0.9224	0.6427	0.9338	0.6556	0.9281	0.6438	0.9284	0.6444	0.9558	0.6486	0.9556	0.6428	0.9528	0.6447
	150	0.9327	0.6097	0.9229	0.6032	0.9213	0.6111	0.9397	0.6506	0.9338	0.6377	0.9310	0.6362	0.9494	0.6508	0.9460	0.6311	0.9431	0.6272
	200	0.9371	0.6256	0.9265	0.6224	0.9225	0.6292	0.9355	0.6719	0.9281	0.6512	0.9275	0.6566	0.9714	0.6444	0.9699	0.6292	0.9695	0.6273
	250	0.9371	0.6422	0.9265	0.6393	0.9225	0.6469	0.9647	0.6486	0.9596	0.6356	0.9577	0.6386	0.9774	0.6116	0.9773	0.5939	0.9758	0.6020
	300	0.9327	0.6055	0.9230	0.6006	0.9213	0.6091	0.9647	0.6342	0.9596	0.6187	0.9577	0.6238	0.9714	0.6346	0.9699	0.6222	0.9695	0.6231
	350	0.9327	0.6210	0.9230	0.6194	0.9213	0.6262	0.9611	0.6299	0.9544	0.6119	0.9528	0.6163	0.9774	0.6615	0.9773	0.6539	0.9758	0.6569
	400	0.9371	0.6402	0.9265	0.6379	0.9225	0.6457	0.9611	0.6501	0.9544	0.6415	0.9528	0.6418	0.9730	0.6232	0.9738	0.6025	0.9747	0.6041

[1]The best model selected on the basis of best R^2_{cv} value is highlighted in gray.

general CP ANN model (R^2_{avg} = 0.9054, $R^2_{cv\_avg}$ = 0.7098; Table **3**) which subsequently was used for prediction of experimentally determined biological

Table 3: Generalized CP ANN models obtained by averaging R^2 and R^2_{cv} values for each case from Table **2**[1]

Training Set		NN-Architecture					
24 Compounds		5 x 5		6 x 6		7 x 7	
		(48h + 72h + 96h)/3		(48h + 72h + 96h)/3		(48h + 72h + 96h)/3	
		R^2_{avg}	$R^2_{cv\_avg}$	R^2_{avg}	$R^2_{cv\_avg}$	R^2_{avg}	$R^2_{cv\_avg}$
Learning Epochs	10	0.8750	0.6899	0.8802	0.6289	0.8832	0.6828
	15	0.8870	0.6748	0.8900	0.6693	0.8986	0.6625
	20	**0.9054**	**0.7098**	0.9016	0.6840	0.9136	0.6685
	25	0.9084	0.6635	0.9122	0.6860	0.9215	0.6571
	50	0.9259	0.6553	0.9404	0.6512	0.9381	0.6419
	100	0.9286	0.6416	0.9301	0.6479	0.9547	0.6454
	150	0.9256	0.6080	0.9348	0.6415	0.9462	0.6364
	200	0.9287	0.6257	0.9304	0.6599	0.9703	0.6336
	250	0.9287	0.6428	0.9607	0.6409	0.9768	0.6025
	300	0.9257	0.6051	0.9607	0.6256	0.9703	0.6266
	350	0.9257	0.6222	0.9561	0.6194	0.9768	0.6574
	400	0.9287	0.6413	0.9561	0.6445	0.9738	0.6099

[1]The best general CP ANN model was selected according to the best $R^2_{cv\_avg}$ value (highlighted in gray).

toxicity values for the previously excluded test set compounds ($Q^2_{te\_48h}$ = 0.8610, $Q^2_{te\_72h}$ = 0.8226, $Q^2_{te\_96h}$ = 0.8074). The results of the test set predictions, together with the models for each case (48h, 72h, and 96h), respectively, are presented in Table **4** and Fig. (**5**). These results apparently show the good predictive ability of these models for the prediction of toxicity values against *D. tertiolecta* for new toxicants.

Table 4: Experimental *vs.* predicted biological toxicity values (*p*T) for training/test set compounds obtained by CP ANN*

Training Set						
ID	$pT_{48h\_exp\_tr}$	$pT_{72h\_exp\_tr}$	$pT_{96h\_exp\_tr}$	$pT_{48h\_pred\_tr}$	$pT_{72h\_pred\_tr}$	$pT_{96h\_pred\_tr}$
1	-0.240	-0.280	-0.300	0.315	0.242	0.214
2	0.440	0.340	0.330	0.602	0.521	0.493
4	0.810	0.670	0.630	1.029	0.938	0.888
5	0.980	0.920	0.880	1.133	1.014	0.948
6	1.230	1.200	1.140	1.029	0.938	0.888
7	0.980	0.970	0.950	0.602	0.521	0.493
8	0.430	0.300	0.250	0.602	0.521	0.493
10	1.710	1.460	1.390	1.401	1.219	1.158
11	1.880	1.680	1.650	2.138	1.929	1.820
12	1.780	1.780	1.690	1.651	1.584	1.497
13	0.800	0.600	0.580	0.874	0.720	0.686
15	0.750	0.680	0.670	0.830	0.742	0.716
16	2.220	2.150	2.150	2.304	2.203	2.172
17	2.590	2.460	2.410	2.757	2.652	2.598
18	2.290	2.040	1.850	2.138	1.929	1.820
19	2.320	2.110	1.900	2.244	2.043	1.866
20	3.030	2.960	2.910	2.757	2.652	2.598
21	0.590	0.570	0.530	0.713	0.658	0.618
22	2.560	2.520	2.510	2.074	2.042	2.025
24	1.680	1.660	1.640	2.074	2.042	2.025
26	0.860	0.740	0.700	0.874	0.720	0.686
27	1.190	1.030	0.940	1.133	1.014	0.948
28	-0.420	-0.420	-0.430	0.115	0.102	0.095
30	0.510	0.490	0.490	0.115	0.102	0.095
Test Set						
ID	$pT_{48h\_exp\_te}$	$pT_{72h\_exp\_te}$	$pT_{96h\_exp\_te}$	$pT_{48h\_pred\_te}$	$pT_{72h\_pred\_te}$	$pT_{96h\_pred\_te}$
3	0.910	0.620	0.490	1.401	1.219	1.158
9	1.430	1.350	1.350	1.669	1.536	1.463

Table 4: contd….

14	1.990	1.900	1.730	1.709	1.568	1.484
23	2.590	2.590	2.580	2.075	2.042	2.025
25	0.730	0.670	0.600	0.874	0.720	0.686
29	0.160	0.040	0.000	0.115	0.102	0.095

* ID designates the number of the compound in the data set; ($pT_{48h\_exp\_tr}$, $pT_{72h\_exp\_tr}$, $pT_{96h\_exp\_tr}$) and ($pT_{48h\_pred\_tr}$, $pT_{72h\_pred\_tr}$, $pT_{96h\_pred\_tr}$) are the experimental and predicted biological toxicity values for training set compounds for 48h, 72h, and 96h exposures, respectively; while ($pT_{48h\_exp\_te}$, $pT_{72h\_exp\_te}$, $pT_{96h\_exp\_te}$) and ($pT_{48h\_pred\_te}$, $pT_{72h\_pred\_te}$, $pT_{96h\_pred\_te}$) are the experimental and predicted biological toxicity values for test set compounds for 48h, 72h, and 96h exposures, respectively.

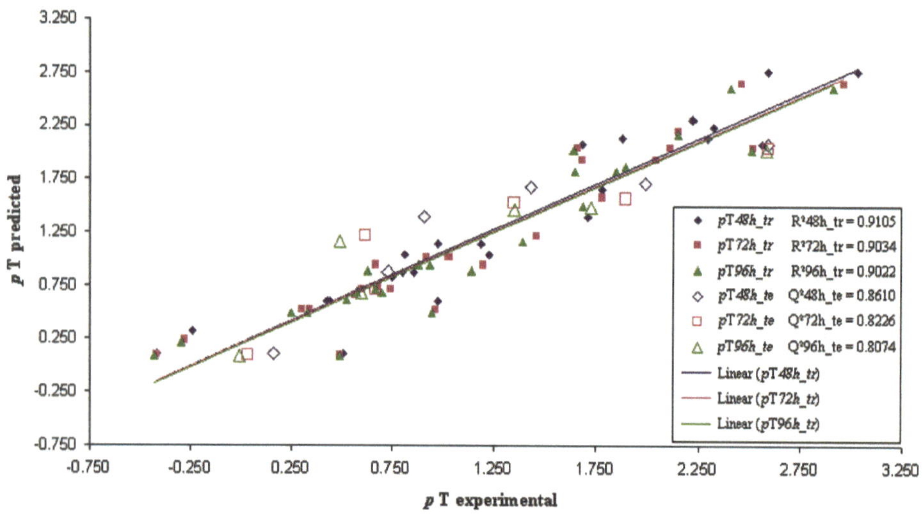

Figure 5: The predicted *vs.* experimental pT values for training/test compounds (predictions from CV LOO method were used for the training set).

Mechanistic Interpretation of the CP ANN Model

Nearly forty years ago, McFarland [70] stated that toxicity is a combination of uptake of chemicals into biological membranes and the interaction of the toxicant with the site of action.

Based on this toxicity approach, several researchers have developed simple models using small number of descriptors (known as the response-surface analyses) as a method to develop QSARs [71 – and references therein]. The main advantage of developing a QSAR that employs small number of descriptors is its ease of interpretation, compared to a model containing large number of descriptors. While the latter might be better at explaining the variance in toxicity compared to the former, the presence of a lot of descriptors makes it often difficult to interpret the results. Therefore, we preferred developing a two descriptor model that would be both good at prediction and enable a sound basis for the interpretation of marine algal toxicity of phenols towards *D. tertiolecta*, Fig. (**6**).

The developed CP ANN model employed Mor24m and C_{ortho} as the descriptor pair. Mor24m is a 3D-MoRSE descriptor (3D-MoRSEsignal 24/weighted by atomic masses) retrieved from the Dragon 5.4. These descriptors are calculated by summing atom weights viewed by a different angular scattering function [64] and provide information of a compound from its three-dimensional structure using a molecular transform derived from an equation used in electron diffraction studies [69, 72]. The other descriptor that was used in the model was C_{ortho}, which indicates the partial charge on the carbon atom located at the (2-) position relative to the OH moiety. It was retrieved from the Codessa Pro software (originally named as MOPAC Partial Charges for atom # 0000002 - Carbon).

As can be seen from Fig. (**6**), the distribution of the weights in the Kohonen layer that account for the Mor24m descriptor (plot **A**) matches very well with the distribution of the toxicity values in plot **C**, **D** and **E**, which confirms a straightforward positive correlation regarding all compounds, except hydroquinones; while C_{ortho} descriptor (plot **B**) follows a slightly different color distribution, which makes a clear separation of hydroquinones (ID = **22**, **23**, **24**) from the rest of the compounds. Indeed, closer inspection of C_{ortho} values (Table **1**) revealed that only the hydroquinones have positive partial charges on their second carbon atom, which in turn supports the color distribution noted above in that this descriptor discriminates hydroquinones from rest of the phenols in the data set.

A combination of the two descriptors, which can be explained with overlapping of plots **A** and **B**, enables an effective grouping of the compounds with similar toxicity in the CP ANN. Moreover, the analysis of the *top-map* superimposed on the output layer for 48h toxicity exposition Fig. (**6C**), shows a good distribution of the test set compounds (black numbers) over the entire network.

The descriptor's values (**A** and **B**) are normalized between 0.00 (blue - minimum value) and 1.00 (red – maximum value). The output layers for all three time expositions ($p\text{T}_{48\text{h}}$, $p\text{T}_{72\text{h}}$, and $p\text{T}_{96\text{h}}$) are presented as **C**, **D**, and **E**, respectively, where minimum toxicity is designated with blue, while maximum toxicity with red. The numbers in the plot (**C**) correspond to the positions of the compounds from the training (white) and test set (black) in the *top-map* of the Kohonen layer, which can be superimposed equally on all the layers **A**-**E**.

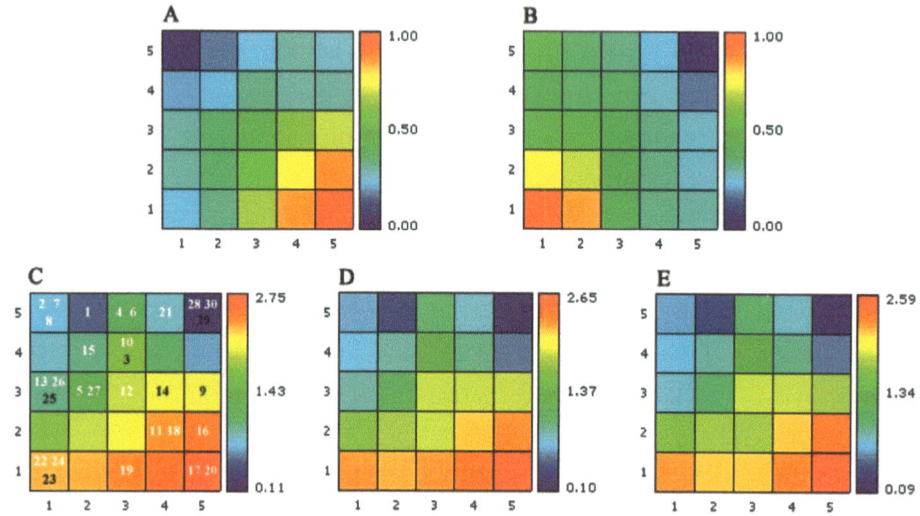

Figure 6: Output layers of the selected CP ANN model. A) Mor24m; B) C_{ortho}; C) *p*T 48h; D) *p*T 72h; E) *p*T 96h.

Both the distribution of the weights in the Kohonen layer and the analysis of the *top-map* superimposed on the output layers provide valuable insights on the role of descriptors in explaining the toxicity of phenols towards *D. tertiolecta*. Closer inspection of Mor24m revealed that chlorophenols containing chlorine atoms at 2- and 6- positions have significantly lower values compared to the congeners with the same number of chlorine atoms. Note that the *ortho* substituted chlorophenols were also found to be less toxic than the *meta* and *para* substituted congeners (Table **1**). As stated by Boyd *et al*. [73], this has been ascribed to the shielding of the OH group by *ortho* substituted chlorine atom(s). Since the 3D-MoRSE descriptors encode three-dimensional information, the conformation of the molecules probably has an impact on their toxicity as well. On the other hand, the inability of this descriptor in explaining the toxicity of hydroquinones suggests that information other than encoded in the three-dimensional structure is required to understand the adverse effects of these compounds on *D. tertiolecta*. Such information seems to be provided by the other descriptor, C_{ortho}, in order to explain the variance inflicted by hydroquinones. From McFarland's toxicity perspective, it is likely that the presence of Mor24m in the developed QSAR lies in its ability to encode relevant information for the bio-uptake process while the presence of C_{ortho} can be said to reflect the reactivity of chemicals, particularly hydroquinones, towards *D. tertiolecta*.

The reactive toxicity is an established concept in toxicological sciences and a detailed discussion on this subject was provided by Schultz *et al.* [74]. In summary, the majority of adverse effects due to the reactivity of chemicals are attributed to the reactions between an electrophilic toxicant and a biological nucleophile (*e.g.* protein and amino acid residues rich in electrons). The reactions often result in the formation of an irreversible covalent bond between the electrophile and the nucleophile; and this leads to strong interaction between the biological organism and the toxicant [74].

The toxicity of hydroquinones to *D. tertiolecta* can be attributed to the transformation of these compounds into their more electrophilic, hence more reactive, quinone forms. Satoh *et al.* [75] indicated that hydroquinones are not themselves electrophilic and require oxidative conversion to their quinone forms (*e.g.*, semi-quinone, benzoquinone) to be involved in electrophilic processes. The transformation of hydroquinones into more electrophilic quinone forms is well established. As stated by Aptula *et al.* [76], hydroquinones can undergo one-electron oxidation to semi-quinone, and being a radical anion the semi-quinone can be further oxidized to benzoquinone. Alternatively, hydroquinone can be oxidized directly to benzoquinone in a two-electron transfer reaction [76].

Experimental toxicological data for hydroquinone [30] indicate high toxicity to aquatic organisms (acute LC_{50}/EC_{50} values < 1 mg/L for fish and water flea, and as low as 0.044 mg/L for fish). The 72h IC_{50} value of hydroquinone using freshwater algae *Selenastrum capricornutum* was found to be 0.34 mg/L (0.0031 mM) [77], which is very close to 72h IC_{50} value determined in this study (0.33 mg/L – 0.0030 mM). Almost similar IC_{50} values determined for two different algal species suggest that toxicity mechanism of hydroquinone towards algae might be similar and it is very reactive in algal test systems.

CONCLUSION

D. tertiolecta has significant applications in aquatic toxicity testing ranging from detecting toxicants in raw and treated effluents to the toxicity of metals and organic compounds. As a representative species of marine ecosystems and a primary producer, *Dunaliella* has been regarded as a model test species.

Using a novel dataset containing the toxicity of 30 phenols to *D. tertiolecta*, we developed a two-descriptor QSAR using CP ANN technique considering the non-linear nature of toxicity. The developed model not only produced acceptable statistics to predict the toxicity of compounds not included in the training set, but also allowed a mechanistic interpretation regarding the toxicity of phenols towards *D. tertiolecta*. Using the relatively small data set, evidence is provided in the case study that QSAR methodology can be successfully applied to fill the data gap in marine ecotoxicity.

ACKNOWLEDGEMENTS

The support of this study by TUBITAK-ARRS (Project No: 108Y119) is gratefully acknowledged. The financial support of Boğaziçi University Research Fund (Project Number: 5564) and of Slovenian Ministry for Higher Education, Science and Technology (Grant P1-0017) are gratefully appreciated. We would also like to thank to the Slovenian Research Agency for their financial support (BI-TR/09-11-003).

CONFLICT OF INTEREST

The authors confirm that this chapter contents have no conflict of interest.

REFERENCES

[1] Boyce, D.G.; Lewis, M.R.; Worm, B. Global phytoplankton decline over the past century. *Nature*, **2010**, *466*, 591-596.

[2] DeLorenzo, M.E. Utility of *Dunaliella* in Ecotoxicity Testing. In: *The Alga Dunaliella: Biodiversity, Physiology, Genomics and Biotechnology*; Ben-Amotz, A., Polle, J.E.W., Rao, D.V.S., Eds.; Science Publishers: Enfield, New Hampshire, **2009**; pp. 495-512.

[3] Pereira, M.J.; Resende, P.; Azeiteiro, U.M.; Oliviera, J.; de Figueiredo, D.R. Differences in the effects of metals on growth of two freshwater green algae (*Pseudokirchneriella subcapitata* (Korshikov) Hindak and *Gonium pectorale* Müller). *Bull. Environ. Contam. Toxicol.*, **2005**, *75*, 515-522.

[4] Saçan, M.T.; Balcıoğlu, I.A. A case study on algal response to raw and treated effluents from an aluminum plating plant and a pharmaceutical plant. *Ecotoxicol. Environ. Saf.*, **2006**, *64*, 234-243.

[5] Chang, G-H.; Chen, H.; Chen, H-L.; Li, W.; Pan, G. Effect of sorbed and desorbed Zn(II) on the growth of a green alga (*Chlorella pyrenoidosa*). *J. Environ. Sci.*, **2007**, *19*, 1028-1031.

[6] Franz, S.; Altenburger, R.; Heilmeier, H.; Schmitt-Jansen, M. What contributes to the sensitivity of microalgae to triclosan? *Aquat. Toxicol.*, **2008**, *90*, 102-108.

[7] Scragg, A.H.; Spiller, L.; Morrison, J. The effect of 2,4-dichlorophenol on the microalga *Chlorella* VT-1. *Enzyme Microb. Technol.*, **2003**, *32*, 616–622.

[8] Olivier, S.; Scragg, A.H.; Morrison, J. The effect of chlorophenols on the growth of *Chlorella* VT-1. *Enzyme Microb. Technol.*, **2003**, *32*, 837–842.

[9] Novotny, C.; Dias, N., Kapanen, A.; Malachova, K.; Vandrovcova, M.; Itavaare, M.; Lima, N. Comparative use of bacterial, algal and protozoan tests to study toxicity of azo- and anthraquinone dyes. *Chemosphere*, **2006**, *63*, 1436-1442.

[10] Kulacki, K.J.,Lamberti, G.A. Toxicity of imidazolium ionic liquids to freshwater algae, *Green Chem.*, **2008**, *10*(1), 104–110.

[11] Cai, X.Y.; Ye, J.; Sheng, G.; Liu, W. Time-dependent degradation and toxicity of diclofop-methyl in algal suspensions. *Environ. Sci. Pollut. Res.*, **2009**, *16*, 459-465.

[12] Murkovski, A.; Skórska, E. Effect of $(C_6H_5)_3PbCl$ and $(C_6H_5)_3SnCl$ on delayed luminescence intensity, evolving oxygen and electron transport rate in photosystem II of *Chlorella vulgaris*. *Bull. Environ. Contam. Toxicol.*, **2010**, *84*, 157-160.

[13] American Public Health Association, American Water Works Association, and Water Environment Federation. *Standard Methods for the Examination of Water and Wastewater*, 20th ed.; Washington, DC, **1998**.

[14] United States Environmental Protection Agency. *Methods for Measuring the Acute Toxicity of Effluents and Receiving Waters to Freshwater and Marine Organisms*. Washington DC, **2002**.

[15] European Centre for Ecotoxicology and Toxicology of Chemicals. *Workshop on the probabilistic approaches for marine hazard assessment*, Workshop Report No. 15. Brussels, **2008**.

[16] Environmental Chemicals Bureau. *European Union Technical Guidance Document on risk assessment in support of the commission directive 93/67/EEC on risk assessment for new notified substances Commission Regulation (EC) No 1488/94 on risk assessment for existing substances directive 98/8/EC of the European Parliament and of the council concerning the placing of biocidal products on the market*. Ispra, **2003**.

[17] Ben-Amotz, A.; Polle, J.E.W.; Rao, D.V.S. (Eds.) *The Alga Dunaliella: Biodiversity, Physiology, Genomics and Biotechnology*; Science Publishers: Enfield, New Hampshire, **2009**.

[18] Lin, C.F.; Hao, O.J.; Jeng, F.T. Microtox evaluation of industrial wastewaters. *Water Sci. Technol.*, **1994**, *30*(10), 97-106.

[19] Ternes, T.A. Occurrence of drugs in German sewage treatment plants and rivers. *Water Res.*, **1998**, *32*, 3245-3260.

[20] Balcıoğlu, I.A.; Arslan, İ.; Saçan**,** M.T. Homogenous and heterogenous advanced oxidation of two commercial reactive dyes, *Environ. Technol.*, **2001**, *22*, 813-822.

[21] Balcıoğlu, I.A., Tarlan, E.; Kıvılcımdan,Ç.; Sacan, M.T. Merits of ozonation and catalytic ozonation pre-treatment to algal treatment of pulp and paper mill effluent. *Environ. Manage.*, **2007**, *85*, 918-926.

[22] Ali, M., Sreekrishnan, T.R. Aquatic toxicity from pulp and paper mill effluents: a review. *Adv. Environ. Res.*, **2001**, *5*(2), 175-196.

[23] Sponza, D.S. Application of toxicity tests into discharges of the pulp-paper industry in Turkey. *Ecotoxicol. Environ. Saf.*, **2003**, *54*(1), 74-86.

[24] Wong, S.L.; Nakamoto, J.; Wainwright, F. Detection of toxic organometalic complexes in wastewaters using algal assays. *Arch. Environ. Contam. Toxicol.*, **1997**, *32*, 358-366.

[25] Saçan, M.T.; Özkovalak, N. Interaction between synthetic dye bath and selenium in their toxicity to *Dunaliella tertiolecta* under two light intensities. *Bull. Environ. Contam. Toxicol.*, **2007**, *78*, 132-136.

[26] Saçan, M.T.; Öztay, F.; Bolkent Ş. Exposure of *Dunaliella tertiolecta* to lead and aluminum: Toxicity and effects on ultrastructure. *Biol. Trace Elem. Res.,* **2007**, *120*(1-3), 264-272.

[27] Saçan, M.T.; Balcıoğlu, I.A.; Ercan, C. Laboratory bioaccumulation of copper, lead and selenium in the marine alga *Dunaliella tertiolecta.* Metal pair situation. *Toxicol. Environ. Chem.,* **2000**, *76*, 17-27.

[28] Saçan, M.T.; Balcıoğlu, I.A. Bioaccumulation of aluminium in *Dunaliella tertiolecta* in natural seawater: Aluminium – metal (Cu, Pb, Se) interactions and influence of *p*H. *Bull. Environ. Contam. Toxicol.*, **2001**, *66*(2), 214-221.

[29] TerraTox™ database, version 3.0, Terrabase, Inc., **2006**

[30] United States Environmental Protection Agency. ECOTOXicology Database. Last update in June, 2011. www.epa.gov/ecotox [Accessed July 2011].

[31] Esposito, E.X.; Hopfinger A.J.; Madura J.D. Methods for applying the quantitative structure-activity relationship paradigm. *Methods Mol. Biol.*, **2004**, *275*, 131-213.

[32] REACH Legislation Commission of the European Communities. 2006.Regulation (EC) No 1907/2006 of the European Parliament and of the Council of 18 December 2006concerning the Registration, Evaluation, Authorisation and Restriction of Chemicals (REACH), establishing a European Chemicals Agency, amending Directive1999/45/EC and repealing Council Regulation (EEC) No793/93 and Commission Regulation (EC) No 1488/94 as well as Council Directive 76/769/EEC and Commission Directives 91/155/EEC, 93/67/EEC, 93/105/EC and2000/21/EC. *Off J Eur Union*, L 396/1 of 30.12.2006. http://eurlex.europa.eu/LexUriServ/site/en/oj/2006/l_396/l_39620061230en00010849.pdf [Accessed July 2011].

[33] http://www.oecd.org/document/2/0,3746,en_2649_34379_42926338_1_1_1_1,00.html. [Accessed July 2011].

[34] Bosque, R.; Sales, J.; Bosch, E.; Rosés, M.; García-Alvarez-Coque, M.C.; Torres-Lapasió, J.R. A QSPR study of the *p* solute polarity parameter to estimate retention in HPLC. *J. Chem. Inf. Comput. Sci.*, **2003**, *43*(4), 1240-1247.

[35] Minovski, N.; Vračko, M.; Šolmajer T. Quantitative structure-activity relationship of antitubercular fluoroquinolones. *Mol. Divers.*, **2011**, *15*(2), 417-426.

[36] Siraki, A.G.; Chan, T.S.; O'Brien, P.J. Application of quantitative structure-toxicity relationship for the comparison of the cytotoxicity of 14 *p*-benzoquinone congeners in primary cultured rat hepatocytes *Versus* PC12 Cells. *Toxicol. Sci.*, **2004**, *81*, 148-159.

[37] Free, S. J.; Wilson, J. A mathematical contribution to structure-activity studies. *J. Med. Chem.*, **1964**, *7*, 395-399.

[38] Hansch C.J., Fujita T. ρ-σ-π Analysis. A method for the correlation of biological activity and chemical structure. *J. Am. Chem. Soc.*, **1964**, *86*, 1616-1626.

[39] Lill, M.A. Multi-dimensional QSAR in drug discovery. *Drug Discov. Today*, **2007**, *12*(23-24), 1013-1017.

[40] Vedani, A.; Dobler, M.; Lill, M. A. Combining protein modeling and 6D-QSAR. Simulating the binding of structurally diverse ligands to the estrogen receptor. *J. Med. Chem.*, **2005**, *48*(11), 3700-3703.

[41] Minovski, N.; Šolmajer, T. Chemometrical exploration of combinatorially generated drug-like space of 6-fluoroquinolone analogs: a QSAR study. *Acta Chim. Slov.*, **2010**, *57*, 529-591.

[42] Consonni, V.; Ballabio, D.; Todeschini, R. Comments on the definition of the Q^2 parameter for QSAR validation. *J. Chem. Inf. Model.*, **2009**, *49*, 1669-1678.

[43] Lacy, M. E. Neural network technology and its application in chemical research. *Tetrahedron. Comput. Meth.*, **1990**, *3*(3-4), 119-128.

[44] Zupan, J.; Novič, M., Ruisánchez, I. Kohonen and counter propagation artificial neural networks in analytical chemistry. *Chemometr. Intell. Lab.*, **1997**, *38*, 1-23.

[45] Zupan, J.; Gasteiger, J. Neural Networks in Chemistry and Drug Design, Wiley-VCH, Weinheim, **1999**, p.380.

[46] Abraham, A. Meta-learning evolutionary artificial neural networks. *Neurocomput. J.*, **2004**, *56c*, 1-38.

[47] Fausett, L. Fundamentals of Neural Networks. Prentice Hall: USA, **2004**.

[48] Kohonen, T. Self-Organization and Associative Memory. Springer-Verlag: New-York, **1988**.

[49] Heicht-Nielsen, R. Counter propagation networks. *Appl. Optics.*, **1987**, *26*, 4979-4984.

[50] Duchowicz, P.; Mercader,A.G.; Fernandez, F. M., Castro, E.A. Prediction of aqueous toxicity for heterogeneous phenol derivatives by QSAR. *Chemom. Intell. Lab. Syst.*, **2008**, *90*(2), 97-107.

[51] Rayne, S.; Forest, K.; Friesen, K. J. Mechanistic aspects regarding the direct aqueous environmental photochemistry of phenol and its simple halogenated derivatives. A review. *Environ. Int.*, **2009**, *35*(2), 425-437.

[52] Jensen, J. Chlorophenols in the terrestrial environment. *Rev. Environ. Contam. Toxicol.*, **1996**, *146*, 25-51.

[53] World Health Organization. *Environmental health criteria 161: Phenol*. Geneva, **1994**.

[54] Leung, K. M.Y.; Morritt, D.; Wheeler, J. R.; Whitehouse, P.; Sorokin, N.; Toy, R.; Holt, M.; Crane, M. Can saltwater toxicity be predicted from freshwater data? *Mar. Pollut. Bull.*, **2001**, *42*(11), 1007-1013.

[55] European Centre for Ecotoxicology and Toxicology of Chemicals. *Risk assessment in marine environments*. Technical Report No. 82. Brussels, **2001**.

[56] Hutchinson, T.H.; Scholz, N.; Guhl, W. Analysis of the ECETOC aquatic toxicity (EAT) database IV- comparative toxicity of chemical substances to freshwater *versus* saltwater organisms. *Chemosphere*, **1998**, *36*, 143-153.

[57] Wheeler, J.R.; Leung, K.M.; Morritt, D.; Sorokin, N.; Rogers, H.; Toy, R.; Martin, H.; Whitehouse, P.; Crane, M. Freshwater to saltwater toxicity extrapolation using species sensitivity distributions. *Environ. Toxicol. Chem.*, **2002**, *21*, 2459-2467.

[58] Erturk, M.D. Personal communication. Institute of Environmental Sciences, part of ongoing Ph.D. thesis (supervised by Melek Türker Saçan), Boğaziçi University, Istanbul, Turkey. July, 2011.

[59] Organization for Economic Cooperation and Development. *Guidelines for testing of chemicals: Freshwater Alga and Cyanobacteria Growth Inhibition Test*. Paris, **2006**.

[60] Okay, O.; Gaines, A. Toxicity of 2,4-D acid to phytoplankton. *Water Res*., **1996**, *30*, 688–696.

[61] Statistical Package for Social Sciences (SPSS), version 18.0, **2009**.

[62] Scientific Workplace software, Version 3.0., MacKichan software, Inc. **1998**.

[63] Spartan software, Version 06, Wavefunction Inc., 2006.

[64] Dragon software, Version 5.4., Talete srl, Milan, Italy, 2007.

[65] Codessa Pro software, University of Florida, 2002.

[66] Codessa software, Version 2.2., University of Florida, 1996.

[67] Schuur, J.; Gasteiger, J. Software Development in Chemistry -Vol. 10 (Gasteiger, J., ed.), Fachgruppe Chemie-lnformation-Computer (CIC), Frankfurt am Main, **1996**.

[68] Schuur, J.;Gasteiger J.Infrared Spectra Simulation of Substituted Benzene Derivatives on the Basis of a Novel 3D Structure Representation. *Anal.Chem*., **1997**, *83*, 2398-2405.

[69] Soltzberg L.J.; Wilkins C.L. Molecular transforms: a potential tool for structure-activity studies. *J. Am. Chem. Soc*., **1977**, *99*, 439-443.

[70] McFarland, J. W. On the parabolic relationship between drug potency and hydrophobicity. *J. Med. Chem*., **1970**, *13*, 1092-1196.

[71] Cronin, M. T. D.; Aptula, A. O.; Duffy, J. C.; Netzeva, T. I.; Rowe, P. H.; Valkova, I. V.; Schultz, T.W. Comparative assessment of methods to develop QSARs for the prediction of the toxicity of phenols to *Tetrahymena pyriformis*. *Chemosphere*, **2002**, *49*(10), 1201-1221.

[72] Mercader, A. G.; Pomilio, A. B. QSAR study of flavonoids and biflavonoids as influenza H1N1 virus neuraminidase inhibitors. *Eur. J. Med. Chem*., **2010**, *45*(5), 1724-1730.

[73] Boyd, E.M.; Killham, K; Meharg, A.A. Toxicity of mono-, di- and tri-chlorophenols to lux marked terrestrial bacteria, Burkholderia species Rasc c2 and *Pseudomonas fluorescens*. *Chemosphere*, **2001**, *43*, 157-166.

[74] Schultz, T. W.; Carlson, R. E.; Cronin, M. T. D.; Hermens, J. L. M.; Johnson, R.; O'Brien, P. J.; Roberts, D.W.; Siraki, A.; Wallace, K.B.; Veith, G.D. A conceptual framework for predicting the toxicity of reactive chemicals: Modeling soft electrophilicity. *SAR QSAR Environ. Res*., **2006**, *17*(4), 413-428.

[75] Satoh, T.; Saitoh, S.; Hosaka, M.; Kosaka, K. Simple ortho- and para-hydroquinones as compounds neuroprotective against oxidative stress in a manner associated with specific transcriptional activation. *Biochem. Biophys. Res. Commun*., **2009**, *379*(2), 537-541.

[76] Aptula A.O.; Roberts, D.W.; Cronin, M.T.D.; Schultz, T.W. Chemistry-toxicity relationships for the effects of di- and trihydroxybenzenes to *Tetrahymena pyriformis*. *Chem. Res. Toxicol*., **2005**, *18*(5), 844-854.

[77] Devillers, J; Boule, P.; Vasseur, P.; Prevot, P.; Steiman, R.; Seigle-Murandi, F.; Benoit-Guyod, J.L.; Nendza, M.; Grioni, C.; Dive, D.; Chambon, P. Environmental health risks of hydroquinone. *Ecotoxicol. Environ. Saf*., **1990**, *19*(3), 327-354.

CHAPTER 8

Anti-Tubercular Drug Designing Using Structural Descriptors

Manish C. Bagchi[1,*] and Payel Ghosh[2]

[1]School of Bioscience & Engineering, Jadavpur University, Kolkata 700032, India and [2]Department of Biotechnology, University of Pune, Pune-411007, India

Abstract: As *Mycobacterium tuberculosis* has become drug resistant, we need to design new anti-tuberculosis lead compounds. This review has focused on the application of various chemoinformatics methods that can be used in an attempt to search for potent and selective inhibitors against *M. tuberculosis*. One of the "rational" approaches towards designing such novel anti-bacterials is to develop and deploy QSAR of similar drug-like compounds, that helps in implementing and improvising the techniques and shares some newly identified potential anti-TB drug candidates. Here, we have also mentioned about some of the QSAR models developed and validated by various groups as well as our team for several derivatives showing anti-tuberculosis activity *viz.* fluoroquinolones, quinoxaline and nitrofuranyl amide derivatives *etc.* As the calculation of diverse physicochemical properties for such huge number of compounds is time consuming and also not cost-effective, we have utilized molecular descriptors for regression modeling. Among different types of descriptors, the study has also been extended to understand the influence of each class of molecular descriptor for predicting structure-activity relationships, and the results indicate the preeminence of topological descriptors over other descriptor lessons. The methodologies described in this review are non specific and applicable to other syndromes also.

Keywords: Fluoroquinolone, molecular descriptors, molecular docking, molecular similarity, nitrofuranyl amide, QSAR, quinoxaline, statistical regression models, tuberculosis, virtual screening.

Computer-Aided Drug Design against Tuberculosis

Traditionally, coincidence and trial and error have played a major role in the discovery of drugs. In addition to having many uncertain phases, drug discovery

***Corresponding author Manish C. Bagchi:** School of Bioscience & Engineering, Jadavpur University, Kolkata 700032, India; E-mail: mcbagchi@hotmail.com

Subhash C. Basak, Guillermo Restrepo & José L. Villaveces (Eds)
Copyright © 2015 Bentham Science Publishers Ltd. Published by Elsevier Inc. All rights reserved.
10.1016/B978-1-68108-053-6.50008-9

involves a huge amount of man power and monetary expenses together with a minimum time of around fifteen years for a successful completion. None can ignore the importance of scientific process of screening in modern drug discovery programs [1]. Among the wide gamut of medicinal chemistry programs, latest techniques of organic synthesis and pharmacological testing methods for lead identification dominate the current drug discovery. But the whole system of such discovery processes is not limited within the science of medicinal chemistry only [2, 3]. High throughput screening, protein-ligand docking, protein-protein interactions, drug-protein interactions are several other important areas playing key roles leading to drug like molecules [4, 5]. In computational modeling, one is mainly concerned with design of new lead compounds using computer programs relating to chemotherapeutic and other causative agents which ultimately helps in predicting important biological properties at all stages of drug discovery [6, 7]. Such *in silico* design will definitely boost up the opportunity of predicting novel compounds with useful properties and will reduce the time as well as high cost of experimental research in the drug development strategy. Rational drug design with an aim of arriving at some lead compound is an ever expanding area of research [8-10]. The canvass of the entire gamut of studies in molecular design being too wide, the key feature restricts itself to the development of some leading aspects including molecular similarity analysis, virtual library design and most importantly, a rationalized and systematized approach, better known as structure-activity-property relationship. Although theoretical /computer-aided drug design (CADD) approach cannot replace laboratory experiment with actual molecule, it can give a clear insight by generating and analyzing massive amounts of data in relatively short period of time. This chapter explores those facets of computational aspects used in molecular modeling that have appreciable interfaces with corresponding extant studies in anti tubercular drug design. The focus is mainly on the applications of computational biology in a novel way with a view to drug design and, particularly emphasising on tuberculosis (TB), which is one of most critical global health concerns, among infectious diseases

Application of Chemoinformatics

Tuberculosis (TB) is a devastating infectious disease worldwide, and one of the most common causes of serious respiratory disability. If it is left untreated or not properly treated for a long time, it becomes fatal [11]. More than a century ago, tuberculosis (TB), was ranked among the most feared and dreaded of the numerous diseases afflicting mankind. The discovery of the causative agent of tuberculosis by Robert Koch in 1882 gave great hope that this 'white plague' would soon be vanquished. During the 20th century new anti-tuberculosis drugs were discovered. Chemotherapy of tuberculosis was in progress with the detection of some active anti-TB agents and plans and programs to combat the disease were formulated only in recent years [12]. By 1980's the great majority of patients could be cured by treatment with short-course curative chemotherapy that was possible as a consequence of several newly introduced anti-tuberculosis drugs. Only one decade later, an outbreak of HIV-related multidrug-resistant tuberculosis in New York made tuberculosis suddenly the centre of attention. More than one century after Koch's discovery, far from celebrating the final conquest of tuberculosis, it was declared as a disaster by WHO. Both academia and drug industries joined hand in hand for developing efficient computational methods with a view to accurate predictions and less animal sacrifice [13,14]. It is expected that such computer assisted design programs will help in the planning of chemical synthesis and biological testing in wet labs with a sharper focus. Optimization of ADME/TOX profile and accurate methods for lead selection are also some important aspects of such *in silico* drug design approaches [15]. QSAR was first described by Hansch and Fujita [16] as well as Free and Wilson [17] and are widely used in CADD. These are mainly predictive approaches employing statistical methods. In general, the term QSPR refers to the case where we are considering physico chemical properties alone and QSAR refers to the situation where structural properties are considered for prediction of biological activities. The main theme of QSAR is the relationship between chemical structure and biological response. In order to develop QSAR models, structural parameters are defined numerically at first [18]. Next, the proper chemical structural descriptors have to be chosen which might be responsible for the biological function of molecules. It is worth to mention that physic-chemical constants were only used in the early hours of QSAR research. The formulation of QSAR models against anti-tubercular drugs was started long back, when Seydel *et al.* [19] proposed a model for 2-substituted INH derivatives. After two years, an endeavour was made by Kiritsy *et al.* [20] to synthesize and develop QSAR for a set of piperazinoacethydrazones with their hydrophobic, electronic and steric parameters, but unfortunately no statistically significant correlation was found with the activity of Mtb. After almost two decades, Bergmann *et al.* [21] attempted to understand the relationship between the activity of the pyrazinoic acid esters based on QSAR studies. At the start of 21st century, the QSAR study became very popular and different labs across the world started working on this topic against a number of tubercular agents and this leads to the development of QSAR models for potential anti-tubercular derivatives like pyrroles [22], dihydropteridine [23], dihydropyridines [24] and diarylpyrroles [25]. QSAR models against various steroids like progesterone, cholesterol, ergosterol, betulin *etc.* were also developed by Rugutt & Rugutt against Mtb H37Rv and their study showed that antimycobacterial activity depends on hydrophobicity and type of substituents on the phynyl moeity on steroid backbone [26]. An application of topological indices (TI) was first integrated in the QSAR of anti-tubercular compounds by our group in 2004 and biological activity prediction was done for isoniazid derivatives, potent agents as anti-tb drugs [27, 28]. In the first study [27], theoretical structural properties of the INH derivatives were calculated using software packages and the QSAR models were built together with the physicochemical descriptors. It was seen that role of topological descriptors was dominant in the regression models. In the next piece of work [28], antibacterial activity against *Mycobacterium tuberculosis* was predicted by using 223 mathematical descriptors and 7 geometrical descriptors. A special type of linear statistical analysis like ridge regression was used to build up the models and it was seen again that the topostructural and topochemical indices were able to predict biological activities in a far better way.

One can find the use of QSAR while screening a virtual library of potent anti-tubercular compounds [29] and for extraction of active fragment in databases [30]. 3-dimensional QSAR including 3-D pharmacophore mapping has been applied for analysing ring substituted quinolines [31] and diaryloxy-methano-phenanthrene derivatives as potent antitubercular agents [32]. Besides these, some spectacular applications of 3-D QSAR concerning dTMP derivatives, thymidine analogues and nitroaromatic compounds against tuberculosis [33-35] deserves worth mentioning.

Molecular Similarity Analysis

The theory of intermolecular similarity between two compounds based on atom pairs was first proposed by Carhart [36] and in 2003 our lab [37] used Carhart atom pair method and Carhart similarity in the development of a mathematical model for the prediction of biological activities relative to a reference anti tubercular drug. In the model, emphasis was given on heteroatom atom pairs because of the fact that such atom pairs constitute heterocyclic compounds which are usually more biologically active compared to carbocyclic compounds. We have also implemented the atom-pair based similarity analysis on the second-line anti-tb agents *viz.*, fluoroquinolone derivatives [38]. Considering the efficiency spectrum, it appears that sparfloxacin is perhaps the most important agent against mycobacterial infections [39, 40]. Renau & co-workers measured the biological activities of N-1, C-7 and 8 substituted quinolone antibacterials in laboratory against two mycobacteria *viz.*, *M. fortuitum* and *M. smegmatis* [41, 42]. It is observed that substituent of N-1 greatly influences contribution of the 8 position in respect of anti-mycobacterial activity [see Fig. **1**]. To predict the influence of different sets of molecular descriptors in QSAR models based on structural modifications and activities of 69 quinolone compounds, it was necessary to subgroup the data structural similarity wise. The three categories that represent our subdivision are mainly all the 69 quinolone compounds, and compounds having more than 50% and 60% structural similarities with Sparfoxacin. Finally, from the regression summary for QSARs of Quinolone compounds, it is evident that the computed graphical invariants have a clear edge over the physico chemical data measured in the laboratory and lead to more significant predictions for N-1, C-7 and 8 substituted quinolone derivatives.

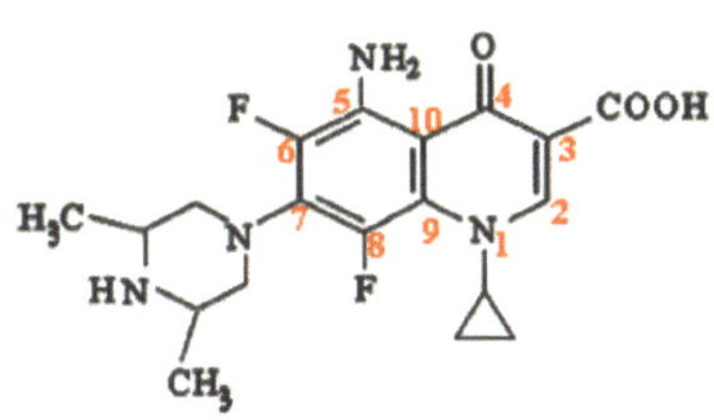

Figure 1: Structure of Sparfloxacin having activity values 0.06 & 0.13 against *M. fortuitum* and *M. smegmatis* respectively. Red marks indicate quinolone rings. Variations are at N1, C7 and 8th positions.

QSAR Modeling Employing Feature Selection

While developing QSAR models, it may be pointed out that if we consider all the computed descriptors in the model, we may obtain a very high correlation coefficient but at the same time the model becomes so complex that it may not be possible to interpret the outcome. So our goal should be to screen the significant descriptors by eliminating multicollinearity and chance correlation factors. Feature selection [43] monitors this task which acts as an important basis on a model building exercise. Genetic algorithm [44] and simulated annealing [45] have become popular optimization methods in QSAR model development and can give an effective solution with limited number of variables. Feature selection methods in conjunction with regression analysis to develop more robust models seems to be very appropriate in our study with different anti-tubercular agents like fluoroquinolones [46], quinoxaline [47] and nitrofuranyl amide derivatives [48]. In all the cases, it has been shown that application of feature selection methods appreciably improves the model quality. In one of our recent studies [46], a large number of fluoroquinolone compounds were considered for building QSAR models against two mycobacterial species. While developing the QSARs, the datasets were divided into training and test sets which were used for internal and external validation. Again, inadequacy of data sets becomes a major problem for building robust QSAR models. In such situations, merging of similar type of data sets may be the solution and therefore an appropriate method should be devised. To develop QSAR models for quinoxaline anti bacterials [49], merging of two small data sets was required. To investigate whether they can be treated as a single set of compounds, we have applied linear statistical techniques like cluster analysis and non-linear methods such as counter propagation neural network (CPNN) over the same set of theoretical structural parameters with a view to arriving at a decision regarding the unification of the two sets of data. Because of the large number of molecular descriptors and complexity of the chemical compounds due to the presence of diverse functional groups, it is essential to consider cluster analysis involving various classes of molecular descriptors. As the total number of calculated molecular descriptors for all the quinoxaline group of compounds is huge in number, it is proposed to carry out PCA prior to cluster analysis to reduce the dimensionality of the descriptors.

Since the clustering result motivates us to combine the two datasets, we have also applied non-linear technique like Kohonen analysis in the next step to verify the same. This analysis finally drives us to take decision for combining the two datasets and developing QSAR models with the merged one. It is worth to mention here that Monge and colleagues have also developed QSAR models for quinoxaline derivatives [50] against *Mycobacterium tuberculosis*.

In our investigation with nitrofuranyl amide derivatives [48], we have also applied different feature selection methods like stepwise method (SW) [51], genetic algorithms (GA) [52] and simulated annealing (SA) [45] to develop 2D QSAR models separately with the compounds having ionizable functional groups and non-ionizable functional groups and a comparative study was performed among different feature selection criteria. A total of 83 of nitrofuranyl amide derivatives were used to formulate 3D QSAR model. The q^2 value signifying the model quality based on training set molecules was obtained as 0.6471 and the model quality based on test set molecules *i.e.*, $pred\_r^2$ value was 0.6849 that clearly indicates the model's acceptability. Interaction pattern of steric and electrostatic descriptors have been discussed and the most potent molecule was identified. From the QSAR model, it is evident that the positive and negative coefficients of steric descriptors are responsible for the enhancement and damaging inhibitory activities respectively.

Molecular Docking & Combinatorial Library

Docking involves the placement of a ligand within a binding site and the prediction of the free energy of binding for such poses. The goal is to find the global energy minimum of the complex, and numerous programs have been developed to solve this non-trivial problem. One of the first docking programs was DOCK, developed by Kuntz *et al.* [53], which treated both the ligand and target as rigid bodies. Today, most algorithms deal with a flexible ligand and a rigid target, although significant effort has been devoted to also include protein flexibility, and the modeling of structural waters.

In molecular docking, the most important aspect is the calculation of binding energy so as to fit a ligand in a binding site. The binding softwares like Gold [54] and AutoDock [55] are frequently used to compute the binding affinities and scoring functions of ligands. The descriptor based QSAR models are often very useful in predicting biological activities of molecules and thus enriches the concept of virtual screening. The fast expanding protein bank information coupled with molecular descriptor based virtual screening methods supplement the whole drug design process by identifying lead molecules.

In the treatment of tuberculosis, fluoroquinolones play a major role as these are considered to be second line anti-TB agents and are very effective when the disease becomes drug resistant [56]. DNA gyrase protein is the main target of such fluoroquinolone derivatives [57, 58] and interaction pattern of fluoroquinolones and DNA gyrase is the key feature in molecular docking. Combinatorial library generation program is employed with fluoroquinolone template for obtaining a virtual library which consists of permissible substituents at particular positions. The virtual library is again screened by applying Lipinski's rule of five criteria and the remaining molecules are then tested with QSAR models for activity prediction. Molecules with high predicted activities are subjected to docking studies to scrutinize the dock score and interaction patterns. A selected number of molecules with high activity profiles, minimum dock scores and desired interaction patterns are recommended for further chemical synthesis and testing for lead identification.

CONCLUSIVE REMARKS

Genomic methods, *in silico* screening and receptor based drug design programs in the invention of new and effective anti TB drug development are being studied rapidly and a large number of new promising compounds from existing drugs are coming out as a result. Such a pattern will definitely boost up for a better perception of the biology of *Mycobacterium tuberculosis* and validation of targets in the near future.

ACKNOWLEDGEMENTS

MCB acknowledges the Council of Scientific and Industrial Research (C.S.I.R), New Delhi, India for the grant of an Emeritus Scientistship to him.

CONFLICT OF INTEREST

The authors confirm that this chapter contents have no conflict of interest.

REFERENCES

[1] Chandra, N., Computational systems approach for drug target discovery. *Expert Opinion on Drug Discovery* **2009,** *4* (12), 1221-1236.

[2] Reiss, T., Drug discovery of the future: the implications of the human genome project. *Trends Biotechnol* **2001,** *19* (12), 496-9.

[3] Searls, D. B., Data integration: challenges for drug discovery. *Nat Rev Drug Discov* **2005,** *4* (1), 45-58.

[4] Apic, G.; Ignjatovic, T.; Boyer, S.; Russell, R. B., Illuminating drug discovery with biological pathways. *FEBS Lett* **2005,** *579* (8), 1872-7.

[5] Claus, B. L.; Underwood, D. J., Discovery informatics: its evolving role in drug discovery. *Drug Discov Today* **2002,** *7* (18), 957-66.

[6] Jorgensen, W. L., The many roles of computation in drug discovery. *Science* **2004,** *303* (5665), 1813-8.

[7] Pirard, B., Computational methods for the identification and optimisation of high quality leads. *Comb Chem High Throughput Screen* **2004,** *7* (4), 271-80.

[8] Bohacek, R.; McMartin, C.; Guida, W., The art and practice of structure-based drug design: a molecular modeling perspective. *Medicinal Research Reviews* **1996,** *16* (1), 3-50.

[9] Kaye, S. B., New drug development. *Eur J Cancer* **1991,** *27* (3), 377-80.

[10] Veselovsky, A.; Ivanov, A., Strategy of computer-aided drug design. *Current Drug Targets-Infectious Disorders* **2003,** *3* (1), 33-40.

[11] Dye, C.; Watt, C.; Bleed, D.; Hosseini, S.; Raviglione, M., Evolution of tuberculosis control and prospects for reducing tuberculosis incidence, prevalence, and deaths globally. *JAMA* **2005,** *293* (22), 2767.

[12] Tripathi, R.; Tewari, N.; Dwivedi, N.; Tiwari, V., Fighting tuberculosis: an old disease with new challenges. *Medicinal Research Reviews* **2005,** *25* (1), 93-131.

[13] Kumar, M.; Verma, R.; Raghava, G. P., Prediction of mitochondrial proteins using support vector machine and hidden Markov model. *J Biol Chem* **2006,** *281* (9), 5357-63.

[14] Stahl, M.; Guba, W.; Kansy, M., Integrating molecular design resources within modern drug discovery research: the Roche experience. *Drug Discov Today* **2006,** *11* (7-8), 326-33.

[15] Kapetanovic, I. M., Computer-aided drug discovery and development (CADDD): *in silico*-chemico-biological approach. *Chem Biol Interact* **2008,** *171* (2), 165-76.

[16] Hansch, C.; Fujita, T., p- - Analysis. A Method for the Correlation of Biological Activity and Chemical Structure. *Journal of the American Chemical Society* **1964,** *86* (8), 1616-1626.

[17] Free, S.; Wilson, J., A mathematical contribution to structure-activity studies. *J Med Chem* **1964,** *7* (4), 395-399.

[18] Hansch, C.; Leo, A.; Hoekman, D., *Exploring QSAR.:. Fundamentals and applications in chemistry and biology*. An American Chemical Society Publication: 1995.

[19] Seydel, J.; Schaper, K.; Wempe, E.; Cordes, H., Mode of action and quantitative structure-activity correlations of tuberculostatic drugs of the isonicotinic acid hydrazide type. *J Med Chem* **1976,** *19* (4), 483-492.

[20] Kiritsy, J.; Yung, D.; Mahony, D., Synthesis and quantitative structure-activity relationships of some antibacterial 3-formylrifamycin SV N-(4-substituted phenyl) piperazinoacethydrazones. *J Med Chem* **1978,** *21* (12), 1301-1307.

[21] Bergmann, K.; Cynamon, M.; Welch, J., Quantitative Structure- Activity Relationships for the *in Vitro* Antimycobacterial Activity of Pyrazinoic Acid Esters. *J. Med. Chem* **1996,** *39* (17), 3394-3400.

[22] Ragno, R.; Marshall, G.; Di Santo, R.; Costi, R.; Massa, S.; Rompei, R.; Artico, M., Antimycobacterial pyrroles: synthesis, anti-Mycobacterium tuberculosis activity and QSAR studies. *Bioorganic & Medicinal Chemistry* **2000,** *8* (6), 1423-1432.

[23] Suling, W.; Maddry, J., Antimycobacterial activity of 1-deaza-7, 8-dihydropteridine derivatives against Mycobacterium tuberculosis and Mycobacterium avium complex *in vitro*. *J. Antimicrob Chemotherapy* **2001,** *47* (4), 451.

[24] Desai, B.; Sureja, D.; Naliapara, Y.; Shah, A.; Saxena, A., Synthesis and QSAR Studies of 4-Substituted phenyl-2, 6-dimethyl-3, 5-bis-N-(substituted phenyl) carbamoyl-1, 4-dihydropyridines as potential antitubercular agents. *Bioorg Med Chem* **2001,** *9* (8), 1993-1998.

[25] Biava, M.; Porretta, G.; Poce, G.; Supino, S.; Deidda, D.; Pompei, R.; Molicotti, P.; Manetti, F.; Botta, M., Antimycobacterial agents. Novel diarylpyrrole derivatives of BM212 endowed with high activity toward Mycobacterium tuberculosis and low cytotoxicity. *J. Med. Chem* **2006,** *49* (16), 4946-4952.

[26] Rugutt, J.; Rugutt, K., Relationships between molecular properties and antimycobacterial activities of steroids. *Natural Product Research* **2002,** *16* (2), 107-113.

[27] Bagchi, M. C.; Maiti, B. C.; Bose, S., QSAR of anti tuberculosis drugs of INH type using graphical invariants. *J Mol Struct-Theochem* **2004,** *679* (3), 179-186.

[28] Bagchi, M. C.; Maiti, B. C.; Mills, D.; Basak, S. C., Usefulness of graphical invariants in quantitative structure-activity correlations of tuberculostatic drugs of the isonicotinic acid hydrazide type. *J Mol Model* **2004,** *10* (2), 102-111.

[29] Besalu, E.; Ponec, R.; Vicente de Julian-Ortiz, J., Virtual generation of agents against Mycobacterium tuberculosis. A QSAR study. *Molecular Diversity* **2003,** *6* (2), 107-120.

[30] Prakash, O.; Ghosh, I., Developing an antituberculosis compounds database and data mining in the search of a motif responsible for the activity of a diverse class of antituberculosis agents. *J. Chem. Inf. Model* **2006,** *46* (1), 17-23.

[31] Nayyar, A.; Malde, A.; Jain, R.; Coutinho, E., 3D-QSAR study of ring-substituted quinoline class of anti-tuberculosis agents. *Bioorg Med Chem* **2006,** *14* (3), 847-856.

[32] Kumar, A.; Panda, G.; Siddiqi, M., CoMFA and CoMSIA 3D-QSAR analysis of diaryloxy-methano-phenanthrene derivatives as anti-tubercular agents. *J Mol Model* **2007,** *13* (1), 99-109.

[33] Aparna, V.; Jeevan, J.; Ravi, M.; Desiraju, G.; Gopalakrishnan, B., 3D-QSAR studies on antitubercular thymidine monophosphate kinase inhibitors based on different alignment methods. *Bioorg Med Chem Lett* **2006,** *16* (4), 1014-1020.

[34] Andrade, C.; Pasqualoto, K.; Ferreira, E.; Hopfinger, A., 3D-Pharmacophore mapping of thymidine-based inhibitors of TMPK as potential antituberculosis agents. *J Comp-Aided Mol Des* **2010,** *24* (2), 157-172.

[35] Tawari, N.; Degani, M., Pharmacophore mapping and electronic feature analysis for a series of nitroaromatic compounds with antitubercular activity. *J Comp Chem* **2010,** *31* (4), 739-751.

[36] Carhart, R.; Smith, D.; Venkataraghavan, R., Atom pairs as molecular features in structure-activity studies: definition and applications. *J. Chem. Inf. Comput. Sci* **1985,** *25* (2), 64-73.

[37] Bagchi, M. C.; Maiti, B. C., On application of atom pairs in drug design. *J Mol Struct-Theochem* **2003,** *623*, 31-37.

[38] Ghosh, P.; Thanadath, M.; Bagchi, M. C., On an aspect of calculated molecular descriptors in QSAR studies of quinolone antibacterials. *Mol Divers* **2006,** *10* (3), 415-27.

[39] Nakamura, S.; Minami, A.; Nakata, K.; Kurobe, N.; Kouno, K.; Sakaguchi, Y.; Kashimoto, S.; Yoshida, H.; Kojima, T.; Ohue, T.; *et al. In vitro* and *in vivo* antibacterial activities of AT-4140, a new broad-spectrum quinolone. *Antimicrob Agents Chemother* **1989,** *33* (8), 1167-73.

[40] Rastogi, N.; Labrousse, V.; Goh, K. S.; De Sousa, J. P., Antimycobacterial spectrum of sparfloxacin and its activities alone and in association with other drugs against Mycobacterium avium complex growing extracellularly and intracellularly in murine and human macrophages. *Antimicrob Agents Chemother* **1991,** *35* (12), 2473-80.

[41] Renau, T. E.; Gage, J. W.; Dever, J. A.; Roland, G. E.; Joannides, E. T.; Shapiro, M. A.; Sanchez, J. P.; Gracheck, S. J.; Domagala, J. M.; Jacobs, M. R.; Reynolds, R. C., Structure-activity relationships of quinolone agents against mycobacteria: effect of structural modifications at the 8 position. *Antimicrob Agents Chemother* **1996,** *40* (10), 2363-8.

[42] Renau, T. E.; Sanchez, J. P.; Gage, J. W.; Dever, J. A.; Shapiro, M. A.; Gracheck, S. J.; Domagala, J. M., Structure-activity relationships of the quinolone antibacterials against mycobacteria: effect of structural changes at N-1 and C-7. *J Med Chem* **1996,** *39* (3), 729-35.

[43] Guyon, I.; Elisseeff, A., An introduction to variable and feature selection. *The Journal of Machine Learning Research* **2003,** *3*, 1157-1182.

[44] Hasegawa, K.; Kimura, T.; Funatsu, K., GA strategy for variable selection in QSAR studies: enhancement of comparative molecular binding energy analysis by GA-based PLS method. *Quantitative Structure-Activity Relationships* **1999,** *18* (3), 262-272.

[45] Zheng, W.; Tropsha, A., Novel variable selection quantitative structure--property relationship approach based on the k-nearest-neighbor principle. *J Chem Inf Comput Sci* **2000,** *40* (1), 185-94.

[46] Ghosh, P.; Bagchi, M. C., Anti-tubercular drug designing by structure based screening of combinatorial libraries. *J Mol Model 17* (7), 1607-20.

[47] Ghosh, P.; Bagchi, M. C., QSAR modeling for quinoxaline derivatives using genetic algorithm and simulated annealing based feature selection. *Curr Med Chem* **2009,** *16* (30), 4032-48.

[48] Ghosh, P.; Bagchi, M. C., Comparative QSAR studies of nitrofuranyl amide derivatives using theoretical structural properties. *Molecular Simulation* **2009,** *35* (14), 1185-1200.

[49] Ghosh, P.; Vracko, M.; Chattopadhyay, A. K.; Bagchi, M. C., On application of constitutional descriptors for merging of quinoxaline data sets using linear statistical methods. *Chem Biol Drug Des* **2008,** *72* (2), 155-62.

[50] Vicente, E.; Duchowicz, P. R.; Castro, E. A.; Monge, A., QSAR analysis for quinoxaline-2-carboxylate 1,4-di-N-oxides as anti-mycobacterial agents. *J Mol Graph Model* **2009,** *28* (1), 28-36.

[51] Darlington, R., *Regression and linear models*. McGraw-Hill College: 1990.

[52] Hasegawa, K.; Kimura, T.; Funatsu, K., GA strategy for variable selection in QSAR studies: application of GA-based region selection to a 3D-QSAR study of acetylcholinesterase inhibitors. *J Chem Inf Comput Sci* **1999,** *39* (1), 112-20.

[53] Kuntz, I.; Blaney, J.; Oatley, S.; Langridge, R.; Ferrin, T., A geometric approach to macromolecule-ligand interactions. *J Mol Biol* **1982,** *161* (2), 269-288.

[54] Jones, G.; Willett, P.; Glen, R.; Leach, A.; Taylor, R., Development and validation of a genetic algorithm for flexible docking. *J Mol Biol* **1997,** *267* (3), 727-748.

[55] Morris, G.; Goodsell, D.; Huey, R.; Olson, A., Distributed automated docking of flexible ligands to proteins: parallel applications of AutoDock 2.4. *J Comp-Aided Mol Des* **1996,** *10* (4), 293-304.

[56] Blumberg, H. M.; Burman, W. J.; Chaisson, R. E.; Daley, C. L.; Etkind, S. C.; Friedman, L. N.; Fujiwara, P.; Grzemska, M.; Hopewell, P. C.; Iseman, M. D.; Jasmer, R. M.; Koppaka, V.; Menzies, R. I.; O'Brien, R. J.; Reves, R. R.; Reichman, L. B.; Simone, P. M.; Starke, J. R.; Vernon, A. A., American Thoracic Society/Centers for Disease Control and Prevention/Infectious Diseases Society of America: treatment of tuberculosis. *Am J Respir Crit Care Med* **2003,** *167* (4), 603-62.

[57] Madurga, S.; Sánchez-Céspedes, J.; Belda, I.; Vila, J.; Giralt, E., Mechanism of binding of fluoroquinolones to the quinolone resistance-determining region of DNA gyrase: towards an understanding of the molecular basis of quinolone resistance. *ChemBioChem* **2008,** *9* (13), 2081-2086.

[58] Maxwell, A., The molecular basis of quinolone action. *J Antimicrob Chemotherapy* **1992,** *30* (4), 409.

CHAPTER 9

Integrating Bioinformatics and Systems Biology for Exploring Novel Lipid Pathways in Infectious Diseases

Sonali Shinde, Vineetha Mandlik and Shailza Singh*

Computational and Systems Biology Laboratory, Lab 9, New Building, NCCS, Pune -411007, India

Abstract: Systems Biology aims to define biological problems using the language of mathematics. With the advancement of high throughput technologies and their ever expanding capabilities to generate large scale "-omics" data, the basic goal of systems biology would be to integrate global data sets and develop a coherent understanding to the biological system under study. Biological interactions are highly complex where the components of the systems are connected in a highly intricate manner. Mathematical modeling plays a major role in capturing the dynamics of each and every component in the system, simplifying complex biological networks. In the field of infectious diseases, system level understanding is essential to gain valuable insights into the pathogenic processes. A thorough understanding of the perturbations in biological networks may aid in prioritizing of the drug targets. Schistosomiasis and Trypanosomiasis remain as the two neglected tropical diseases affecting human population worldwide. The central theme of this work revolves around developing a systems level understanding of the lipid metabolism of these two parasites, abstracting complex biological processes as a collection of interacting functions driven in time by a set of discrete biological events. An insight has been laid into the importance and application of systems biology which is emerging as an amalgamation of two important sciences "Mathematics" and "Biology".

Keywords: Drug repositioning, gene regulatory network, graph theory, kinetic laws, kinetic modeling, mathematical modeling, metabolic pathway, neglected tropical diseases, schistosomiasis, stochastic simulation, systems biology, trypanosomiasis.

INTRODUCTION

In the 1960's and 1970's, several approaches were developed to study complex molecular systems, such as the metabolic control analysis and the biochemical

***Corresponding author Shailza Singh:** Computational and Systems Biology Laboratory, Lab 9, New Building, NCCS, Pune -411007; Tel: +91-20-25708296; Fax: +91-20-25692259; E-mails: shailza_iitd@yahoo.com; singhs@nccs.res.in

Subhash C. Basak, Guillermo Restrepo & José L. Villaveces (Eds)
Copyright © 2015 Bentham Science Publishers Ltd. Published by Elsevier Inc. All rights reserved.
10.1016/B978-1-68108-053-6.50009-0

systems theory. However, despite the success of molecular biology in the 1980's skepticism toward theoretical biology continued to grow and the importance of quantitative modeling of biological processes remained somewhat oblivious. With the birth of functional genomics in the 1990's, large quantities of high throughput data started becoming available with the exploding computing power, increasing the possibility of making more realistic models that would address various biological problems by defining the biological systems in a more meaningful manner. For example, in 1997, the first quantitative model of the metabolism of a whole (hypothetical) cell was published by Tomita [1]. Around the year 2000, systems biology emerged as a movement in its own right, spurred on by the completion of various genomic projects, the large increase in data from the omics and the accompanying advances in high-throughput experiments and bioinformatics. Since then, various research institutes dedicated to only systems biology have started developing in many parts of the world.

The availability of genome sequences of various organisms, alongside with accelerated development of high-throughput techniques for the systematic analysis of metabolism and gene function is posing new challenges on data analysis and interpretation [2]. The proliferation of genomics platform data is providing new insights into the interplay between human and pathogen genes during infection; the so-called 'host–pathogen interactome'. Exploiting the interactome for novel human drug targets could provide new therapeutic avenues towards the treatment of infectious disease, which could ameliorate the growing clinical challenge of drug-resistant infections [3]. Over the past few decades, many new and re-emerging diseases have posed a continuous and increasing threat to human health and welfare. Despite the availability of antibiotics and vaccines against many of the causative pathogens, the mortality rates remain high. Due to emerging pathogens and multiple drug resistance (MDR), infectious diseases continue to remain a major concern where systems biology plays a pivotal role in novel drug designing against neglected tropical diseases (NTDs).

Systems biology is an emerging field in life sciences research that studies the underlying network structure and dynamics of cells or organisms as an integral whole. It aims to investigate all interacting components simultaneously and requires data from high-throughput techniques, namely proteomics, transcriptomics and metabolomics [2]. Last couple of decades has seen remarkable advancements in the field of molecular biology which led to a better understanding of biological systems at the cellular level. High throughput technologies have enabled sequencing of the whole genomes of organisms, gathering data on temporal gene expression patterns and uncovering new relationships in metabolic and signal transduction pathways. Traditionally biology dominated by a reductionist approach focuses on specific cellular components like a single gene or single protein. Such low-throughput studies have generated a wealth of information but in a confined cellular space. Due to emergence of powerful sequencing technologies, there is an extensive parts-list of the cell and a general idea of the interactions among genes, proteins, RNA and small molecules. One of the theorists who can be seen as a pioneer of systems biology is Ludwig von Bertalanffy with his general systems theory. Metabolic and gene regulatory pathways playing crucial roles in processes like cell division, response to external stimuli and development have been identified. However, the high degree of cross-talk between these pathways has resulted in an increase of informational complexity of the system. The challenge is to assemble all the 'bits and pieces' together to create an integrated view of whole cell transactions. Mathematical modeling and computer simulation techniques have proved useful for understanding the topology and dynamics of such networks. An important aspect of modeling cellular networks is the occurrence of stochastic or random events. One of the first numerical simulations in biology was published in 1952 by the British neurophysiologists and Nobel Prize Winners Alan Lloyd Hodgkin and Andrew Fielding Huxley who constructed a mathematical model that explained the action potential propagating along the axon of a neuronal cell. Their model described a cellular function emerging from the interaction between two different molecular components, potassium and a sodium channel, and was referred as the beginning of computational systems biology. The goal of modeling in systems biology is to provide a framework for hypothesis generation and prediction based on *in silico* simulation of human disease biology across the multiple distance and time scales of an organism [4, 5]. Biological processes are highly connected and are therefore better described as a network rather than as a set of separated pathways. Thus, interactome modeling is employed for targeting, identifying and prioritizing potential targets in biological systems through network perturbation analysis [6]. Reconstruction of metabolic pathways with bottom-up approach is another step which includes different enzymatic reactions and is further needed for discrete-stochastic modeling and simulation, which allow to study the stochastic effects occurring in a system containing all the essential elements responsible for regulation of cellular metabolism [7]. Although considerable progress has been made so far in the molecular, biological characterization of several cellular metabolic pathways, but many more genes and their products still require detailed genetic and biochemical validation including detailed enzymatic studies to differentiate the pathways between human and parasites towards the development of drugs for neglected tropical diseases (NTDs).

In recent years, Schistosomiasis and Trypanosomiasis are two notable neglected tropical diseases (NTD) of concern worldwide. Schistosomiasis, also known as snail fever continues to rank second after malaria in the world's parasitic diseases in terms of prevalence, chronicity, morbidity and mortality rates [8]. Currently, over 210 million individuals in 76 countries have been affected, while close to 800 million people are at risk of contracting this disease [9]. At least 200,000 people die each year of schistosomiasis in Africa (WHO, 1993).The causative organisms for this disease are schistosomes, the blood-dwelling trematode parasites. *Schistosoma mansoni*, one of the prevalent species associated with this disease, is found to occur across much of Africa, Middle East and South America. The trypanosomiasis consists of a group of important animal and human diseases caused by parasitic protozoa of the genus *Trypanosoma*. In sub-Saharan Africa, the final decade of the 20th century witnessed an alarming resurgence in sleeping sickness or Human African Tryapanosomiasis [10]. Human African Trypanosomiasis or HAT is caused by Trypanosoma brucei rhodensiense in Eastern Africa and *Trypanosoma brucei* gambiense in Western Africa. Both forms of sleeping sickness affect the central nervous system. More than 12000 new infections are reported to the WHO each year (http//www.parasitesinhumans.org). The two subspecies do not overlap in geographic distribution. They infect humans and tsetse flies (Glossina genus) in the woodlands, savannah and the dense vegetation between Kalahari and Sahara deserts (www.parasitesinhumans.org.). *Schistosomes* exhibit complex life cycle comprising of morphologically distinct phenotypes in intermediate snail host (molluscs) and definitive human host [11].

Cercariae are the larvae of schistosomes, capable of infecting mammals (human).On the other hand; *trypanosomes* are parasites with a 2-host life cycle: mammalian and arthropod. The metacyclic trypomastigotes infect humans. According to clinical manifestations, two different conditions – acute and chronic schistosomiasis are associated with *S.mansoni* infection. Consequences of chronic infection may range from granuloma formation to fibrosis to severe mal-function of the liver and spleen [12]. HAT causes symptoms including lymphadenopathy, organomegaly, CNS damage, stupor and coma (giving rise to the name sleeping sickness), sensory disorders *etc.* (http://www.medscape.com).With the advent of safe and effective antischistosomal drugs in the 1960s and 1970s [13], morbidity control became the mainstay of schistosomiasis control (WHO, 1985, 1993).Besides antimony based drugs, Praziquantel (PZQ) has widely been used to combat this disease. But, recent reports of treatment failure after PZQ use in Egypt and Senegal [14] and in returned travelers from endemic areas [15, 16] have raised concerns about the development of drug resistance. 'Drug repositioning', an alternative approach against schistosomiasis also failed due to drug resistance (WHO, 2004b) [17]. Recent studies showed that, miltefosine, a membrane active synthetic ether lipid analogue, initially developed for cancer treatment [18] and also approved for leishmaniasis control, can act as antischistosomal drug through oral administration whenever resistance to praziquantel is reported [19]. For trypanosomiasis, suramin has been used for first-stage *T b rhodesiense* disease, whereas pentamidine is the drug of choice for treatment of first-stage disease caused by *T b gambiense*. Various combinations of eflornithine, melasoprol and nifurtimox for the treatment of second stage of *T b gambiense* showed safety, comparatively high cure rates and a good intermediary effectiveness [19]. But still there are arguments because some researchers suggest that due to unacceptable side effects number of drug-resistant strains of *T. brucei spp* has increased.

Drug resistance mechanism is related to different metabolism. Lipid metabolism is highly relevant as it plays a central role in a number of human diseases. Due to the highly interactive structure of lipid metabolism and its regulation, it is necessary to apply a holistic approach, and systems biology is therefore well suited for integrated analysis of lipid metabolism. Lipids play a crucial role in host-parasite interaction and pathogenicity in infectious diseases. *Schistosoma mansoni* and *Trypanosoma brucei* both depend on host to obtain precursors for their metabolism. Lipids are not only major constituents of biomembranes compartmentalizing the cell, but they also have a role in host recognition, immune response modulation and evasion [20]. Thus due to complex life cycles and changing environments, both organisms develop dramatic morphological and metabolic changes, including adaptation of their lipid and energy metabolism [21]. As the lipid metabolism of these two organisms differs in many aspects from that of its mammalian host, it offers potential and promising targets for the development of urgently needed, new chemotherapeutic drugs to combat these parasites [22].

Cellular metabolism is characterized by interwoven networks of metabolic compounds, biochemical fluxes and regulatory interactions [23]. After analyzing metabolic pathways, the gene regulatory network have been built which makes it possible to make an *in silico* simulation of gene and protein expression over time. A common mathematical framework for discrete-stochastic simulation of gene regulatory network (GRN) has been developed. Through GRN it is possible to identify drug targets for infectious diseases *viz*., schistosomiasis and trypanosomiasis. The strongest assumption on discrete regulation of gene or protein expression also serves as a focal point of this chapter as modeling can be done based on pure qualitative knowledge about interaction between genes or proteins. Simulation mechanisms are mainly based on certain defined kinetic laws such as Convenience Kinetics, Generalized Mass Action Kinetics and the Hill-Hinze equation. The estimation of kinetic parameters for the mathematical modeling provides a basis for iterative manipulation of biochemical pathways. The 'scale-free' structure of biological networks implies that a drug that targets the most highly connected nodes will be the most effective at disrupting cellular metabolism whereas inhibiting processes that are not highly connected to other metabolic pathways or functions may not perturb cellular metabolism sufficiently to result in antimicrobial activity [2]. 'Metabolic flux control analysis' can be used to predict the essential core reactions in the network that are required for maintaining net cellular growth under defined conditions. A drug target needs to be specific for the pathogen or, if not absent from the host; the protein needs to be sufficiently divergent from the host in sequence or mechanism of catalysis to avoid interfering with host metabolism [2]. 'Choke point analysis' is a proposed approach in this direction. The concept of choke points is used to find enzymes (edges) which uniquely consume or produce a particular metabolite (nodes) [24]. Several crucial enzymes involved in glycerolipid and glycerophospholipid metabolism in trypanosomes and also for glycerophospholipid and sphingolipid metabolism in schistosomes identified as novel drug targets are reported. Further, kinetic modeling of these enzymes based on explicit enzyme-kinetic rate equations helps in elucidation and understanding regulatory properties, physiological functions and biochemical significance of important metabolic pathways [25].

Schistosomes and *trypanosomes* owe many of their unique qualities to lipids, which play important role for their growth, survival and pathogenicity. A comprehensive model of the whole lipid metabolic pathway in these organisms has been constructed based on a detailed survey of literature and pathway databases.

Enlistment of Pathways Involved in Whole Lipid Metabolism

- Fatty acid biosynthesis.
- Fatty acid elongation in mitochondria.
- Fatty acid metabolism.
- Synthesis and degradation of ketone bodies.
- Steroid biosynthesis.
- Glycerolipid metabolism.
- Glycerophospholipid metabolism.
- Ether lipid metabolism.
- Sphingolipid metabolism.

- Arachidonic acid metabolism.
- Alpha-Linolenic acid metabolism.
- Biosynthesis of unsaturated fatty acids.

The above mentioned pathways were taken into consideration for pathway analysis of lipid metabolism of trypanosomes and schistosomes (Arachidonic acid metabolism for trypanosomes are excluded due to lack of knowledge of genes required to be incorporated in building up gene regulatory network).

Construction of Gene Regulatory Network

Genetic regulatory networks (GRN) model is a collection of all of the molecular species and their regulatory interactions that regulate the gene expression patterns. GRN's can be extremely complex and highly dynamic in nature. To unravel this complexity of the network, computational tools are essential. Predictive computational models of regulatory network can be used to describe a disease which is nothing but a perturbation in the biochemical network. Network analysis can shed light on several questions pertaining to the behavior of system under certain conditions. It describes the dynamic nature of network giving an idea about the robustness of the system to external perturbations along with the information about the essentiality of various components involved in the system. This allows us to identify the key regulators of the system which could have the potential to act as good drug targets which would facilitate the design of effective containment strategies against infectious diseases.

The basic mathematical concept used to construct a GRN is a graph. Transcriptional regulatory networks are usually modeled as directed graphs where the nodes represent the genes and the edges define the interactions between the nodes as shown in Fig. (**1**). A finite directed graph, G, consists of a set of vertices or nodes,$V(G)$ where $V(G) = v1, v2 \ldots vn)$ along with an edge set$\varepsilon\ (G) \subseteq V(G) X\ V\ (G)$.

Each edge $(u, v)\ \epsilon\ \varepsilon\ (G)$ can be thought of a connecting link between the starting node u to the terminal node v. The in-degree, $\deg_{in}(u)$ (out-degree, degout(u)) of a vertex is given by the number of edges that terminate (start) at u.

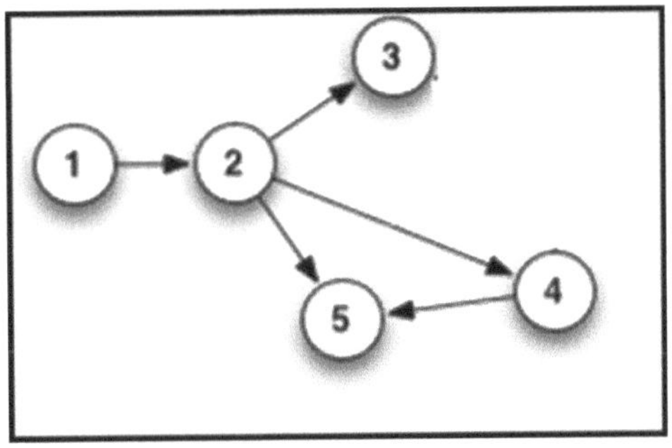

Figure 1: Example of a directed graph with 5 vertices $(\mathbf{1}, \mathbf{2}, \mathbf{3}, \mathbf{4}, \mathbf{5})$.

Adjacency matrices are constructed to understand the interactions within the GRN; the adjacency matrix for a directed graph is generally non- symmetrical. The distance between any two vertices (nodes) in the network is the number of edges in a shortest path connecting the two vertices. This is also referred to as the geodesic distance or path length. If there is some path between the two vertices then they are said to be connected, however if there is no path between the two vertices (nodes) then such nodes belong to entirely different connected components and the distance between them is infinite. For *e.g.* the distance between vertices $v2$ *and* $v5$ is 1. The network diameter is the largest distance between two nodes. If a network is disconnected, its diameter is the maximum of all diameters of its connected components. The clustering coefficient is a measure of degree to which nodes in a graph tend to cluster together. To gain further insights into the network structure and to study the clustering and modular properties of the network, clustering coefficient is calculated. The clustering coefficient C_i, defined as the ratio between the number of edges linking nodes adjacent to i and the total possible number of edges among them. In a scale free network, some nodes have low connectivity and removal of such nodes would not affect the network connectivity. However some nodes also called as 'hub nodes' are highly connected and upon the removal of such nodes will disconnect the network easily. Measures like degree centrality, closeness, betweeness and eigenvector centrality have to be computed to determine the importance of a particular node within the network. For a directed graph G, the degree centrality of a node $u \in V\ (G)$ is given by $Cdeg\ (u) = \deg(u)$ and there are there are two notions of degree centrality: one based on in-degree and the other on out-degree. Closeness centrality on the other hand ranks the nodes based on its closeness and its ability to communicate with other nodes easily in the networks. Betweeness centrality measures the number of paths a node lies within the network. The eigenvector centrality on the other hand looks at a particular node with respect to its neighbors and their relative importance in the network. Together with the centrality measures, the essentiality of a particular node within the network decides the relative importance of that node and the effect upon its removal. The most prominent nodes are the one with the highest information flow scores and are considered as the centers of the network. Therefore the main of reconstructing a GRN's would be therefore to identify and target those 'hubs' or nodes that are essential for the basic functioning of the network, that play an important role in the parasite metabolism [26, 27].

One such example for construction of GRN for targeting the lipid metabolism of *Schistosoma mansoni* and *Trypanosoma brucei* was made as shown in Fig. (**2**)

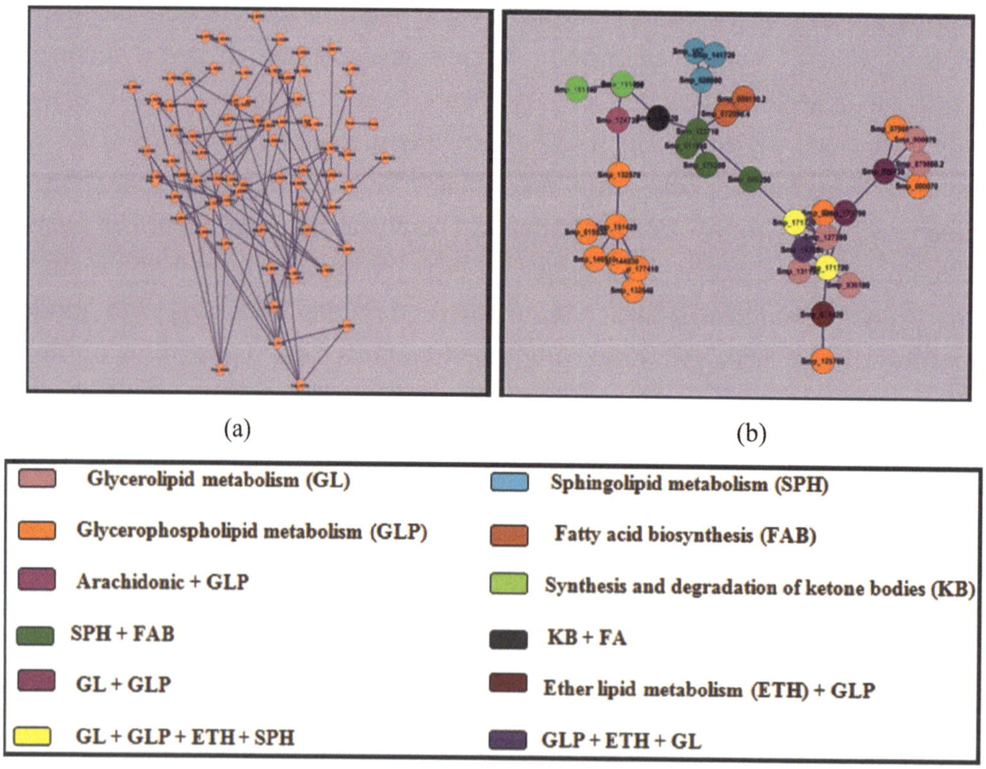

Figure 2: (a) Gene regulatory network of (*Schistosoma masoni*) visualized through Cytoscape. **(b)** Gene cluster showing maximum number of genes in GLP, GL and SPH metabolism.

GRNs of lipid metabolism of these two model organisms were built using Cytoscape (www. cytoscape.org). Spring embedded layout is being the most widely used automatic network layout algorithm models a mechanical system in which edges correspond to springs, creating an attractive force between nodes (*i.e.* one substrate and one product) that are far apart and a repulsive force between nodes that are close together [28]. The spring embedded layout for the network is shown in Fig. (**S1**) (Supplementary material). The highly connected nodes formed in GRN were taken into consideration to lay emphasis upon clustering of genes into metabolic pathways of related functions. The total numbers of nodes and edges in GRNs were also identified.

From the built-up GRNs, it's seen that maximum number of genes exclusive for *S. mansoni* were involved in glycerophosholipid (part of glycerolipid pathway) and sphingolipid metabolism shown in Fig. (**2**) Likewise, for *T.brucei*, genes found in GRN were prevalent in glycerolipid and glycerophospholipid metabolism. GRNs thus give an overview of gene-gene interaction and important lipid metabolic pathways responsible for the survival of pathogens in host system.

Reconstruction of Metabolic Pathway

Reconstruction of the metabolic pathway enables the identification of gaps in our understanding of parasite metabolism, facilitates the computational interrogation of the overall properties of the prebuilt gene regulatory network and provides context for analysis of "-omics" data sets.

Reconstruction of the global metabolic network in a standardized, quality-controlled and bottom-up manner. Manual literature-based reconstruction ensured that the network components and their interactions were based on direct physical evidence and reflected the current knowledge of parasite metabolism. Thus reconstruction represents a versatile and effective integration tool enabling visualization and analysis of a genome-wide metabolic network in a highly curated manner [29].

The metabolic reconstruction of the glycerophospholipid and sphingolipid pathway for ***S. mansoni*** and ***T .brucei*** has been presented in Fig. (**3a, 3b**) while the metabolic reconstruction of the Glycerolipid pathway for ***T.brucei*** has been included in the supplementary file (Fig. (**S2**)).

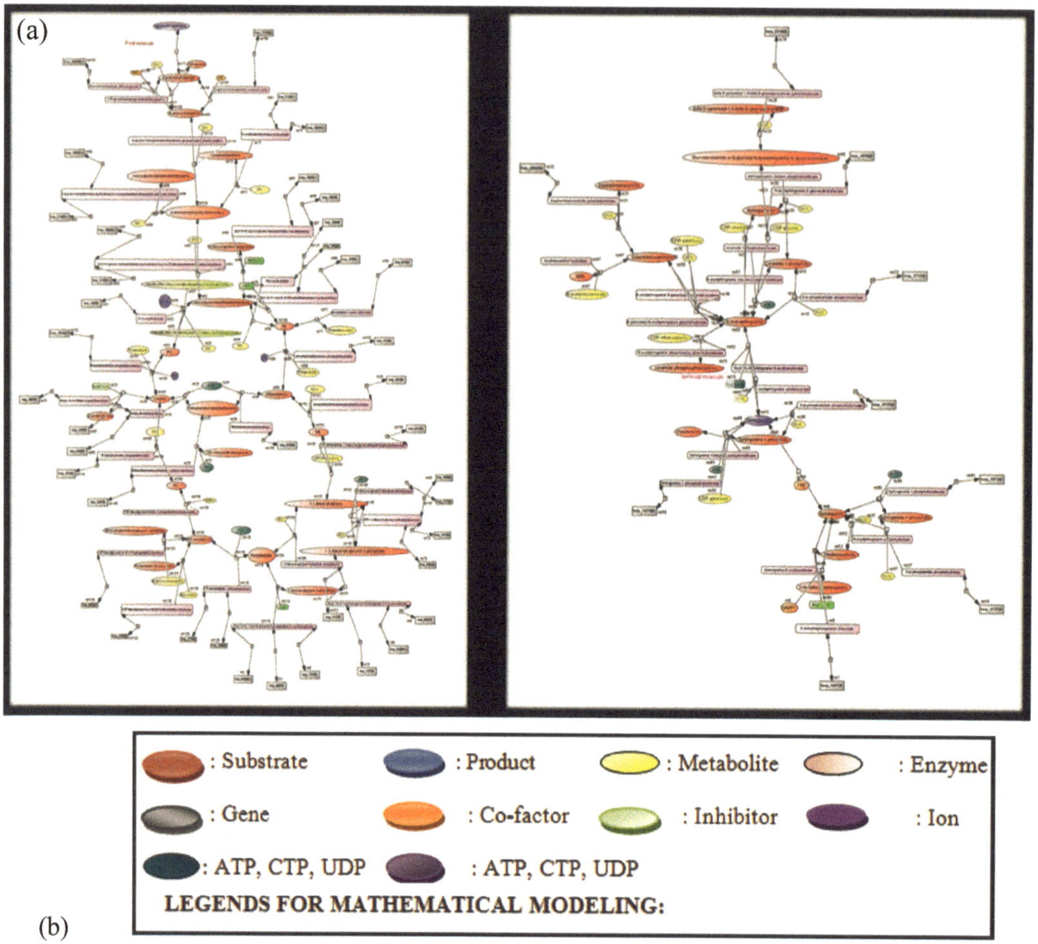

Figure 3: (a) Reconstructed glycerophospholipid metabolic pathway of *S.mansoni* and *T.brucei* (b) Reconstructed Sphingolipid metabolic pathway in *S.mansoni.*

Compartmentalization and Sub-Cellular Localization

Initial analysis of the network structure and the formulation of an *in silico* model from the reconstruction highlighted the importance of intracellular compartmentalization [29]. Sub-cellular localization provides spatial information of proteins in the cell; proteins target sub-cellular localizations to interact with appropriate partners and form functional complexes in metabolic processes [30]. Measurement of sub-cellular localization is invaluable for assessing a protein as a potential drug target (www.molecular-cytomics.com) against infectious diseases. Compartmentalizations were determined from protein databases *viz*. Uniprot, gene ontology, literature data, and indirect physiological evidence. Subcellular localization of enzymes present in

GLP metabolism in *S. mansoni* and *T.brucei* shown in Table **1**. The subcellular localization of enzymes present in *S. mansoni* Sphingolipid metabolism shown in Table **2** whereas in Table **3** subcellular localization of enzymes present in *T. brucei* Glycerolipid metabolism is shown. If these data are unavailable, reactions are modeled as cytoplasmic [29]. To refine knowledge, transmembrane domain for each protein is also predicted.

Subcellular Localization of Enzymes

1. Glycerophospholipid Metabolism

Table 1: Subcellular localization of enzymes present in GLP metabolim in *S. mansoni* and *T. brucei*

Serial No.	Enzyme name	Gene	TMHMM domains	Location
1	sn-glycerol-3-phosphate:NAD+ 2-oxidoreductase	Smp_030500.1	-	Cytosol
2	sn-glycerol-3-dehydrogenase(FAD dependent)	Smp_121990	-	Mitochondria
3	acyl-CoA:1-acyl-sn-glycerol-3-phosphate-2-O-acyltransferase	Smp_000070 Smp_079860.2 Smp_173790	2 7 10	Integral to membrane
4	1,2-Diacyl-sn-glycerol-3-phosphate phosphohydrolase	Smp_171720	6	Cell membrane
5	ATP:1,2-diacyl-sn-glycerol-3-phosphotransferase	Smp_036180 Smp_131190 Smp_163080	- - -	Cytosol
6	Phosphatidylcholine/ethanolamine phosphatidohydrolase	Smp_151420	-	Golgi apparatus
7	Phosphatidylcholine/ethanolamine 2-acylhydrolase	Smp_133290	-	Cytosol
8	Acyl-CoA:1-acyl-sn-glycero-3-phosphocholine/ethanolamine O-acyltransferase	Smp_132850.1 Smp_169040	1 3	Cytosol Integral to membrane
9	2-lysophosphatidylcholine acylhydrolase	Smp_166530.2 Smp_171820.1	- 1	Cell inner membrane
10	Acetyl-CoA:choline O-acetyltransferase	Smp_146910	-	Nucleus
11	Choline/Ethanolamine phosphotransferase	Smp_015030 Smp_132570	- -	Cytosol
12	a) CTP:choline-phosphate cytidylyltransferase b) CTP:ethanolamine-phosphate cytidylyltransferase	Smp_124730 Smp_124730	- -	Cytosol Cytosol

Table 1: contd…

13	CDP-ethanolamine:1,2-diacyl-sn-glycerol ethanolamine phosphotransferase	Smp_071020	9	ER membrane
14	ATP:ethanolamine-O-phosphotransferase	Smp_015050	-	Cytosol
15	CTP:phosphatidate cytidylyltransferase	Smp_144030 Smp_177410	7 2	Integral to membrane
16	Phosphatidylserine decarboxylase	Smp_021830	-	Mitochondria(inner)
17	CDP-diacylglycerol:sn-glycerol-3-phosphate 3-phosphatidyltransferase	Smp_090820	-	Cytosol
18	1-Acyl-sn-glycero-3-phosphoethanolamine aldehydohydrolase	Smp_166530.2 Smp_171820.1	- 1	Nucleus
19	CDP-diacylglycerol: myo-inositol 3-phosphatidyltransferase	Smp_132640	4	Membrane

2. *S. Mansoni* Sphingolipid Metabolism

Table 2: Subcellular localization of enzymes present in *S. mansoni* Sphingolipid metabolism

Serial No.	Enzyme Name	Gene	TMHMM Domains	Location
1	3-dehydrosphinganine reductase	Smp_141720	4	Cytosol
2	3-sn-phosphatidate phosphohydrolase	Smp_171720	6	Integral to membrane
3	a) Sphinganine-1-phosphotransferase b) Sphingosine-1-phosphotransferase	Smp_157100	-	Cytosol
4	N-acylsphingosine-D-glucosyltransferase	Smp_157080	3	Cytosol
5	Beta-D-galactosyl-1,4-beta-D-glucosylceramide galactohydrolase	Smp_011990	-	Cytosol
6	Digalactosylceramide galactohydrolase	Smp_089290	1	Cytosol

3. *T. Brucei* Glycerolipid Metabolism

Kinetic Modeling and Stochastic Simulation

Mathematical modeling and numerical simulation aid in the reduction of the inherent complexity of the large-scale biological systems. Kinetic modeling is based on certain rate-equations and parameters involved in enzymatic reactions. This depends mainly on the following rate laws:

1) Convenience Kinetics.

2) Generalized Mass Action Kinetics.

3) Hill-Hinze equation.

Table 3: Subcellular localization of enzymes present in *T. brucei* Glycerolipid metabolism

Serial No.	Enzyme Name	Gene	TMHMM Domains	Location
1	ATP:Glycerol-3-phosphotransferase	Tb09.211.3540 Tb09.211.3550 Tb09.211.3560 Tb09.211.3570 Tb09.211.3590	- - - - -	Glycosome
2	Acyl Co-A:sn-G-3-P-O-acyltransferase	Tb10.70.4400	3	Cytosol
3	Diacylglycerol-3-phosphate-phosphohydrolase	Tb10.389.0020 Tb10.61.2970	1 6	Cytosol Integral to membrane
4	Phospholipid:1,2-diacylglycrol-sn-glycerol-O-acyltransferase	Tb11.01.4790	1	Cytosol
5	Triacylglycerol acylhydrolase	Tb927.1.2740	1	Cytosol
6	Glycerol-ester-acylhydrolase	Tb927.4.4360 Tb927.8.8020	- -	Cytosol

Convenience Kinetics is inferred from molecular enzyme mechanism as like many established rate laws. It exemplifies a rapid equilibrium random order enzyme mechanism. It transforms a biochemical network into a dynamical model either manually or automatically with believable biological properties. It implements enzyme saturation and regulation by activators and inhibitors, addressing all possible reaction stoichiometry that is specified by a small number of parameters [31]. The mathematical form makes it desirable for parameter estimation and optimization. The parameters are estimated using least-square fit to Michaelis-Menten values, equilibrium constants, the turn-over rates and the quantities that are routinely assessed in enzyme assay and stored in kinetic databases. Convenience kinetics takes into account the enzyme catalyzed reaction that is from substrate to ES complex formation. It constitutes a simple molecular reaction in which a substrate binds rapidly and in random order to the enzyme [32]. The equation for convenience kinetics is:

$$\frac{E}{default} \times \frac{Kcat_{re} \times \frac{\frac{R}{default}}{Kmc_{re_R}} \times \frac{\frac{M}{default}}{Kmc_{re_M}}}{1 + \frac{\frac{R}{default}}{Kmc_{re_R}} \times 1 + \frac{\frac{M}{default}}{Kmc_{re_M}}}$$

where E-Enzyme concentration, Kcat_re - catalysis constant, R-Reactant concentration, M- Metabolite constant, Kmc_re_R and Kmc_re_M -Michels constant for reactant and metabolite, *default* indicates the initial default values used by the user.

Generalized mass-action kinetics accounts modeling of reactions with any number of reactant and product molecules especially used for non-enzyme catalyzed reaction, where the product is formed by the ES complex. This approach has been successfully utilized for optimization of valine and leucine biosynthesis [33]. The equation is given by:

$$Kass\_re \times \frac{ES - complex}{default}$$

Where, Kass_re- association constant, ES-complex-enzyme substrate complex, *default* indicates the initial default values used by the user.

Hill-Hinze equation is used to quantify the binding of a ligand to a macromolecule [34]. It describes the macromolecule saturated by the ligand, which is used in determining the degree of cooperatively of the ligand binding to the enzyme or receptor. Gene regulation can be best studied using Hill's equation, Hill Hinze kinetics are particularly used for the formation of gene products. It is known that transition to a protein from gene is "transition omitted". The Hill equation modified by SBML squeezer is given as vre2:

$$Vre2 = vmax\_re2 \times \frac{[R1]^{h1c\_re2\_s2}}{[R1]^{h1c\_re2\_s2} + KSP\_re2\_S2^{h1c\_re2\_s2}}$$

Where, *vmax* is the maximum velocity, R1 is the Reactant concentration. As the d/dt [mRNA] is zero, the concentration of mRNA reaches the steady state quickly as compared to the protein concentration.

Based on these kinetic rate laws, discrete stochastic simulations are carried out by changing the initial concentrations of molecular species and varying the kinetic parameters. The focus is laid on study of time-dependent behavior of the complex biological system by the discretization of events. Numerical simulation helps to study the metabolic fluxes involved in each and every reaction of the system thus aiding the process of identification of the key enzymes involved in the pathway regulation.

In the graphs presented in Fig. (**4**), it is postulated that there was a constant decrease in the concentration of substrate, as the concentration of the product falls. When going downstream the reaction, there is a potential fall in flux. The concentration of gene almost remains unchanged during the process as the transcription rate does not change the state of the gene. The graphs for rest of the reactions of the metabolic network have been included in the supplementary material (Fig. (**S3-S12**))

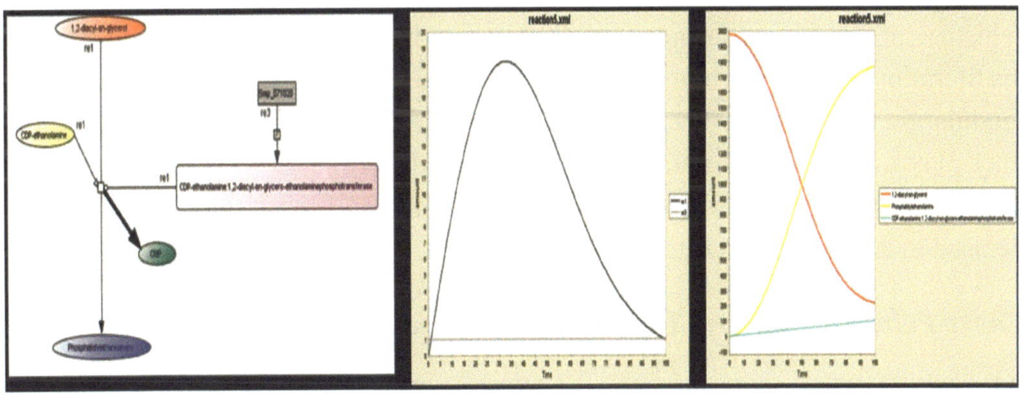

Figure 4: Mathematical modeling of single reaction, flux and product formation.

By definition, an ideal target should be playing a key role in the metabolic pathway under study and concentration of the target should be low. By kinetic modeling and simulation study sharp flux for each reaction was obtained by changing the concentration of the enzyme in a minimal range. The enzymes hypothesized to be potential drug targets were found to be located at the crucial junction of these metabolic pathways either a bifurcation or a bottle-neck or where the flux was operating.

Choke-Point Analysis

The completion of the respective genome projects for schistosomes and trypanosomes has enabled *in silico* to deduce a general metabolic pathway map for these organisms; however this strategy alone misses to determine the presence and importance of single metabolic steps in each complex life cycle stages. Network analysis and reconstruction of these complex metabolic pathways require kinetic modeling and discrete-stochastic simulation study to understand the basis of normal and abnormal cellular function. Henceforth, systems biology adopts an integrated approach to study dynamic behavior of a biological system analyzing the response of the system to perturbations. Detailed analysis of important enzymes taking part in biochemical reaction of the pathways holds promise for target identification to control the parasites in infectious diseases. Emergence of drug resistance seeks urgent need of novel chemotherapeutic drugs against neglected tropical diseases *viz.* Schistosomiasis and Trypanosomiasis. Characterizing fundamental differences in the biochemical pathways and enzymes therein of the parasite and its mammalian host is an important first step towards this goal. Drug resistance is related to different metabolic pathways. In schistosomes and trypanosomes lipids play role in cellular growth, multiplication, proliferation, virulence, homeostasis control, host recognition, immune response modulation and evasion. The crucial enzymes present in these metabolic pathways may play a vital role in parasite metabolism. Metabolic choke-point analysis of lipid metabolic pathways reconstructed from genome of these two organisms may be used to identify drug targets.

Choke Point Enzymes and Their Function in Parasite Metabolism:

A) *Schistosoma mansoni* and *Trypanosoma brucei* glycerophospholipid metabolism:

1) Ethanolamine/choline phosphotransferase.

2) Choline/ethanolamine phosphate cytidylyl transferase.

3) 3)CDP-ethanolamine or choline: 1, 2-diacyl-sn-glyceroethanolamine or choline phosphotransferase.

4) Phosphatidyl-L-serine carboxylase or Phosphatidyl serine decarboxylase.

5) CDP-diacylglycerol: sn-glycerol-3-phosphate-3-phosphatidyltransferase.

6) CTP: phosphatidate cytidylyl transferase.

7) Phosphatidylcholine-2-acylhydrolase.

8) Phosphatidylethanolamine-2-acylhydrolase.

(Note: First three enzymes are involved in Kennedy pathway).

The major phospholipids present in the phospholipid extract of *S.mansoni* are PC (28%), PE (25%), PS (15%) and phosphatidylglycerol (8%) [35]. *T.brucei* bloodstream and procyclic forms contain all major phospholipid classes known in other eukaryotes, accounting for ~80% of membrane lipids [36]. PC and PE represent the most abundant glycerophospholipid classes, comprising 45-60% and 10-20% respectively. Kennedy pathway involves major phospholipid biosynthesis in all eukaryotes.

In schistosomes, during life cycle the adult parasite can survive for many years in close contact with the immune system of the final host. One of the adaptations of the parasite to deal with this hostile environment is the outer tegumental membrane complex that surrounds the parasite. This membrane complex is composed of two stacked lipid bilayers (referred to as the outer tegumental membranes) of which the main constituents are phospholipids [37]. It has been established that incorporated fatty acids can be modified by chain elongation, resulting in a fatty acid profile significantly different from that of the host. For a better comprehension of the functions of the (modified) fatty acids in schistosomes, analysis of the molecular species of the phospholipids in schistosomes, and in particular those of the outer tegumental membranes is needed. Experimental evidences have shown PC molecular species that are major constituents in the body of the schistosome, like (20:1–16:0)PtdCho, are absent from the blood of the host and nearly absent from the outer tegumental membranes. In general, the PC species from the schistosome and those of the outer tegumental membranes in particular, are more saturated than the species from the blood of the host. The high amount of (16:0–16:0)PtdCho, which comprises nearly 50% of the tegumental membrane PC fraction, can make the outer tegumental membranes more resistant to the damage caused by reactive oxygen species secreted by neutrophils and macrophages. The PC composition of the outer tegumental membranes therefore seems to reflect another adaptation of the parasite to the intravascular habitat and the exposure to the hosts immune system. Noteworthy was the high amount of plasmalogens present in schistosomes, which constituted for 20-25% of the total amount of PE. The most prominent plasmalogen species contained 20:1 linked to the sn-2 position. This particular species was not found in the blood of the host. The outer tegumental membranes were highly enriched in plasmalogens (44%) compared to the blood of the host (14%) [20]. The synthesis of PC in *S.mansoni* adults occurred by the choline to PC or Kennedy pathway [35].

Similar to higher eukaryotes, genes are present in *T.brucei* for *de novo* biosynthesis of phospholipids PC and PE *via* PA and DAG (Kennedy pathway) and pulse-chase experiments using radioactively labelled ethanolamine demonstrated indeed significant activity of this pathway in bloodstream form trypanosomes [38]. Trypanosomes do not obtain intact phospholipids from their hosts, but instead synthesize their own phospholipids using head groups (such as choline, ethanolamine) acquired from the host (and the often modified) [22]. In *T. brucei* bloodstream and procyclic forms PE is generated *via* its CDP-activated intermediate CDP-ethanolamine through this pathway. It involves the phosphorylation of ethanolamine kinase, followed by activation of ethanolamine phosphate to CDP-ethanolamine *via* ethanolamine-phosphate cytidyltransferase. Finally the activated head group is transferred to diradylglycerol by ethanolamine phosphotransferase. The first two reactions are catalyzed by cytosolic enzymes, whereas the third step is mediated by an integral membrane protein of the ER [39]. All enzymes involved in PE formation by the ‘Kennedy pathway’ in *T. brucei* have been identified and experimentally confirmed [40]. Disruption of the CDP-ethanolamine branch in procyclic forms using RNA interference (RNAi) resulted in severe growth phenotypes and revealed dramatic changes in cellular PE, PC and PS levels [41]. In addition, a block in PE synthesis caused alterations in mitochondrial morphology and the formation of multinucleate parasites [42]. Similar observations were also made in *T. brucei* bloodstream forms after knocking out ethanolamine-phosphate cytidyltransferase, which validates the Kennedy pathway as a potential drug target [43]. A similar pathway, involving CDP-activated choline, also leads to the formation of PC, the most abundant phospholipid class in eukaryotic cells, including *T. brucei*. Candidate trypanosome genes for all three enzymes of the CDP-choline branch of the 'Kennedy pathway' have been identified. The first and third enzymes, choline kinase [40] and choline phosphotransferase [42], have been characterized experimentally and were found to show dual specificities for choline and ethanolamine, and CDP-choline and CDP-ethanolamine, respectively. Choline kinase is essential in *T. brucei* bloodstream forms [21], indicating the importance of this pathway in choline metabolism of the parasite. The second enzyme, choline-phosphate cytidyltransferase, which, in analogy to its homologues in mammalian cells or to *T. brucei* ethanolamine-phosphate cytidyltransferase in the CDP-ethanolamine branch, likely represents the rate-limiting reaction. Experiments involving RNAi-mediated gene silencing of ethanolamine phosphotransferase revealed that a block in PE synthesis *via* the CDP-ethanolamine branch led to increased PE formation *via* choline phosphotransferase, which shows dual specificity for the substrates of both branches(CDP-choline and CDP-ethanolamine), thereby changing the molecular species composition of PE from mostly alk-1-enyl-acyl-type species to mostly diacyl-type species [15]. Phosphatidylethanolamine in *Trypanosoma brucei* is organized in two separate pools and is synthesized exclusively by the Kennedy pathway. Thus, parasites try to compensate for a lack of one subclass of PE by up-regulating another subclass, indicating that the PE content of *T. brucei* is tightly regulated, and essential, for its survival. In *T. brucei*, decarboxylation of PS to PE has been shown to occur in procyclic forms [41]. A gene encoding a putative PS decarboxylase-I have been identified in the *T. brucei* genome and was found to be expressed in both main life cycle stages. In addition, recombinant *T. brucei* PS decarboxylase-I has been shown to be correctly processed and catalytically active, implying its activity may be highly regulated or restricted in the parasite [20]. All together these results show that trypanosomes can synthesize phospholipids by Kennedy pathway suggesting that enzymes involved in reactions are potential drug targets against HAT.

CDP-Diacylglycerol: Sn-Glycerol-3-Phosphate-3-Phosphatidyltransferase CTP: Phosphatidate Cytidylyl Transferase

These two enzymes are involved in the formation of DAG which in turn is essential for PC or PE synthesis.

Phosphatidylcholine-2-Acylhydrolase and Phosphatidylethanolamine-2-Acylhydrolase

PC or PE acylhydrolases are phospholipase A2.The (16:0-16:0) AlkCho species of PC is the only alkylacyl species found in the schistosome but it was absent in the blood of the host. This species was particularly present in the outer tegumental membranes and hypothetically it could be converted to PAF by substitution of the palmitoyl chain by acetate. PAF is important to the process of homeostasis and mediator of bronchoconstriction. Although the schistosome might profit from the increased supply of blood resulting from the vasodialating properties of PAF, the strong pro-inflammatory effects mediated by PAF make it questionable whether this putative conversion would be beneficial to the parasite [20] and seek experimental evidences. Thus, the enzymes of the synthetic pathways for phospholipids in these parasites may offer unique sites for drug action.

B) *Schistosoma mansoni* Sphingolipid metabolism:

1) 3-sn-phosphatidate phosphohydrolase.

2) N-acylsphingosine D-glucosyltransferase.

Sphingolipids are a diverse and ubiquitous group of lipids in cell membranes and are presumed to play a vital role in membrane function. Sphingolipids are found in all eukaryotic and some prokaryotic organisms and participate in the regulation of cell growth, differentiation, cell-cell communication, cell-substrate interactions and intracellular signal transduction [45]. They are complex lipids which have as their backbone sphingosine (a long chain aliphatic amine) or a related base. The "sphingosin" backbone of sphingolipids was so named by J. L. W. Thudichum in 1884 for its enigmatic ("Sphinx-like") properties. In addition to the newly discovered role of ceramide as an intracellular second messenger for tumor necrosis factor-alpha, IL-1beta, and other cytokines, sphingosine, sphingosine-1-

phosphate, and other sphingolipid metabolites have recently been demonstrated to modulate cellular calcium homeostasis and cell proliferation. Perturbation of sphingolipid metabolism using synthetic and naturally occurring inhibitors of key enzymes of the biosynthetic pathways is aiding the characterization of these processes; Ceramide and Sphingosine are highly bioactive compounds that affect cellular regulatory pathways [45]. 3-sn-phosphatidate phosphohydrolase plays pivotal role in synthesizing ceramide (N-acylsphingosine),a key metabolite involved in sphingolipid metabolism.

Glycosphingolipids (GSLs) are glycosylated derivatives of ceramide in the lipid bilayer. Their ubiquitous distribution and complexity suggest that they have important functions. N-acylsphingosine D-glucosyltransferases are essential for glycosphingolipid synthesis [46]. *Schistosoma mansoni* synthesize novel glycosphingolipids [47, 48]. Glycosphingolipids are ubiquitous membrane components that play important roles in signal transduction events thereby affecting many cellular functions, including modulation of the immune response. Previous studies have shown that the glycoprotein oligosaccharides synthesized by adult *Schistosoma mansoni*, the organism responsible for human schistosomiasis, are unusual in that they contain terminal beta-GalNAc residues and lack sialic acid. These observations and other studies indicating that schistosome glycoproteins and glycolipids are antigenic in infected animals led further investigation to the structures of the glycosphingolipids synthesized by these organisms and to determine whether they are structurally related to those synthesized by their vertebrate hosts [47]. Many different glycosphingolipids of the human parasitic helminth *Schistosoma mansoni* have been structurally characterized [49]. Thus inhibition of key enzymes 3-sn phosphatidate phosphohydrolase and N-acylsphingosine D-glucosyltransferase involved in sphingolipid metabolism may pave a way in chemotherapeutic discovery.

C) *Trypanosoma brucei* Glycerolipid metabolism:

1) ATP: glycerol 3-phosphotransferase:

Ethanolamine and choline are major components of the trypanosome membrane phospholipids, in the form of glycophosphoethanolamine and choline.

Ethanolamine is also found as an integral component of the GPI anchor that is required for membrane attachment of cell-surface proteins, most notably variant surface glycoproteins or VSG. The *de novo* synthesis of glycerophosphoethanolamine and glycerophosphocholine starts with the generation of phosphoethanolamine and phosphocholine by ethanolamine and choline kinase *via* the kennedy pathway [44]. The Kennedy pathways initiate from exogenous ethanolamine and choline involving ethanolamine kinase and choline kinase followed by the ethanolamine-phosphate cytidylyltransferase (ECT) and choline-phosphate cytidylyltransferase (CCT) that catalyse the formation of CDP-ethanolamine and CDP-choline [50].

Phosphotransferases are mainly involved in crucial processes such as the maintenance of ATP balance in the cell. Slight disruption on this equilibrium is usually lethal for all living organisms; in consequence it is considered that for trypanosomatids phosphotransferases are promising targets for rational drug design. In trypanosome energy metabolism the key role of glycosomes become evident with the fact that the bloodstream form of *T. brucei* depends exclusively on glycolysis for ATP generation. The end metabolite of this pathway of hexose sugars corresponds to excreted pyruvate. Intraglycosomal redox balance is maintained using a glycerol-3-phosphate dehydrogenase shuttle. The reoxidation of the glycolysis-derived NADH coenzyme is accomplished inside this organelle through an NAD-linked glycerol-3-phosphate dehydrogenase which reduces dihydroxyacetone phosphate to glycerol-3-phosphate.After exportation from the glycosome this molecule is directed to the mitochondria in the presence of molecular oxygen,where it becomes reoxidized to dihydroxyacetone phosphate by the cyanide-insensitive trypanosome alternative oxidase which then returns to the glycosome. It is observed that in half of the glycolytic pathway the final product that leaves the glycosome is 3-phosphoglycerate.In the cytosol the remaining three glycolytic steps take place producing pyruvate as end-product,thus the net ATP yield corresponds to two molecules per glucose. However in anaerobic conditions the yield is halved due to the inability to reoxidize glycerol-3-phosphate and glycerol becomes an end-product equimolar to pyruvate. The essential role of glycolysis in trypanosome energy metabolism and its particular and divergent strategy of glycosomal confinement,which distinguishes them from other eukaryotes, constitute a clear and plausible target for chemotherapeutic molecules. In this regard, glycerol phosphotransferases involved in trypanosma glycerolipid metabolism play an important role in ATP balance and may be targeted for novel drug discovery.

To construct biochemical network models that are relevant to cellular physiology it is important to determine the kinetic properties of the enzyme in conditions as close as possible to cellular milieu. The principal result of this chapter is laid onto the idea that it shows a simply identifiable class of kinetic expressions, including the familiar detailed balanced kinetics as a proper subclass which ensures consistency with the extended thermodynamic conditions. The rate laws applied provided a very good fit to the data. With the network model it is identified which steps have strong effect on other parts of the network and those steps that do indeed have high levels of control are then chosen further for kinetic analysis. This means that Glycerolphospholipid (in *S. mansoni* and *T. brucei*) sphingolipid metabolism (*S. mansoni*) and glycerolipid metabolism (*T. brucei*) could be examined further in case it has a strong effect on the rest of the network model. The most important feature for building a bottom-up biochemical network model is that the relation between the concentration of the effectors and the rate be accurate. The objective is to learn more about how cells work *via* the means of computational models and not all about the mechanisms of catalysis, except when they are found themselves of importance to cellular function. The concept of "gene function" becomes synonym with the kinetics of its protein product embedded in the cellular biochemical network [51]. It can further be added that differential network analysis is a successful mode of genomic analysis which can be attributed towards high-throughput experimental methods. Further insights need to be drawn from differential mapping which may allow us to explore a previously unexplored interactome providing a deeper understanding of complex biological phenomena.

In nutshell, this chapter focuses onto the discrete modeling of GRNs. The central theme revolves around abstracting a complex biological process as a collection of interacting functions driven in time by a set of discrete biological events. Analyzing the system at a molecular level, the temporal dynamics of the system are revealed by the interaction of these events. The stochastic behavior of the interactions is captured through the mathematical formalism characterizing the time associated with each of the biological events. The discrete event models built helps to capture the interaction to create the dynamics of the biological system.

ACKNOWLEDGEMENTS

The author would like to thank the reviewers for their valuable suggestions and constructive guidance which went a long way in the improvisation of the said chapter. The authors are also thankful to Department of Biotechnology, Government of India for financial support (BT/PR/3140/BID/7/379/2011). Ms. Sonali Shinde acknowledges the financial support as Junior Research Fellow (BT/PR/3140/BID/7/379/2011) to Department of Biotechnology, Government of India. We would also like to thank Dr. S.C. Mande, Director, National Centre for Cell Science, Pune for supporting the Bioinformatics and High Performance Computing Facility.

CONFLICT OF INTEREST

The authors confirm that this chapter contents have no conflict of interest.

ABBREVIATIONS

MDR = Multi-drug resistance

NTD = Neglected tropical disease

PZQ = Praziquantel

GLP = Glycerophospholipid

GL = Glycerolipid

SPH = Sphingolipid

GRN = Gene regulatory network

PC = Phosphatidylcholine

PE = Phosphatidylethanolamine

DAG = Diacylglycerol

multi = Scale modeling

REFERENCES

[1] Tomita, M.; Hashimoto, K.; Takahashi, K.; Shimizu, T.; Matsuzaki, Y.; Miyoshi, F.; Saito, K.; Tanida, S.; Yugi, K.; Venter, J.C.; Hutchison, C.A. E-CELL: Software Environment for Whole Cell Simulation. *Genome Inform Ser Workshop Genome Inform.,* **1997**, (8), 147-155.

[2] Boshoff, H.I.; Manjunatha, U.H. The impact of genomics on discovering drugs against infectious diseases. *Microbes Infect.*, **2006**, 8(6), 1654-61.

[3] Brown, J.R.; Magid-Slav, M.; Sanseau, P.; Rajpal, D.K. Computational biology approaches for selecting host–pathogen drug targets. *Drug Discovery Today.,* **2011**, 16(5-6), 229-236.

[4] Ideker, T.; Galitski, T.; Hood, L. A new approach to decoding life: systems biology. Annu. Rev. Genomics Hum. Genet., 2001, 2, 343-72.

[5] Hunter, P.J.; Borg, T.K. Integration from proteins to organs: the physiome project. *Nat.Rev.Mol.Cell Biol.*, **2003**, (4), 237-243.

[6] Azuaje, F.; Devaux, Y.; Wagner, D.R. Identification of potential targets in biological signaling systems through network perturbation analysis. *Biosystems.,* **2010**,100(1), 55-64.Puchalka, J.; Kierzek, A.M. Bridging the gap between stochastic and deterministic regimes in the kinetic simulations of the biochemical reaction networks. *Biophysical Journal.*, **2004**, 86(3),1357-72.

[7] Eissaa, M.M.; El-Azzounia, M. Z.; Amera, E. I.; Baddourb N. M. Miltefosine, a promising novel agent for schistosomiasis mansoni. *International Journal for Parasitology.,***2011**,41(2),235-242

[8] Peter,S.; Jennifer,; Robert,B.; Marcel Tanner.; Jürg,U. Schistosomiasis and water resources development: systematic review, meta-analysis, and estimates of people at risk. *The Lancet Infec. Dis.,* **2006**, 6(7), 411–425.

[9] Barrett, M.P.; Burchmore, R.J.; Stich, A., Lazzari, O.J., Frasch, A.C.; Cazzulo, J.J.; Krishna, S. The trypanosomiases. *Lancet.,* **2003**; 362(9394), 1469–80.

[10] Leiper, R.T. Report of the results of the bilharzia mission in Egypt. Part I: transmission. *Journal of Royal Army Med. Corps.,* **1915**, 25(1), 1-55.

[11] Gryseels, B.; Mbaye, D.; De vlas, S.J.; Stelma, F.F.; Guisse, F.; Van, Lieshout, L.V.; Faye, D.; Diop, M.; Ly, A.; Tchuem-Tchuente, L.A.; Engels, D.; Polman, K. Are poor responses to praziquantel for the treatment of *Schistosoma mansoni* infections in Senegal due to resistance? An overview of the evidence. *Trop. Med. Int. Health.*, **2001**, 6(11), 864–873.

[12] Davis,A.; Wegner,D.H. Multicentre trials of praziquantel in human schistosomiasis: design and techniques. *Bull World Health Organ.,* **1979**, 57(5), 767-71.

[13] Ismail, M.; Botros, S.; Metwally, A.; William, S.; Farghally, A.; Tao, L.; Day, T.A.; Bennett, J.L.; Resistance to praziquantel: direct evidence from Schistosoma mansoni isolated from Egyptian villagers. *Am. J. Trop. Med. Hyg.* **1999**, 60 (6), 932–935.

[14] Lawn, S.D.; Lucas, S.B.; Chiodini, P.L. Case report: *Schistosoma mansoni* infection: failure of standard treatment with praziquantel in a returned traveler. *Trans. R. Soc. Trop. Med. Hyg.*, **2003**, 97(1), 100–1.

[15] Alonso, D.; Muñoz, J., Gascòn, J.; Valls, M.E.; Corachan, M. Failure of standard treatment with praziquantel in two returned travelers with *Schistosoma haematobium* infection. *Am. J. Trop. Med. Hyg.,* **2006**, 74(2), 342–4.

[16] Jennifer, K.; Jacques, C.; Shu-Hua, X.; Jin-Yan, M.; Jiao, P.Y.; Utzinger, J.; Marcel, T. Mefloquine – an aminoalcohol with promising antischistosomal properties in mice. *PLoS NeglectedTrop. Dis*., **2009**, 3(1), e350.

[17] Hilgard, P.; Klenner, T.; Stekar, J.; Unger, C. Alkylphosphocholines, a new class of membrane active anticancer agents. *Cancer Chemother Pharmacol*., **1993**, 32(2), 90–5.

[18] Priotto, G.; Fogg, C.; Balasegaram, M.; Erphas, O.; Louga, A.; Checchi, F.; Ghabri, S.; Piola, P. Three drug combinations for late-stage *Trypanosoma brucei gambiense* sleeping sickness: a randomized clinical trial in Uganda. *PLoS Clin Trials*., **2006**, 1(8), e39.

[19] Brouwers, J.F.; Van Hellemond, J.J.; Van Golde, L.M.; Tielens, A.G. Ether lipids and their possible physiological function in adult *Schistosoma mansoni. Mol Biochem Parasitol.*, **1998**, 96(1-2), 49-58.

[20] Smith. T.K.; Bütikofer, P. Lipid metabolism in *Trypanosoma brucei. Mol Biochem Parasitol.,* **2010**, 172(2), 66-79.

[21] Van Hellemond, J.J.; Tielens, A.G. Adaptation in the lipid metabolism of the protozoan parasite *Trypanosoma brucei. FEBS Letters*., **2006**, 580(23),5552-5558.

[22] Steuer, R. Computational approaches to the topology, stability and dynamics of metabolic networks. *Phytochemistry.,* **2007**, 68(16-18),2139-51.

[23] Iwei, Y.; Theodor, H.; Sophia T.; Peter, D. K.; Russ, B. A. Computational Analysis of Plasmodium falciparum Metabolism: Organizing Genomic Information to Facilitate Drug Discovery. *Genome Res.* **2004,** 14(5), 917–924.

[24] Fell, D. *Understanding the control of metabolism*, Portland press: London, **1997**, vol.41 (3), pp. 187-188.

[25] Mason, O.; Verwoerd, M. Graph theory and networks in biology. *IET Syst Biol*., **2007** 1(2), 89-119.

[26] Xiaowei, Z.; Mark, G.; Michael, S. Getting connected: analysis and principles of biological networks. Genes & Dev., **2007**, 21, 1010-1024.

[27] Shannon, P.; Markiel, A.; Ozier, O.; Baliga, N.S.; Wang, J.T.; Ramage, D.; Amin, N.; Schwikowski, B.; Ideker, T. Cytoscape: A Software Environment for Integrated Models of Biomolecular Interaction Networks. *Genome research.,* **2003**,13(11),2498-2504.

[28] Duarte, N.C.; Becker, S.A.; Jamshidi, N.; Thiele, I.; Mo, M.L.; Vo, T.D.; Srivas, R.; Palsson, B.Ø.Global reconstruction of the human metabolic network based on genomic and bibilomic data. *Appl Biol.Sci.,***2007**, 104(6),1777-82.

[29] Au, C.E.; Bell, A.W.; Gilchrist, A.; Hiding, J.; Nilsson, T.; Bergeron, J.J. Organellar proteomics to create the cell map. Curr Opin Cell Biol.,**2007**, 19(4),376–385.

[30] Liebermeister, W.; Klipp, E. Bringing metabolic networks to life: convenience rate law and thermodynamic constraints. Theoretical Biology and Medical Modelling **2006**, 3(41),1-13.

[31] Walsh, R.; Martin, E.; Darvesh, S. A method to describe enzyme-catalyzed reactions by combining steady state and time course enzyme kinetic parameters. *Biochim Biophys Acta.,* **2010**,1800(1),1-5.

[32] Dräger, A.; Supper, J.; Planatscher, H.; Magnus, J.B.; Oldiges, M.; Zell, A. Comparing Various Evolutionary Algorithms on the Parameter Optimization of the Valine and Leucine Biosynthesis in *Corynebacterium glutamicum. IEEE Congress on Evolutionary Computation.,* **2007**, 620–627.

[33] Mestl, T.; Plahte, E.; Omholt, S.W. A Mathematical Framework for Describing and Analysing Gene Regulatory Networks. *J. Theor. Biol.,* **1995**, 176(2),291-300.

[34] Young, B.W.; Podesta, R.B. Major Phospholipids and phosphatidylcholine synthesis in adult *Schistosoma mansoni. Mol Biochem Parasitol.*, **1982**, 5(3),165-72

[35] Patnaik, P.K.; Field, M.C.; Menon, A.K.; Cross, G.A.; Yee, M.C.; Butikofer, P. Molecular species analysis of phospholipids from *Trypanosoma Brucei* bloodstream and procyclic forms. *Mol Biochem Parasitol.,* **1993**, 58(1):97-105.

[36] Allan, D.; Payares, G.; Evans, W.H. The phospholipid and fatty acid composition of Schistosoma mansoni and of its purified tegumental membranes. *Mol Biochem Parasitol.*, 1987, 23(2) 123–128.

[37] Rifkin, M.R.; Strobos, C.A.; Fairlamb, A.H. Specificity of ethanolamine transport and its further metabolism in *Trypanosoma brucei. J.Biol.Chem.,* **1995**, 270(27), 16160-6.

[38] Vance, J.E.; Vance, D.E. Phospholipid biosynthesis in mammalian cells. *Biochem Cell Biol.*, **2004,** 82(1), 113-28.

[39] Gibellini, F.; Hunter, W.N.; Smith, T.K. Biochemical characterization of the initial steps of the Kennedy pathway in *Trypanosoma brucei*: the ethanolamine and choline kinases. *Biochem J.,* **2008**, 415(1), 135-44.

[40] Signorell, A.; Jelk, J.; Rauch, M.; Butikofer, P. Phosphatidylethanolamine in *Trypanosoma brucei* is organized in two separate pools and is synthesized exclusively by the Kennedy pathway. *J Biol Chem.,* **2008***,* 283(35), 23636-44.

[41] Signorell, A.; Gluenz, E.; Rettig, J. Perturbation of phosphatidylethanolamine synthesis affects mitochondrial morphology and cell-cycle progression in procyclic-form *Trypanosoma brucei. Mol Microbiol.,* **2009***,* 72(4), 1068-79.

[42] Gibellini, F.; Hunter, W.N.; Smith, T.K. The ethanolamine branch of the Kennedy pathway is essential in the bloodstream form of *Trypanosoma brucei. Mol Microbiol.*, **2009,** 73(5), 826-43.

[43] Signorell, A.; Jelk, J.; Rauch, M.; Butikofer, P. Phosphatidylethanolamine is the precursor of the ethanolamine phopshoglycerol moiety bound to eukaryotic elongation factor 1A. *J Biol Chem.,* **2008** 283(9), 20320-9.

[44] Merrill, A.H.; Schmelz, E.M.; Wang, E.; Schroeder, J.; Dillehay, D.L.; Riliy, R.T. Role of dietary sphingolipids and inhibitors of sphingolipid metabolism in cancer and other diseases. *J. Nutr.,* **1995,** 125(6 Suppl), 1677S-1682S.

[45] Marza, E.; Simonsen, K.T.; Faergeman, N.J.; Lesa, G.M. Expression of ceramide glucosyltransferases, which are essential for glycosphingolipid synthesis, is only required in a small subset of C. elegans cells. *J Cell Sci.*, **2009**, 122(6), 822–833.

[46] Makaaru, C.K.; Damian, R.T.; Smith, D.F.; Cummings, R.D. The human blood fluke *Schistosoma mansoni* synthesizes a novel type of glycosphingolipid. *J. Biol. Chem.*, **1992**, 267(4), 2251-7.

[47] Persat, F.; Bouhours, J.F.; Mojon, M.; Petavy, A.F. Glycosphingolipids with Gal beta 1-6 Gal sequences in metacestodes of the parasite Echinococcus multilocularis. *J. Biol. Chem.*, **1992**, 267(13), 8764-9.

[48] Irma van, D.; Carolina, M.W.; Van, S.; Hildegard, G.; Rudolf, G. Structural and functional analysis of glycosphingolipids of *Schistosoma mansoni*. *Methods in Enzymol.*, **2010**, (480), 117-140.

[49] Déchamps, S.; Wengelnik, K.; Berry-Sterkers, L.; Cerdan, R.; Vial, H.J.; Gannoun-Zaki, L.The Kennedy phospholipid biosynthesisT pathways are refractory to genetic disruption in *Plasmodium berghei* and therefore appear essential in blood stages. *Mol Biochem Parasitol.,* **2010**, 173(2), 69-80.

[50] Mendes, P.; Messiha, H.; Malys, N.; Hoops, S. Enzyme kinetics and computational modeling for systems biology. *Methods Enzymol.*, **2009**, (467), 583-599.

[51] Integrating Bioinformatics and Systems Biology for Exploring Novel Lipid Pathways in Infectious Diseases, *Vol. 2,* 2014*,* 161-190.

WEBSITE REFERENCES

1) World Health Organization, Media Center, Leishmania. http://www.who.int/mediacenter/factsheets/fe375/en/

2) Network Data Integration, Analysis, and Visualization in a Box. www.cytoscape.org

3) Parasites In Humans. Find the Nastiest parasites In Humans. http://www.parasitesinhumans.org/Website references:

SUPPLEMENTARY FILES

http://www.nccs.res.in/download-docs/Advances%20in%20Mathematical%20Chemistry%20Supplementary%20files.pd

CHAPTER 10

Applications of Molecular Docking and Molecular Dynamics on the Inhibition of *Quorum Sensing* Systems

Santiago Medina[1], Susana Casas[1], Mariana Restrepo[2], Alejandro Alvarez[1,2], Adriana J. Bernal[2] and Andrés Fernando González Barrios[1,*]

[1]Grupo de Diseño de Productos y Procesos (GDPP), Department of Chemical Engineering, Universidad de los Andes, Bogotá, Colombia and [2]Mycology and Plant Disease Laboratory (LAMFU), Department of Biological Sciences, Universidad de los Andes, Bogotá, Colombia

Abstract: Owing to the evolution of resistant cells using standard antimicrobial methods, it is necessary for us to find alternative strategies to block the attack without exerting pressures on their duplicated structures and in some cases degradation of their *Quorum Sensing* signals is suggested to be more efficient. In this chapter, we give a brief introduction of *quorum sensing* basics. Then, we describe the computational approaches that are used to find molecules to inhibit the chemical signals. After that, we present the case studies using these computational methods. In one of the cases performed in *E. coli* with the autoinducer molecule indole, we elaborate how to evaluate and find molecules with the ability to degrade the indole. We present another case where biological molecules with potential ability to degrade the autoinducer 3-OH-PAME *Quorum Sensing* system in *R. solanacearum* were evaluated as potential quenchers.

Keywords: 3-OH-PAME, AHLs, autoinducer, bacterial signaling, biofilm, *Escherichia coli*, indole, indole oxidase, lipase, molecular docking, molecular dynamics, plant infection, Quorum quenching, *Quorum sensing*, *Ralstonia solanacearum*.

INTRODUCTION

Networks of signaling cascades that respond to environmental stimuli, control the expression of virulence or pathogenicity genes in some bacteria. A subset of these networks control gene expression in a population-dependent manner in a

***Corresponding author Andrés Fernando González Barrios:** Grupo de Diseño de Productos y Procesos (GDPP), Department of Chemical Engineering, Universidad de los Andes, Carrera 1 Este No 19 A 40 Edificio Mario Laserna, Oficina ML 779, Bogota, Colombia; Tel: (571) 3394949 Ext. 3094; Fax: (571) 3324334; E-mail: andgonza@uniandes.edu.co

Subhash C. Basak, Guillermo Restrepo & José L. Villaveces (Eds)
Copyright © 2015 Bentham Science Publishers Ltd. Published by Elsevier Inc. All rights reserved.
10.1016/B978-1-68108-053-6.50010-7

phenomenon known as *Quorum Sensing* (QS). Only when bacteria sense high levels of populations with these systems do they express certain types of genes, which include virulence and pathogenicity, as well as genes that improve their chances of survival in hostile environments. Bacteria achieve this through the production of signaling molecules or autoinducers and the concentration of this signaling molecule is directly proportional to the "census" of the population.

Quorum Sensing is a form of bacterial communication through chemical signals that allows the regulation of a variety of physiological functions and group behavior. Through these autoinducers, they are able to count each other by measuring the density of these signals in the environment, and when a certain population level is reached, they are able to induce or prevent the production of certain proteins and modify their group behavior by changing gene expression [1]. This behavior could be ineffective and unproductive when started by a single bacterium, but if these activities are performed when the right cell density is achieved, they constitute the key for survival. This kind of interaction based on communication can explain the fast adaptation of these organisms to the environment [2].

Quorum Sensing: Historical Perspectives

QS responses were discovered in the early 1970s by Nelson *et al.* (1972). They discovered that the bacterium *Vibrio fischeri* only displayed bioluminescence when bacterial numbers were high and they were confined, but not when they were low, such as in free marine water [1]. Moreover, they found that these bacteria respond to a specific concentration of a secreted N-acyl homoserine lactone (AHL) that acts as the autoinducer. When a threshold concentration is reached, the metabolic path for the production of luciferase is activated [3, 4] and this results in luminescence.

There are different types of signaling molecules, which vary according to the type of organism. In most Gram-negative bacteria the AHL is part of the QS regulation by target genes, which allows bacteria to survive environmental changes Fig. (**1**). Some important proteins in QS signaling are Lux I, which is part of the AHL synthase family, and LuxR, a protein that plays the role of transcriptional activator and is activated by the AHL. At high levels of population, LuxI from different bacteria produce the autoinducer and it reaches high levels in the medium. Threshold concentrations of the autoinducer induce the activation of LuxR, which in turn activates target genes controlled by QS. In other words, Lux I proteins are in charge of synthesizing the AHL signal molecules and LuxR proteins sense and transduce the AHL signal [5].

Although AHLs are the classical signals or autoinducers in QS systems, they are not the only QS signals used by Gram-negative bacteria. Other QS signal molecules include Quinol signals and cyclip dipeptides produced by Pseudomonas [6], as well as small peptides [7] and fatty acids or their ester derivatives [8]. Gram-positive bacteria use mainly peptides as their QS signals [9].

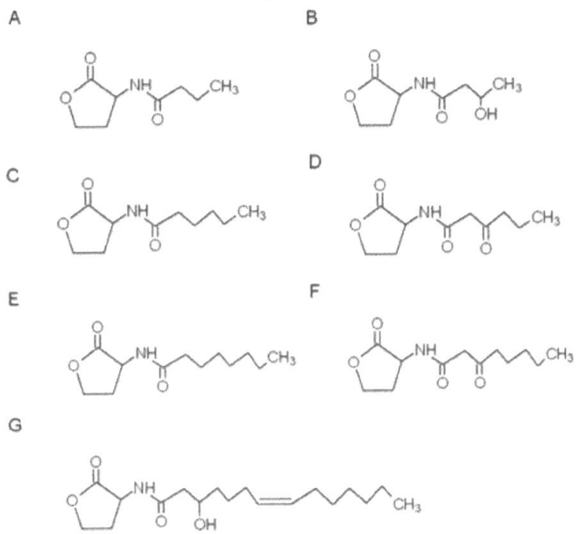

Figure 1: Some structures of N acyl homoserines lactones: (A) N-butanonyl-L-homoserine lactone (BHL); (B) N-(3-hydroxybutanonyl)-L-homoserine lactone (HBHL; (C) N-hexanonyl-L-homoserine lactone (HHL); (D) N-(3-oxohexanoyl)-L-homoserine lactone (OHHL); (E) N-octanonyl-L-Homoserine lactone (OHL); (F) N-(3oxooctanonyl)-L homoserine lactone (OOHL); (G) N-(3hydroxy-7-*cis*-tetradecenoyl)-L-homoserine lactone (HtdeDHL) [10].

Although the mechanisms for QS in gram negative bacteria differ from those in gram positive bacteria, for example, in the structure of their signaling molecules, they can both regulate many activities such as expression of virulence factors, biofilm formation, sporulation, and mating.

Diversity of QS Systems

QS systems are classified according to the type of transcriptional regulator with which the signal molecule interacts to activate the specific phenotype of the organism. The autoinducer activates the regulator by direct binding, which leads the induction or inhibition of the transcription of genes regulated by QS. This, in turn, results in the presence of certain phenotypes. The mechanisms of transcriptional induction occur differently in each organism [11]. Table **1** shows the different types of signal molecules, the regulatory proteins and their respective organism.

Table 1: *Quorum sensing* signals in Gram Negative Bacteria [9]

Organism	Signal Molecule	Regulatory Proteins	Phenotype
Vibrio fischeri	3-Oxo-C6-HSL	LuxI/LuxR	Bioluminescense
Vibrio harvey	3-Hydroxy-C4-HSL	LuxLM/LuxN	Bioluminescense
Pseudomonas aeruginosa	3-Oxo-C12-HSL	LasI/LasR	RhIR, Xcp, biofilms formation
	C4-HSL	RhII/RhIR	Rhamnolipid, Rpods, secundary metabolites
Pseudomonas aeurofaciens	C6-HSL	PhzI/PhzR	Phenazine antibiotics
Agrobacterium tumefaciens	3-Oxo-C8-HSL	TraI/TraR	Ti plasmid conjugation
Erwinia carotovora	3-Oxo-C6-HSL	ExpI/ExpR	Exoenzymes
Erwinia chrysanthemi	3-Oxo-C6-HSL	ExpI/ExpR	Pectate lyases
	C6-HSL		
Erwinia stewartii	3-Oxo-C6-HSL	EsaI/EsaR	Virulence factors, Exopolysaccharide
Rhizobium leguminosarum	C6-HSL	RhII/RhIR	RhiABC rhizosphere expressed gene
Aeromonas hydrophila	C4-HSL	AhyI/AhyR	Exoprotease production
Ralstonia solanacearum	C8-HSL	SolI/SolR	Exopolysaccharide, Virulence factors
Escherichia coli	?	SdiA	Cell division, attachment and effacing lesion formation

QS plays an important role in bacterial pathogenicity, for example in the case of the well studied *P. aeruginosa,* which uses more than one QS system for the genetic regulation. Each of these systems plays a different role and is activated by different autoinducers, for example, the AHL or homoserine lactone (3-oxo-C12-HSL) regulates the expression of virulence genes; another autoinducer known as N-butyryl homoserine lactone-L (C4-HSL) regulates the production of rhamnolipids and secondary metabolites, or the case of PQS or Pseudomonas Quinol Signal (2-heptyl-3-hydroxy-4-quinolone) that is related to the expression of *lasB,* which encodes for elastase [10].

Ralstonia solanacearum uses 3-OH palmytic acid methyl ester (3-OH-PAME) as one of the signal molecules [12], which interacts with the transcription regulator known as *PchA* and forms a regulatory complex. The 3-OH-PAME acts as a signal in the post-transcriptional regulatory system that directs the activity of *PchA*. This system consists of a kinase, PhcS, which phosphorylates the PhcR, a regulator with a kinase domain in its C- terminus. The regulatory system Pch in *R. solanacearum* acts as an on-off control behavior off the survival strategies of bacteria, biofilm formation and pathogenesis itself. Other signal molecules synthesized by this bacterium are known as C8HL and C6HL [13-15]. However, these molecules do not seem to control virulence in this species.

Quorum Sensing in Plant Associated Bacterial Infections

The QS system is used in several phytopathogenic bacteria as their global regulator of pathogenicity factors, examples of phytopathogenic bacteria of great importance and that require QS are: *Agrobacterium tumefaciens, Ralstonia solanacearum, Pseudomonas aeruginosa, Erwinia carotovora, Xanthomonas campestris* among others [15].

Most plant pathogenic bacteria are directly dependent on QS signals to infect the host and become virulent. Mutant bacteria, which lose the ability to produce these signals, also lose their ability to colonize the host. However, there are responses to QS signals by plants. Plants secrete compounds similar in structure to the AHLs, deceiving bacteria that use these as QS signals; so they can interact with the AHL protein receptors and cause them to be degraded Table **2**. The best known inhibitory compounds in plants are halogenated furanones, produced by a red *algae* known as *Delisea pulchra* Fig. (**2**) [16, 17].

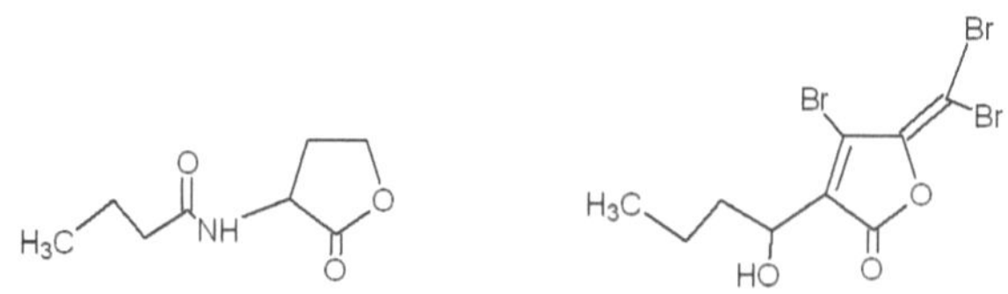

Figure 2: Molecular structures of a typical bacterial AHL (left) and a halogenated furanone AHL mimic from *Delisea pulchra* (right) [16].

In addition to these substances that inhibit the AHLs QS regulation, plants can respond to QS signals perceiving some changes that AHLs produce in proteins of the root, some of them related to defense, directly affecting molecular processes within the plant. This allows the plant to recognize the AHL and produce a systemic response. There are other types of compounds produced by plants that affect other QS signals, therefore affecting processes regulated by QS, for example, it was found that certain compounds significantly affect the production of biofilms in *P. aeruginosa* by inactivating AHL directly, this being one of the key objectives in the QS interference in this bacterium [16, 18].

Higher plants not only have compounds that inhibit the QS, but also compounds that promote QS-regulated behaviors in getting reporters to stimulate certain AHL production. However, not much is known about these substances. This type of stimulation may often be beneficial to the host by stimulation of virulence genes in the bacterium, leading to a faster defense response by the host [18].

Quorum Quenchers

The identification of molecules that interfere with QS could have potential applicability for disease control in diverse hosts. There exist in nature molecules that inhibit the QS signals and they are called Quorum Quenchers [17].

Considering that the major group of Gram negatives has been found to use AHLs, one should expect the presence in nature of a variety of quenchers with the capability of degrading lactone rings. Indeed the first report of an enzyme with such ability was reported in 2000 in the genus *Bacillus*, but in 2002 it was shown that the use of other bacteria (*Bacillus thuringiensis, Bacillus cereus, Bacillus mycoides*) that produce enzymes such as lactonase, amidohydrolase, and reductase might contribute to the modification and inactivation of the QS [17]. The strains that contribute to disrupt the QS signals are named *Quorum Quenching* bacteria [17, 19, 20]. The QS-inhibiting enzymes have been used as biocontrol of diseases in both plants and animals to be useful in blocking the QS signals [19, 21].

As the major group of autoinducers is AHLs, several enzymes with the ability to degrade AHLs have been used as *quorum quenchers*. *Pectobacterium* is a pathogen that produces blackleg and soft rot diseases in potato, two important diseases in this crop. *Quorum Sensing* controls the pathogenic behavior in this bacterium *via* biosynthesis and perception of an AHL signal. Certain attempts to quench this signal have been made, some with promising results [22].

Proteins in charge of inhibiting *Quorum Sensing* lactones are classified as AiiA, AttM, AhlD and AiiB, which are in charge of the AHL inactivation, by hydrolysis of the ring that conforms to the homoserine lactone [11, 17, 19]. Some of the most common *Quorum Quenchers* are presented in Table **2**.

Table 2: Some Quorum Quenchers found in Eukaryotes [11]

QSI Compound	Structure	R	Source
Agrocinopine	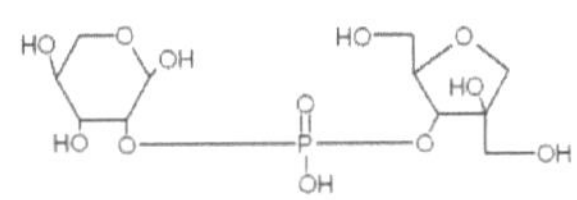		Crow gall cells of host plants
Furanone	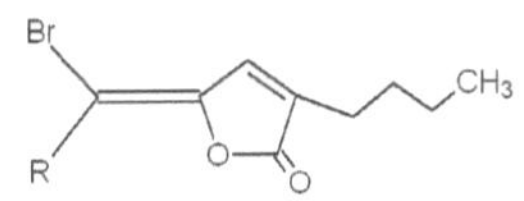	H or Br	*D. pulchra*
L-Canavanine	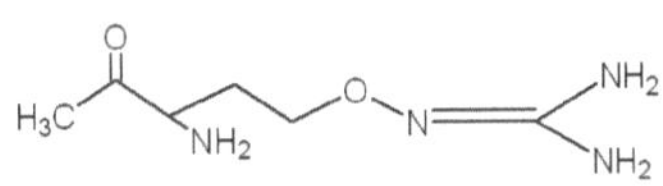		*M. sativa*
Norepinephrine, epinephrine	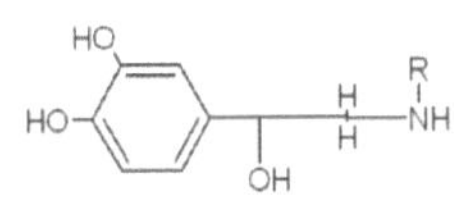	H or CH_3	Human Hormones
Penicilic acid	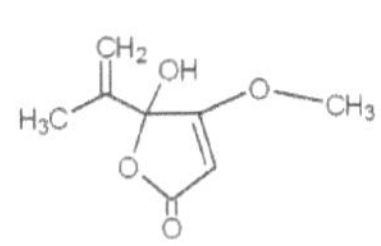		*Penicillium spp.*
Patulin	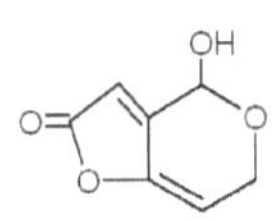		*Penicillium spp.*

RATIONAL DESIGN OF *QUORUM SENSING* INHIBITORS

It has been showed that some pathogenic bacteria can be controlled by inhibiting their QS signals. If these signals are degraded or inhibited, the bacteria behavior can be restricted [23]. Because of the importance of these outcomes the rational design of QS inhibitors could play an important role in the design of new drugs.

One way to carry out rational design of QS inhibitors is to perform virtual screening of potential quenchers by using molecular docking and molecular dynamics. Molecular docking is a method which has the ability to predict the three-dimensional orientation of a molecule (ligand) when bound to a second molecule (substrate) using different scoring functions and to calculate the binding energy required so bonding takes place. Molecular docking algorithms present two main components: the search algorithm that generates different forms of union between ligand and active site and the ranking section where of the different modes of union, through a scoring function based on the complementarity of the ligand to the active site. The definition of a successful binding mode is given by the root mean square deviation (RMSD) of the docked pose compared to the crystal pose at a convention 2- Å cutoff. Both number of rotatable bonds of the ligand and the number of degrees of freedom increase. Firstly, docking programs explored only three degrees of translation and three degrees of rotation (rigid docking). However, with the increased capacity of computers, flexibility in the ligands was introduced and the docking results started to be more accurate [24].

There exist three types of search algorithms: a) the search system where each degree of freedom is explored based on a grid of values, b) stochastic search based where random changes are made for each degree of freedom (Monte Carlo and genetic algorithm) and c) deterministic approach. Regardless of the approach, results provided by docking strongly depend on the initial conditions, as optimization algorithms have to usually deal with local minima. Then different search approaches for the same case is strongly recommended [24].

Some of the interactions that can be taken into account for energy evaluation are Coulomb´s, intermolecular, van der Waals, hydrogen bonding, polar and nonpolar interactions. The sum of these energies is known as the force field. Due to the complexity of this calculation most of the programs disregard some of these interactions and use functions and features based on empirical experience to calculate the energy of interactions. Generally force fields include energy-related intermolecular force, torsion and bond angle, van der Waals interactions, which are calculated using the potential of Lennard-Jones 6-12 and the Coulomb interactions. Then, finding the most probable position for both: substrate and ligand implies to determine the minimum for the force field equation established. Moreover docking algorithms usually ignore entropy and intermolecular terms. A mathematical expression for energy calculation follows [24]:

$$U_t = \sum U_{Etor} + \sum U_{AB} + \sum U_{DA+} + \sum U_{PAP} + \sum U_{Idip} + \sum U_{Icou} \tag{1}$$

where U_t is the total binding energy, $\sum U_{Etor}$ is the atomic interactions term related to torsion, $\sum U_{AB}$ is the atomic interactions term related to the angle bond, $\sum U_{DA}$ is the atomic interactions term related to the dihedral angle, $\sum U_{PAP}$ is the atomic interactions term related to the potential angle outside the plane, $\sum U_{Idip}$ is the energy term related to van der Waals interactions, hydrogen bonding and, polar and non-polar interactions, and $\sum U_{Icou}$ is the atomic interactions term related to the charge generated forces. The summation operator is used because the total energy is the sum of the energy of all possible pairs of atoms in the molecule [24]

Regarding the scoring there exist two approaches: *force field* and empirical data based. Calculation of free energy, taking into account all the variables in the *force field*, is possible but is not computationally efficient. Then, scoring functions based on empirical data and knowledge-based scoring functions could perform better [24].

Molecular Dynamics

Molecular Docking just account for the most probable configuration in equilibrium, nevertheless, one cannot assure if this configuration would be stable, then the evolution of the complex in time is necessary to assess its stability. Molecular dynamics, through numeral integrating second Newton´s law permits to elucidate the evolution of the complex.

Among the applications of molecular dynamics to biomacromolecules, they can either to predict or to refine three dimensional configurations to determine equilibrium averages, structural and motion properties and the thermodynamics of the system and to examine the actual dynamics [25].

Briefly, molecular dynamics requires integrating the equations of motion based on force calculation and this force is estimated assuming conservatives forces so equation 2 becomes useful. Once the initial conditions are established, it is possible to obtain a deterministic behavior of all trajectories for every single molecule either at constant temperature (NVT) or constant pressure (NPT). The equation of motion can be integrated using the leapfrog algorithm or Verlet. The second-order differential equation derived from Newton´s law is equivalent to the following first order differential equations:

$$f_i = \mathrm{m}\frac{dv_i}{dt} \tag{2}$$

$$v_i = \frac{dr_i}{dt} \tag{3}$$

Given the leapfrog method one can expand the equations using the Taylor Expansion for $v_i(t)$,

$$v_i\left(t+\frac{\Delta t}{2}\right) = v_i(t) + \left(\frac{dv_i}{dt}\right)_t \frac{\Delta t}{2} + \frac{1}{2}\left(\frac{d^2 v_i}{dt^2}\right)_t \left(\frac{\Delta t}{2}\right)^2 + \dots \tag{4}$$

$$v_i\left(t-\frac{\Delta t}{2}\right) = v_i(t) - \left(\frac{dv_i}{dt}\right)_t \frac{\Delta t}{2} + \frac{1}{2}\left(\frac{d^2 v_i}{dt^2}\right)_t \left(\frac{\Delta t}{2}\right)^2 + \dots \tag{5}$$

subtracting and rearranging we obtain,

$$v_i\left(t+\frac{\Delta t}{2}\right) = v_i\left(t-\frac{\Delta t}{2}\right) + a_i(t)\Delta t + \dots \tag{6}$$

where a_i is the acceleration of the molecule i and it is calculated from the force. For equation 6 we obtain:

$$r_i\left(t+\frac{\Delta t}{2}\right) = r_i\left(t-\frac{\Delta t}{2}\right) + v_i(t)\Delta t + \dots \tag{7}$$

when using this algorithm the velocities are calculated at time $t+\frac{1}{2}\Delta t$, generally using steps of 10^{-15} s, and calculating the position of the molecules at time $t+\frac{1}{\Delta t}$.

For Verlet´s algorithm the Taylor expansion is carried out for position,

$$r_i(t+\Delta t)=r_i(t)+\left(\frac{dr_i}{dt}\right)_t \Delta t+\frac{1}{2}\left(\frac{d^2r}{dt^2}\right)_t (\Delta t)^2+\ldots \tag{8}$$

$$r_i(t-\Delta t)=r_i(t)-\left(\frac{dr_i}{dt}\right)_t \Delta t+\frac{1}{2}\left(\frac{d^2r_i}{dt^2}\right)_t (\Delta t)^2+\ldots \tag{9}$$

subtracting and rearranging we obtain,

$$r_i(t+\Delta t)=2r_i(t)-r_i(t)(t-\Delta t)+\left(\frac{d^2r_i}{dt^2}\right)_t (\Delta t)^2 \tag{10}$$

we can assume that third order and higher terms are negligible. The acceleration is obtained from the Newton's second law equation at time t and the velocity can be obtained from the finite difference formula

$$v_i(t)=\frac{r_i(t+\Delta t)-r_i(t-\Delta t)}{2\Delta t} \tag{11}$$

This algorithm calculates the position at time $t-\Delta t$ to calculate a new position at time $t+\Delta t$ and the information has to be stored at every iteration [26].

APPLICATIONS OF MOLECULAR DOCKING

The Case of Indole in *Escherichia coli*

Different studies have shown that for bacteria *E. coli*, although generally harmless, some strains are capable of producing extra-intestinal and intestinal infections usually severe in the excretory system as meningitis, peritonitis, mastitis, septicemia and pneumonia [27].

Pathogenic strains are named (ECEI), (ECEH), among others [28] and are capable of forming biofilms and this phenotype enables the bacteria to acquire resistance to antibiotics. Biofilm formation is controlled by several factors including *quorum sensing*. For example, AI-2 has been reported to regulate biofilm formation through regulation of motility [28].

Indole plays a fundamental role for QS in *E. coli*, being one of the signal molecules responsible for the transcription of a variety of genes (*gabT*, and *tnaB* ASTD). It is known that transcription activation of the operon *tnaB* increases in the production of indole. *tnaB* is in charge of encoding the tryptophanase enzyme that synthesizes tryptophan from which is formed indole. *E. coli* does not produce AHLs as signal molecule and for that reason it makes use of some secondary metabolites that accumulate as indole [29]. Moreover, it is known that it controls biofilm formation. However, the role of indole in the cell has not been fully elucidated [1]. Either way indole could be a target for quenchers so the biofilm formation phenotype is controlled in the cell. These quenchers can either be found using a "trial and error" approach in the laboratory or rational design. In order to carry out a rational design, it is necessary to reduce the search space for compounds capable of degrading or affect indole´s functionality. In order to find potential quenchers for indole, our groups carried out a virtual screening for indole-oxidases by calculating indol-oxidase-indol binding energy [29] utilizing molecular docking (Table **3**). This analysis was aimed to find potential indole inhibitors, having as a startup framework *Pseudomonas putida* (*P. putida*) indole oxidases and *in silico* mutations, in the active site, hence the affinity could be improved. Generally, in order to find potential quenchers, it is necessary to take the following steps:

1) Quenchers search space generation.

2) Identification of the active site for each candidate.

3) Calculation of the binding energy based on molecular docking.

4) *In silico* validation based on mutations.

5) Evaluation of the affinity through MD.

Here we explain each step in detail:

Quenchers Search Space Generation

Previous studies have identified that indole-oxidases are present in *P. putida*, whose major ligands are heterocyclic substrates and have an interesting affinity when the substrate is indole. These enzymes oxidize the ring so the substrate turns into Indigo [23]. Due to the inability of Indigo of either inhibit or promote *E. coli* biofilm formation we believe that Pseudomonas enzymes constitute candidates for degrading the signal.

Identification of the Active Site

Owing to the lack of knowledge regarding specific site interactions of the substrate with the ligand for all sequences, it is necessary to identify the catalytic site aminoacids as well as their interaction with indole. Furthermore, once this information is known, we calculate all possible regular nonbonding interactions such as Hydrogen bonds. For the evaluation of the binding energy it is also necessary to distinguish the coordinates of the active site. Hence Swiss PDB Viewer® [30] was utilized to determine these coordinates for the PDB format. Active sites were localized using Ligand Explorer®. They can be downloaded from the protein data bank website (PDB).

Calculation of the Binding Energy Based on Molecular Docking

Molecular docking was carried out for diverse indole-oxidases described in PDB in order to determine the binding energy with indole. AutoDock 4® [31] was utilized thorough all calculations based on its versatility and its capability for parallel computing. AutoDock 4® is a program used to perform computational docking of proteins and small molecules in order to estimate the free energy of binding of a ligand to its target. It uses a scoring function based on the AMBER [31] *force field*. Novel hybrid global-local evolutionary algorithms are used to search the phase space of the ligand-macromolecule system. In these experiments we used the genetic algorithm with 250.000 iterations and it took around 25 minutes each calculation. The results of the binding energy for the 5 different indole-oxidases and indole are shown in Table **3**. The affinity between these two molecules could be increased because they form more hydrogen bonds or the van der Waals interactions are stronger and therefore some molecules have greater affinity to the indole. In this case the indole-oxidase, with PDB code 185L, has the greater affinity with the indole in these *in silico* calculations.

In Silico Based Validation with Mutations and Protein Engineering

We selected *P. putidas* indole-oxidase 107N to perform mutations, as it will be simpler to experimentally evaluate the effects on biofilm formation. Then, we selected amino acids under 3 Å from the ligand, in this case indole, for possible candidates to replace these with amino acids having larger side chains and different characteristics in order to increase the interactions with the indole molecule and obtain an enhance molecule to bind with the indole. The goal of these mutations was to increase the affinity of the ligand-enzyme complex.

Table 3: Binding energy evaluation based on molecular docking calculations for different known indole oxidases structures and indole as ligand

Indole Oxidases PDB ID	Source	Binding Energy (kcal/mol)
185L	*Coliphage*	-19.64
1L4H	*Salmonella entérica*	-13.49
1O7N	*Pseudomona putida*	-9.98
1EG9	*Pseudomona putida*	-6.36
1UUV	*Pseudomona putida*	-0.34

A total of nine different modifications were carried out (Table **4**). We performed mutations in the amino acids of the active site. The replacement was based on hydrophobicity and side chain length: both cases aimed to increase interactions between the ligand and the enzyme. These mutations were carried out in single, pairs or three based change in order to find a synergy. Geometric optimization using HyperChem® [32] was mandatory after carrying out the mutations in order to determine possible changes in the tertiary structure. We used AMBER as force field and the structure was optimized using the Polak-Ribiere [33] or conjugate gradient algorithm [34].

Table 4: Mutations made in the enzyme 1O7N to increase the affinity with the indole

Mutation	Original Amino Acid	Mutated Amino Acid	Binding Energy (kcal/mol)
1	PHE 352	TYR 352	-3.72
	PHE 202	TYR 202	
	ALA 206	MET 206	
2	ASN 297	ARG 297	-16.29
	ASN 201	HIS 201	
	ASP 205	LYS 205	
3	ASN 201	ALA 201	-41.65
	ASP 205	LEU 205	
	ASN 297	LEU 297	
4	ALA 206	VAL 206	-8.12
	HIS 208	ALA 208	
	HIS 295	LEU 295	
5	ALA 206	ALA 206	-6.25
	HIS 208	ALA 208	

Table 4: contd…

	HIS 295	ALA 295	
6	HIS 208	ARG 208	-7.78
	HIS 295	ARG 295	
7	HIS 208	LYS 208	-6.41
	HIS 295	LYS 295	
8	HIS 208	PHE 208	-8.03
	HIS 295	PHE 295	
9	HIS 208	MET 208	-7.71
	HIS 295	MET 295	

Interestingly, after replacing the amino acids ASN201, ASP 205 and ASN 297 for ALA 201, LEU 205 and LEU 297 respectively, the binding displayed a four-fold increase (Table **4**).

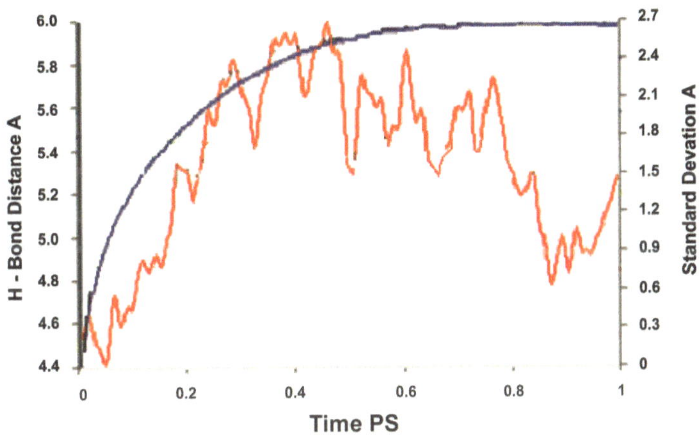

Figure 3: Stability analysis using molecular dynamics (MD) performed in Hyperchem® for 1 ps of the distance between the indole and the amino acid. Red: Distance between the chiral carbon of ALA 205 and indole central carbon (C8); blue: distance standard deviation.

Evaluation of the Stability through Molecular Dynamics

MD studies can be used to determine the stability of the complex. This analysis was aimed to evaluate the bound stability of the mutated amino acids inside the active site and indole, in this case mutation number 3. The simulation was carried out, using AMBER force field with constant and variable temperature in order to have more data in the prediction of the enzyme stability. We measured the distance between the amino acid ALA 205 and indole central carbon (C8) Fig. (**3**), finding a fluctuation in a range between 4.62 and 5.59 Å. The standard deviation is stabilized after 1 psg, so the stability is demonstrated. Similar results were obtained for all mutations (results not shown).

Quenchers Design for Plant Pathogen Ralstonia Solanacearum

R. solanacearum is a plant pathogenic bacterium which causes different diseases in a variety of plants. Its most prevalent hosts are the peanut, potato, tomato, tobacco and banana [35]. It is a soil bacterium which enters the plant through wounds or natural openings invading the xylem and standing fast in the plant's vascular system, making it to prevent the translocation of water and nutrients. As a chemical strategy for penetration and colonization, *R. solanacearum* secrete extracellular polysaccharides (EPS) and extracellular proteins known as (EXPs) essential for its pathogenicity. This type of discharge is part of the type II secretion system, which help to increase the severity of symptoms, in this case the vascular wilt effects, as a result of the condition suffered by the plant's vascular system due to an increase in bacterial density there. Another type of EXPs are involved in the development of the disease by invading the defense response generated by the plant and unlike others are secreted by the type III secretion system that releases proteins directly into the cytoplasm of the cell of the plant. The bacterium secretes effectors proteins that are also released into the cytoplasm of the plant to regulate the survival of the pathogen multiplication and colonization.

Although the pathogenesis of this bacterium is not fully elucidated, molecular and genetic studies have demonstrated the importance of virulence factors which are controlled in response to environmental signals that trigger cascades of target genes regulation associated with disease development. Among the major proteins regulated by this system are proteins for nutrient acquisition, migration *via* the intracellular spaces, penetration of xylem conduits and evasion proteins of the defense system of the plant [8].

The chemical signal that regulates the levels of virulence factors *via* a QS signaling in *R. solanacearum* is 3-hydroxy-palmitic acid methyl ester (3-OH-PAME), a molecule that acts as an autoinducer that is regulated by the gene *phcB* Fig. (**4**) [12].

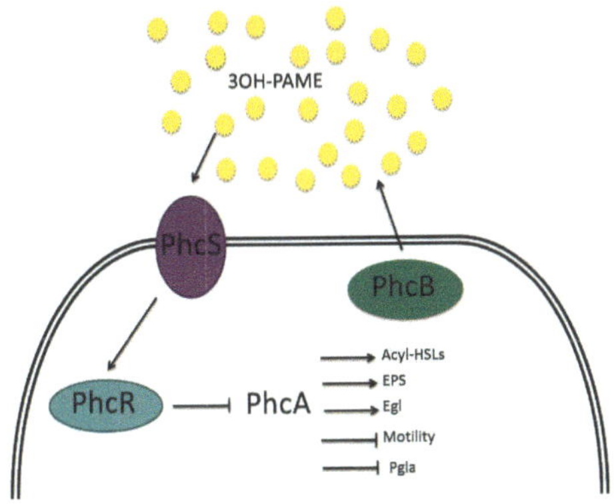

Figure 4: Scheme displaying the regulation mechanism of pathogenicity related genes in *R. solanacearum* by *Quorum Sensing*. PhcB synthesizes the signal 3OH-PAME which interacts with the PhcS (Sensor kinase) in order to phosphorylate the PhcR regulatory protein. PhcR-phosphate is unable to block the activity of the PhcA transcriptional regulator.

Our group sought to develop an algorithm that allows the evaluation of molecules with potential ability to degrade the signal of QS, 3-OH-PAME Fig. (**5**). First we carried out a molecular docking study for a number of candidates and ranked accordingly. This study found lipase B from *Candida antarctica* (*C. antarctica*) as the most suitable quencher for 3-OH-PAME (Table **5** and Fig. (**6**)). Then, we carried out MD studies (Fig. (**7**)) in order to evaluate the stability as the *in silico* analysis identified a binding energy of (-18.375) kcal/mol. These results were verified experimentally using a reporter plasmid that detects the signal presence of *R. solanacearum AW1-3* with which it was shown that in the presence of lipase B from *C. antarctica,* the activity of β-galactosidase is completely reduced.

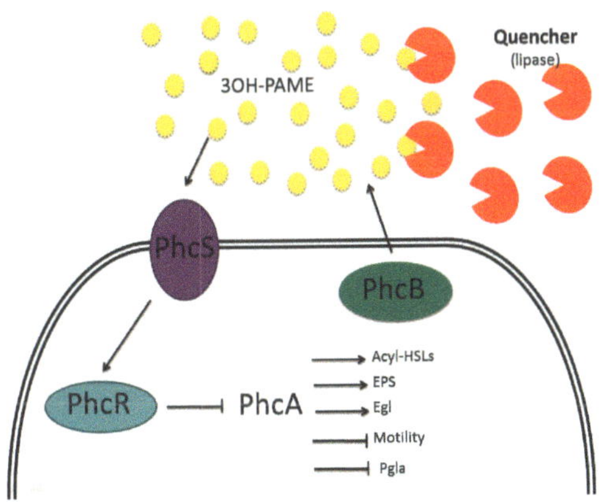

Figure 5: Scheme displaying the inhibition mechanism utilizing lipases in *R. solanacearum Quorum Sensing* system.

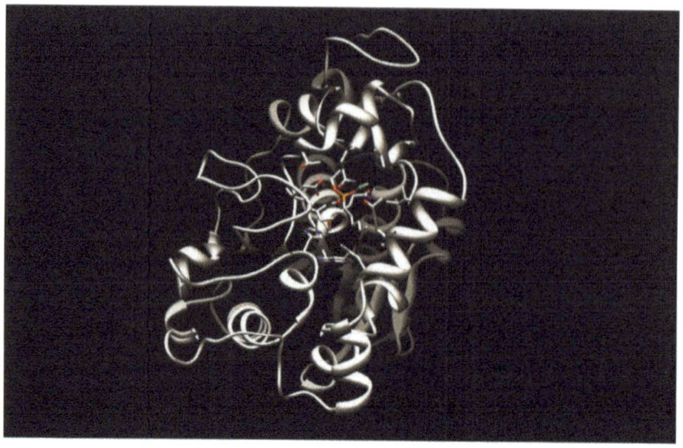

Figure 6: Docked complex of *Candida antarctica's* lipase B with 3-hydroxypalmitic acid methyl ester (3-OH-PAME). The cylindrical color structure represents the ligand.

Table 5: Binding energy evaluation based on molecular docking calculations for different known lipases structures and 3-OH-PAME as ligand

Lipases PDB ID	Source	Binding Energy (kcal/mol)
1LBS	*Candida antartica*	-18.37
1GZ7	*Candida rugosa*	-15.27
1VA4	*Pseudomonas fluresences*	-14.01
1I6W	*Bacillus subtillis*	-13.09
1ZO1	*Pseudomona putida*	-11.75

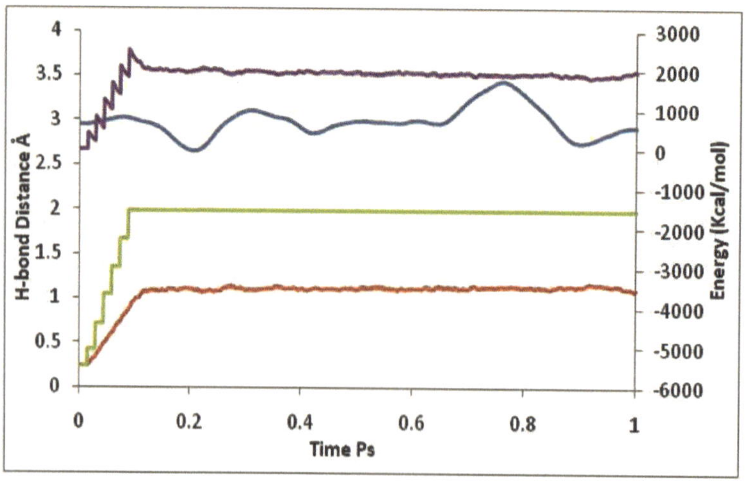

Figure 7: Stability analysis using molecular dynamics (MD) performed in Hyperchem® for 1 ps of the 3-OH-PAME-1LBS complex. Red: Kinetic energy; green: potential energy; purple: total energy; and blue: H-bond distance between serine-105 and 3-OH-PAM.

ACKNOWLEDGEMENTS

Declared None.

CONFLICT OF INTEREST

The authors confirm that this chapter contents have no conflict of interest.

ABBREVIATIONS

QS	=	*Quorum sensing*
AHL	=	N-acyl homoserine lactone
3-Oxo-C6-HSL	=	*N*-3-(oxo-hexanoyl)homoserine lactone
3-Hydroxy-C4-HSL	=	N-(3-hydroxybutanoyl)homoserine lactone
3-Oxo-C12-HSL	=	N-3-oxo-dodecanoyl-L-homoserine lactone
C4-HSL	=	N-butanoyl-L-homoserine lactone
C6-HSL	=	N-hexanoyl-L-homoserine lactones
3-Oxo-C8-HSL	=	N-3-oxo-octanoyl-L-homoserine lactone
3-Oxo-C6-HSL	=	N-(3-oxo-hexanoyl)homoserine lactone
C8-HSL	=	N-octanyl-L-homoserine lactone.
Protein R	=	Protein kinase RNA-activated
PQS	=	*Pseudomonas quinol signals*
SAM	=	S-adenosyl-methionine
DPD	=	4,5 dihydroxy-2,3-pentanedione
SAH	=	S-adenosylhomocysteine

SHR = S-ribosylhomocysteine

EPS = Extracellular polysaccharides

EPS I = Exopolysaccharide I

EXPs = Extracellular proteins

3-OH-PAME = 3-hydroxy-palmitic acid methyl ester

PDB = Protein Data Bank

REFERENCES

[1] Ramsey, M. M.; Korgaonkar, A. K.; Whiteley, M. Quorum-Sensing in Bacteria. In: *Encyclopedia of Microbiology (Third Edition)*; Schaechter,M.,Eds.; Elsevier Inc., San Diego, CA, **2009**, pp. 357-374.

[2] Kroos, L. Who's the Boss? One-Way Conversations between Bacteria. Developmental Cell., **2009,** 17, 155-156.

[3] Bassler, B.L. How bacteria talk to each other: regulation of gene expression by *quorum sensing*. Current Opinion in Microbiology., **1999,** 2(6), 582-587.

[4] Shah, I.M.; Dorwkin, J. Microbial Interactions: Bacteria talk to (some of) their neighbors. Current Biology., **2009,** 19 (16), 689-691.

[5] Engebrecht, J.; Nealson, K.; Silverman, M. Bacterial bioluminescence: isolation and genetic analysis of functions from *Vibrio fischeri.*, Cell., **1983**, 32, 773-781.

[6] Pesci, E.C.; Milbank, J.B.J; Pearson, J.P.; McKnight, S.; Kende, A.S.; Greenberg, E.P.; Iglewski, B.H. Quinolone signaling in the cell-to-cell communication system of Pseudomonas aeruginosa., Proc Natl Acad Sci USA., **1999**, 96, 11229-11234

[7] Han, S.W.; Sriariyanun, M.; Lee, S.W.; Sharma, M.; Bahar, O.; Bower, Z.; Ronald, P. Small protein-mediated *quorum sensing* in a gram-negative bacterium. PloS one., **2011**, 6, e29192.

[8] Rasmussen, T.B.; Givskov, M. *Quorum sensing* inhibitors: a bargain effects. *Microbiology*., **2006**, 152, 895-904.

[9] Kievit, T.R.; Iglewsky, B.H. Bacterial *Quorum Sensing* in Pathogenic Relationships. *Infection and Inmunity*., **2000**, 68 (9), 4839-4849.

[10] Whitehead, N.A.; Barnard, A.M.L.; Slater, H.; Simpson, N.J.L.; Salmond, G.P.C. *Quorum sensing* inGram negative bacteria. FEMS Microbiology Reviews., **2001**, 25, 365- 404.

[11] Gonzalez, J.E; Keshavan, N.D. Messing with Bacterial *Quorum Sensing*. Microbiology and *Molecular Biology Reviews*., **2006**, 70 (4), 859-875.

[12] Flavier, A.B.; Clough, S.J.; Schell, M.A.; Denny, T.P. Identification of 3-hydroxypalmitic acid methyl ester as a novel controlling virulence in *Ralstonia Solanacearum. Molecular Microbiology*., **1997**, 26(2), 251-259.

[13] Denny, T.P. *Ralstonia solanacearum* a plant pathogen in touch with its host. Trends in Microbiology., **2000**, 8 (11), 486-489.

[14] Hikichi, Y.; Yoshomochi, T.; Tsujimoto, S.; Shinojara, R.; Nakaho, K.; Kanda, A.; Kina, A.; Ohnishi, K. Global Regulation of pathogenicy mechanism of *Ralstonia Solanacearum.* Plant Biotechnology., **2007**, 24, 149-154.

[15] Von bodman, S.C.; Diezt Bauer, W.; Coplin, D.L. *Quorum sensing* in plant pathogenic bacteria. Annu. Rev. Phytopathol., **2003**, 41,455-482.

[16] Bauer, W.D.; Mathesius,U. Plant responses to bacterial *quorum sensing* signals. Current Opinion in Biology., **2004,** 7, 429-433.

[17] Zhang, L. Quorum quenching and proactive host defense. *Trends in Plant Science*., **2003,** 8, 238-244.

[18] Bauer, W.D.; Robinson, J.B. Disruption of bacterial *quorum sensing* by other organisms. *Current Opinion in Biotechnology*., **2002**, 13, 234-237.

[19] Chan, K.G.; Atkinson, S.; Mathee, K.; Sam, C.; Chhabra, S.R.; Camara, M.; Koh, C.; Williams, P. Characterization of N- acylhomoserine lactonedegrading bacteria associated with the Zingiber officinale (ginger) rhizosphere: Co-existence of quorum quenching and *quorum sensing* in *Acinetobacter* and *Burkholderia*. BMC Microbiology., **2011**, 11(51), 11-14.

[20] Cirou, A.; Raffoux, A.; Diallo, S.; Latour, X.; Dessaux, Y.; Faure, D. Gamma-caprolactone stimulates growth of quorum-quenching Rhodococcus populations in a large-scale hydroponic system for culturing *Solanum tuberosum*. Research in Microbiology., **2011**, 162(9), 945-960.

[21] Rasmussen, T.B.; Givskov, M. *Quorum sensing* inhibitors: a bargain effects. Microbiology., **2006**, 152, 895-904.

[22] Dong, Y. H., Wang, L. H., Xu, J. L., Zhang, H. B., Zhang, X. F. and Zhang, L. H. Quenching quorum-sensing-dependent bacterial infection by an N-acyl homoserine lactonase. Nature., **2001**, 411, 813-817.

[23] Lee, J.; Bansal, T.; Jayaraman, A.; Bentley, W.; Wood, T. Enterohemorrhagic *Escherichia coli* Biofilms Are Inhibited by 7-Hydroxyindole and Stimulated by Isatin. Applied and Enviromental Microbiology., **2007**, 4100-4109.

[24] Xu, L.; Xu, D.; Liang, J. Computational Methods for Protein Structure Prediction and Modeling; Springer: New York, NY, **2007**, pp. 135-165.

[25] Karplus, M. Molecular Dynamics Simulations of Biomolecules. Accounts of Chemical Research., **2002**, 35(6), 321-323.

[26] Hinchliffe, A. Molecular Dynamics for Beginners; Wiley: Chichester, **2003**, pp. 83-163.

[27] Pace, J. L.; Rupp, M. E.; Finch, R. *Biofilms, Infection, and Antimicrobial Therapy.* Taylor And Francis., Boca Raton, Fl, **2006**.

[28] Nataro, J.; Kaper, J.; Diarrheagenic *Escherichia coli*. Clinical Microbiology Reviews; **1998**, 11, 142-201.

[29] Walters M.; Sperandio,V. *Quorum sensing* in *Escherichia coli* and *Salmonella*. Internal Journal of Medical Microbiology., **2006**, 296,125-131

[30] Swiss PDB Viewer, Version 4.01, **2008**; [Accesed on: January-September 2009].

[31] AutoDock, Version 4, **2007**; [Accesed on: January-September 2009].

[32] HyperChem, Version 8, **2007**; [Accesed on: January-September 2009]

[33] Khoda, K.M.; Liu, Y.; Storey, C.; Generalized Polak Ribiere Algorithm. Journal of Optimization Theory and Applications. **1992,** 75(2), pp 345-354.

[34] Edgar, T.F.; Himmelblau, D.M.; Lasdon, L.S. Optimization of Chemical Processes; Mcgraw-Hill. **2001.**

[35] González, A. F.; Medina, S.; Rational Design of *Quorum Sensing* Inhibitors: The Case of Indole in *Escherichia coli.* **2009,** 2009 AIChE Annual Meeting, Nashville, TN; Available from: http://www.aicheproceedings.org/2009/Fall/data/papers/Paper162607.html; [Accessed 1st September 2011].

CHAPTER 11

Designing Models for Metalloenzymes

James F. Weston*

Departamento de Química, Universidad de los Andes, Carrera 1ª, N° 18-A-12, Bloque Q, 8°Piso, Bogotá, D.C., Colombia, S. A., USA

Abstract: Over the past couple of decades, we have arrived at the point where we are beginning to understand the mode of action of metaloenzymes and to use this knowledge for designing small organometallic catalysts; *i.e.* biomimetics. Quantum chemistry is an essential part of this process. However, due to methodical and technical limitations, quantum mechanical studies are limited to small models of these huge biosystems. This article attempts to bridge the fields of quantum chemistry and biochemistry by illustrating some of the basic mechanisms by which nature accomplishes catalysis. This chapter contains a critical discussion of the limitations of quantum mechanical methods, the pitfalls one can encounter along the way and the necessity for critical model evaluation.

Keywords: Allosteric enzymes, biomimetics, computational models, enzymatic catalysis, metaloenzymes, mode of action, model of enzymatic catalysis, quantum mechanics, sequential metaloenzymes, structure gradient model, superoxide dismutasas, template metaloenzymes, yeast enolase.

INTRODUCTION

Nature's catalysts – the enzymes – perform astonishing tasks. They promote the thousands of chemical reactions that each cell needs to maintain the critical metabolic processes that all life on earth depends upon. Many enzymes contain metal ions in their active sites that are capable of catalyzing processes that every chemist only dreams of. For example, nature uses metalloenzymes to solve its "energy" problem (photosynthesis) and its "food" problem (conversion of atmospheric N_2 into fertilizer). These two fundamental problems are beginning to have a negative impact on our society as earth's finite natural reservoirs of crude oil and nitrates (fertilizers) are rapidly dwindling.

***Corresponding author James Weston:** Departamento de Química, Universidad de los Andes, Carrera 1a No. 18-A-12 Bloque Q 8° Piso, Post Office Box 4976/12340, Bogotá, D.C., Colombia, S. A.; Tel: +57 (1) 33-94-4949; Ext. 3446 or 2786 (Sec.); Fax: +57 (1) 33-24-366; E-mail: jweston@uniandes.edu.co

Subhash C. Basak, Guillermo Restrepo & José L. Villaveces (Eds)
Copyright © 2015 Bentham Science Publishers Ltd. Published by Elsevier Inc. All rights reserved.
10.1016/B978-1-68108-053-6.50011-9

Although we are making continual progress, we are far from fully understanding metalloenzymatic catalysis at a molecular level, not to mention the capability of reproducing such catalytic conversions in the laboratory. In spite of a tremendous amount of effort invested, our ability to *rationally* design and synthesize fully functional biomimetics [1] based on organometallic compounds is still practically nonexistent. If we are truly honest, the few metallocatalysts we have developed can generally be attributed to serendipitous discovery rather than intentional design.

One of the largest hindrances to progress is the paucity of mechanistic details. Enzymes are notorious for their lack of amenity towards direct mechanistic investigations. In cells, they are present in very low concentration, which makes difficult to isolate and study them. A second hurdle is the very characteristic that we want to reproduce; enzymatic rate enhancements are nothing short of astounding. Chemical reactions that have half-lives of around the age of the earth ($\sim 10^9$ years) are accelerated by up to 20 orders of magnitude to very near the diffusion limit ($\sim 10^9$ $s^{-1}M^{-1}$) [2]. This turnover velocity ensures that our tool box of experimental techniques for elucidating mechanisms is rather useless –catalytic turnovers are too fast to be observed. Until recently, almost all of the available mechanistic information on enzymatic action has been obtained from inhibition studies [using a "bad" substrate (inhibitor) to slow the rate down enough to observe the reaction]. The disadvantage with this is that one can never be certain that the mode of action of a true substrate parallels that observed for the inhibitor.

Fortunately, the last ten to fifteen years have seen an astonishing amount of progress. Advances in spectroscopic techniques are making possible to study single molecules [3]. Conformational changes and protein folding dynamics can now be observed with NMR methods [4]. Advances in kinetics are beginning to make feasible direct studies on the slower enzymes [5]. Last, but not least, computational methods have become capable of tying this new information together in a more complete mechanistic picture. As we uncover more details, it is becoming apparent that nature's catalytic strategies are highly variable. The sheer complexity of enzymatic catalysis is slowly being revealed as a quote by Truhlar and co-workers aptly summarize: "*...each enzyme has its unique characteristics* *and enzymes use all possible means to achieve the ultimate objective of reducing the free energy of activation…*" [6].

METHOD SELECTION

Application of quantum modelling techniques in the study of enzymes is becoming commonplace. It has even reached the point that, when experimental studies fail, biochemists approach quantum chemists with "can you calculate this?" However, we pay little attention to designing *in silico* models and just cut out the part of the enzyme which we think is responsible for the chemistry. For a metalloenzyme, this is usually just the metal ion and its ligands. We select a quantum mechanics (QM) method/basis set (usually the combination that works on our computer) and start to work. We do not pay attention to the biochemical characteristics of the enzyme as a whole that predetermine the chemistry occurring in and around the active site to a far greater extent than we realize or would like to admit. Even if we do try to take some of these aspects into account, we have technical limitations as to the size of the QM model we can employ. Even with access to supercomputers, the effective size of QM models remains well below 500 atoms if the system contains a transition metal ion and chemical accuracy is desired. This level of accuracy is, of course, indispensable.

What do we mean by "chemical accuracy"? Well, three basic requirements have to be met. First, the QM method/basis set combination that one employs has to be able to reproduce structural features of all substrate(s), product(s) and within experimental deviations (X-ray and/or spectroscopic data). This means that individual structural parameters need to be accurate to within ±0.1 Å for bond lengths, ±1° for angles and ±3-5° for dihedrals. The second criterion is that the thermodynamics of the system has to be accurate – relative stabilities to ±1 kcal mol^{-1} need to be trusted.

Although tempting due to their applicability, molecular modelling (MM) methods do not possess the accuracy needed for mechanistic studies. However, with scepticism, one can employ these methods for pre-screening, initial structural features and obtaining half-way decent input geometries. With consideration, a combined QM/MM (onium) approach can sometimes be applied. However, these methods are notorious for their unpredictability [7]. If a QM/MM approach is being considered, one needs to verify structural and thermodynamic accuracy.

The third requirement for chemical accuracy is that the method has to be capable of modelling transition structures, transient species and active reaction coordinates within the structural and energetic limits discussed above. This is a more stringent requirement which basically means that a density functional (DFT) method in conjunction with at least a triple-zeta basis set has to be employed. Furthermore, care needs to be taken in the selection of the DFT functional. Some functionals are pretty good for calculating structural and energetic data of stable structures (substrates and products) but are not so good for transition structures. B3LYP, for example, notoriously underestimates reaction barriers [8], which is unacceptable if one wants to calculate a catalytic cycle in an enzyme-based model. Conversely, other functionals like MPW1K [9] or B2PLYP-D [10] yield better reaction barriers. However, using them for structure optimizations is not a good idea; many functionals that yield good energetics are not particularly structurally accurate.

Often, one needs to go with a combined approach –using a structurally accurate DFT method with a medium sized basis set for optimizing the geometries of all species involved and then employing these geometries for single point calculations using a thermodynamically accurate method and a much larger basis set. From a practical point of view, meeting the requirement of chemical accuracy usually limits the model system to between 30 and 500 atoms. Post-HF and higher level methods are usually not under discussion; their very size and the presence of heavy metal ions prohibit their application. Before selecting a method/basis set (or combination thereof), one needs to review the latest QM literature on relevant systems.

Once method selection is out of the way, the PDB file of the enzyme is downloaded; the area of the active site known to be responsible for the chemistry is cut out and these coordinates are used to generate the first input files. In most studies, this cut-out is usually selected on the basis of computational viability. More often than not, the model of enzymatic catalysis that we chemists carry in our subconscious mind has a great deal of influence on the cut-out selection without our realizing it.

GENERAL MODEL OF ENZYMATIC CATALYSIS

Most of us have a picture of enzymatic catalysis that is based on the key-and-lock model proposed by E. Fischer in 1894 [11]. This was modified by the introduction of transition state theory [12, 13] that formed a simple model which we still use today (Fig. **(1)**).

In this model, an enzyme has a small, *localized* active site. If an incoming substrate fits; *i.e.* if it has the proper 3D shape *and* electrostatic surface potential, it docks onto the active site, generating an enzyme-substrate intermediate [ES] called a Michaelis complex [14]. The catalytic activity is assumed to originate from this initial complex and the substrate fit is believed to primarily determine the enzymatic selectivity. Quite some research is based on the often erroneous assumption that substrate-active site interactions in [ES] correlate directly with enzymatic activity. For example, substrate docking models are employed in screening for new substrates (pharmaceutical applications) and QSAR studies for finding and studying structure-function correlations are widely available.

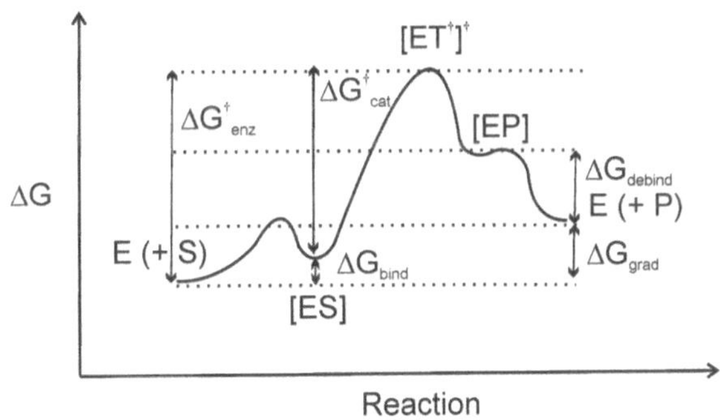

Figure 1: Simple schematic of general enzymatic catalysis.

During docking, we assume that the substrate is positioned and activated, a process which destabilizes or pre-organizes it (a process which can include steric or electronic distortions) [15]. Exactly how enzymes manage this is still under discussion [16]. Some argue that electrostatic effects are the major contributing factor [17]. Others argue that the enzyme forces the substrate to assume a "near attack conformation" that precedes the chemical reaction [18]. The slightest variation in the topology or the electronic attributes of the substrate drastically lowers the ability of the active site to accept it or to promote the subsequent reaction. Sometimes this fact can be exploited; with a suitable inhibitor, an [ES]-analogue is formed and can be studied (substrate analogue inhibition).

As the reaction occurs, the topology of the active site assists by selectively stabilizing the evolving transition structure $[ET^{\dagger}]^{\dagger}$ [16, 17, 19]. The proposal of Pauling that the active site binds the transition structure tighter than the substrate [13] is still widely accepted as one of the major factors in lowering the activation energy. One assumes that the rate-determining step follows the principle of least motion; *i.e.* the transition structure is intermediate between the substrate and product. As a consequence, much effort is devoted to finding substrates that mimic the proposed transition-structure. A familiar example is the use of pentacoordinated phosphorous as S_N2 transition structure analogues. Now and then an article claims to have "isolated" a transition structure when what really happened is that the authors obtained an enzyme complex containing a transition structure analogue. Barrier lowering can also be affected by dynamic movement; the transition structure has evolved to optimize the "reactive" vibrational mode that converts the substrate into the product [20]. The combined effects of substrate destabilization and transition structure stabilization are considered to be the two dominant factors in determining the rate determining barrier $\Delta G^{\dagger}_{cat}$ [16].

The reaction generates a metastable enzyme-product complex [EP]. The evolutional design of the active site is such that the product has a poor fit as compared to the substrate. This guarantees that the product complex [EP] can easily and rapidly dissociate. However, it is possible to find substances which are "product analogues" [21]. These closely resemble the product but have a high affinity for the active site, thus binding quasi-irreversibly. Our basic model (Fig. **(1)**) is idealistic and valid only for one-substrate, one-step enzymes that obey Michaelis-Menten kinetics [14]:

$$E + S \underset{}{\overset{k_{bind}}{\rightleftharpoons}} [ES] \xrightarrow[k_{cat}]{[ET^{\dagger}]^{\dagger}} [EP] \overset{k_{debind}}{\rightleftharpoons} E + P$$

Very few enzymes exhibit Michaelis-Menten behaviour; nature is almost always much more complex. "Reality" involves additional mechanistic steps and the catalytic circle often includes intermediates and transition structures, any one of which may be rate limiting.

TEMPLATE BASED METALOENZYMES

The most simple metalloenzymatic mechanism and that which most closely corresponds to our model is that of a template system. As defined by the author, a fundamental characteristic of these metalloenzymes is that the metal cation (or cations) is/are *irreversibly bound in a small, well defined active site*. The metal ion does *not* participate in the catalytic turnover. Furthermore, *the reaction is localized* on or around the metal ion. Long range and/or dynamic effects are absent and the enzyme functions *via* the "scaffold" [22] or "coordination template" [23] effect. The metal ion recognizes the substrate and activates it through direct metal-substrate interactions and then catalyzes the reaction. Once formed, the product departs from the active site. In this enzymatic category, the metal ion functions as a catalyst. Its activity is modified only by the primary and secondary coordination sphere of the ion with the rest of the enzyme being ballast.

Template features make this category the most attractive one for modelling. Small organometallic complexes and minimal QM models are often thought to provide good biomimetics for metalloenzymes. As a consequence, most of the biomimetic work to date – both of experimental and computational nature – has been based on template models; even when the enzyme does not exhibit template behaviour. True template systems are rare. One should not be surprised when an organometallic complex or QM model that accurately reproduces the enzymatic structure fails spectacularly as a biomimetic. Even when the mode of action is template, successful catalysis often depends critically upon secondary effects that are usually lacking in overly simplistic models. As a consequence, there are still very few *functional* biomimetics available even for this most simple category of mechanistic behaviour.

The superoxide dismutases provide an excellent illustrative example. These metalloenzymes are an indispensable defense against oxidative damage at the cellular level [24]. Oxidative damage contributes to the aging process and is involved in many human diseases ranging from simple infections up to genetic disorders and many age related pathologies such as Alzheimer's and Parkinson's disease [25].

Nickel based superoxide dismutases (NiSOD) contain a short sequence of nine amino acids that fold around the metal ion [26]. Several independent studies have demonstrated that artificial metallocomplexes of this sequence are highly functional SOD biomimetics [27], thus confirming the template nature of this enzyme.

In direct contrast to NiSODs, iron and manganese based superoxide dismutases (Fe- and MnSODs) are not template systems [28]. Small organometallic complexes that reproduce the first ligand sphere of the central metal ion are completely dead in water [29]. QM models containing only the central metal ion and its first sphere ligands are not capable of correctly modelling the chemical catalysis [30]. Enzymatic activity is modulated by a whole series of longer-range interactions that involve residues up to the 5th coordination sphere (Fig. (**2**)) [31].

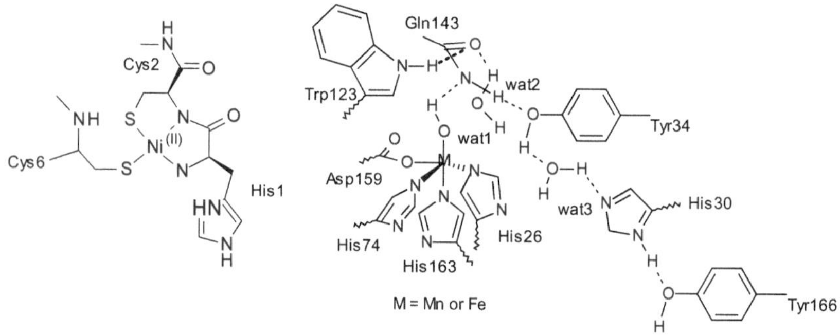

Figure 2: Left: The template nature of NiSOD showing the minimal number (3) of amino acid residues needed to effect chemical catalysis. Right: The extended peptide environment responsible for chemical catalysis in Fe- and MnSODs.

In addition to hydrogen bond networks, there are quite a few other strategies that nature uses to modify modes of action that are predominately template. This introduces additional complications due to medium range interactions in and around the active site. Typical examples include a dynamic component such as a local conformational change between active and inactive conformers. Other modulation possibilities include an active site lid composed of a flexible peptide loop or even an allosteric component that affects the entire enzyme. The list of nature's tricks is long and extremely variable.

DESIGNING TEMPLATE QM MODELS

QM models containing metal ions rely on electrostatics to hold things together. Just remove the metal ion in a QM model and watch the ligands fly away upon optimization. In synthetic chemistry, one encounters the same situation. Synthetic biomimetics are organometallic compounds constructed using the "scaffolding" [22, 23] abilities of the metal ions. One of the greatest problems we face when constructing QM systems models for metalloenzymes is that nature uses a different strategy from what we employ *in silico*. The metal ion in an enzyme is *not* the centre of the universe. Peptide folding gives rise to a scaffold situation with a preformed cavity for the metal ion. For example, the binding sites in both NiSOD and MnSOD form stable preformed binding sites even in the absence of Ni^{2+} or Mn^{2+} [27, 32]. In designing biomimetics, it is apparent that a change of focus is needed in which ligands are specifically designed to provide scaffolds for the metal ion instead of the other way around. A peptide scaffold is capable of stabilizing unusual coordination geometries such as the trigonal bipyrimidal situation found in MnSOD. Simple manganese compounds are usually octahedral. Forcing an unusual coordination geometry is a trick nature uses to potentiate the catalytic capability of a metal ion. QM models based on electrostatics to hold things together return false preferences for the more stable species and will be wrong in many important mechanistic details.

Unfortunately, most QM studies to date have not sufficiently considered scaffolding contributions. The question becomes: how can one begin to include these effects in a QM model that relies on electrostatics? Before getting started, one needs to verify that the enzymatic mode of action is indeed template. Access to kinetic data is therefore essential. Do *functional* synthetic biomimetics based on structurally similar organometallic compounds exist? If so, the mode of action is probably template and simple metal-centred QM models are going to be fairly accurate. However, this is generally not the case.

The scaffolding contributions of the peptide that modify the catalytic capability have to be identified. Moving out from the central metal ion, one needs to identify all residues that may be playing key roles. Selected mutation experiments are a great help. Hydrogen-bond interactions and networks are extremely important as the peptide structure often uses these to impose steric constraints and/or electronic modifications. Is the metal ion located at the bottom of a substrate channel? If so, what residues line the channel? Is the substrate going to be bumping up against anything on its path to the metal ion that is capable of a specific interaction? This may give rise to a very important pre-activation event. Another important point often overlooked is the presence of crystallographic water molecules at positions where they could play a role as proton donors or nucleophilic species. You should get creative. Can important residues be included in the model using simple functionalities? A neat trick is to simply fix the end of a side chain at its PDB coordinates. It is time to move away from free optimizations and include selected fixed coordinates in the model. After all, nature does exactly this using its peptide scaffold. However, very few QM studies even consider this possibility.

SEQUENTIAL SYSTEMS

Many metalloenzymes, especially metabolic enzymes, exhibit a mode of action that differs from the template systems. First described by J. Cowen [33], the author categorizes their mechanistic behaviour as "sequential". Sequential enzymes are still characterized by a *localized* active site. However, an empty resting state is now present because the *metal ion is labile and is an intrinsic part of the catalytic turnover*. A metal ion enters the active site, performs some critical task and then leaves again at some well-defined step in the catalytic circle. A *sequential kinetic mechanism* is present. The relative order in which the metal cofactor and substrate dock onto the active site [and when the metal ion and product leave] is extremely important. The active site of these enzymes is generally a bit larger than that found in a template system. When a metal ion binds to a sequential active site, it usually induces local structural changes (10-50 Å). Things are further complicated by these induced dynamic contributions which can include switching between active and inactive conformations, inducing a loop movement that clamps a lid on the active site during a critical chemical reaction, *etc*. Sometimes the borderline to allosteric behavior is fluxional.

The specific role of the metal ion in a sequential enzyme is varied and can be classified in three general categories:

a) Catalytic function

A solvated metal ion enters the active site, binds to it (loses water ligands) and then performs a catalytic task before leaving again (under uptake of water).

b) Regulatory function

Metal ion docking regulates or even enables some important catalytic aspect. Once this function has been fulfilled, the metal ion departs and the active site returns to its previous state.

c) Cofactor function

The metal ion has little or no direct interaction with the active site. It interacts with the substrate as a metallic cofactor and functions primarily as a Lewis acid, helping in substrate/intermediate/transition structure activation, provides a nucleophile or helps to stabilize a charged product (P_i for example).

Sequential enzymes depend on the bioavailability of the metal ion. Metabolic processes involving sequential enzymes are often crudely regulated by metal ion bioavailability and are then fine-tuned by other, more sensitive, modulation strategies. Due to the ability of water to compete with organic ligands for alkali (Na^+, K^+) and earth alkaline cations (Mg^{2+}, Ca^{2+}), enzymes with a dependency on these ions are predestined for sequential behaviour. Examples of sequential behaviour are the enolases that catalyze the dehydration of 2-phosphoglycerate (2-PGA):

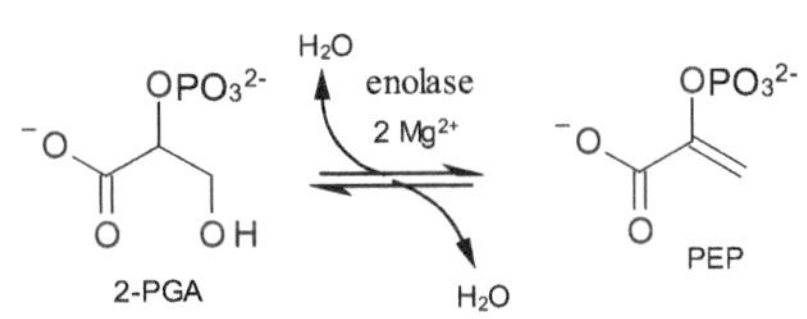

One member of this family, yeast enolase, was one of the very first enzymes to be crystallized (1941) [34]. Since then, a considerable amount of information has been collected and this is the best understood sequential system to date. Yeast enolase requires two Mg^{2+} ions per active site [35] that bind to the active site in a kinetically ordered manner with differential binding affinities [36]. Solid state structures are available for each intermediate kinetically identified [37] and the current understanding of the mode of action is illustrated in Fig. **(3)**.

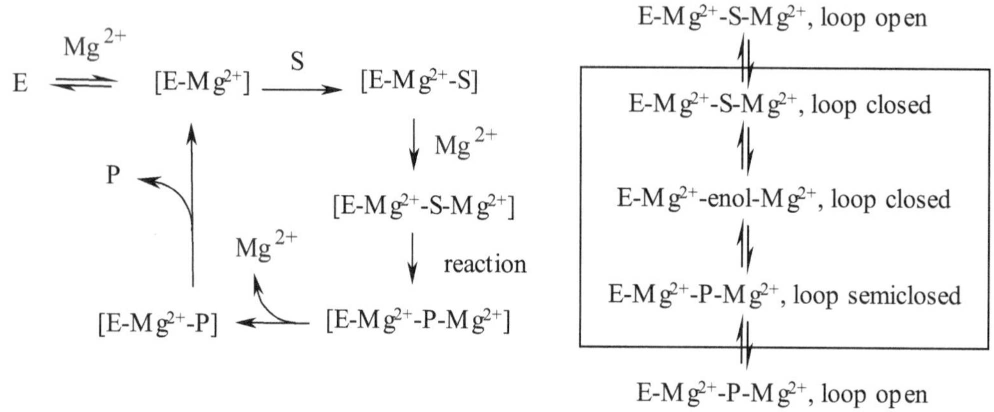

Figure 3: Left: Sequential mode of action of yeast enolase. Right: Conformational changes in the active site during the chemical reaction (enclosed in the box).

The resting state is the metal-free *apo*-form E. Binding the first Mg^{2+} ion to E induces a conformational change in the active site, activating it towards substrate binding [35,36]. This Mg^{2+} ion is therefore termed the "conformational" ion. A [E-Mg^{2+}] intermediate is formed which is in an "open" conformation capable of accepting the substrate (2-PGA) to generate an [E-Mg^{2+}-S] intermediate [38].

The second, "catalytic" Mg^{2+} ion cannot bind until after the substrate [39]. Without it, yeast enolase is inactive [40]. Binding the second Mg^{2+} induces semi-localized dynamic changes. Concerted movements of three short, flexible loops open and close the substrate channel much like opening and closing a lid on a pot [41]. The active intermediate [E-Mg^{2+}-S-Mg^{2+}] is generated in a closed state [39, 42]. As the reaction progresses, dynamic changes cause the lid on the pot to partially (semi-closed state) and then fully open [43]. Product dissociation is also kinetically ordered with the catalytic Mg^{2+} ion leaving first [39] to generate an [E-Mg^{2+}-P] complex [37]. The product then leaves, followed by the conformational Mg^{2+} ion.

Dehydration probably occurs *via* a two-step mechanism over an enol intermediate (Fig. **(4)**) [44]. This is the only intermediate that has not yet been experimentally observed. However, theoretical calculations (all based on template models) are in favor of an enol intermediate [45].

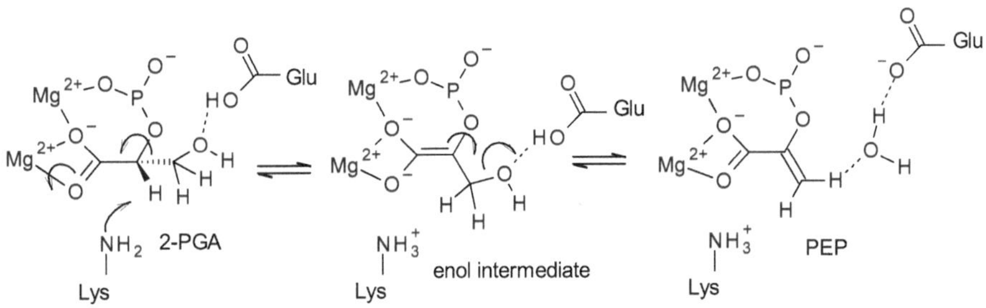

Figure 4: The proposed enol intermediate in the dehydration step of the conversion of 2-PGA into PEP catalyzed by yeast enolase.

DESIGNING SEQUENTIAL QM MODELS

To date, almost all QM studies on sequential enzymes have employed template models that depend on electrostatic interactions between the metal ion(s) and their ligands to hold them together. Often, it is unclear if the theoretician was even aware that the enzyme exhibited a behaviour other than that of a template system.

While a template strategy may suffice for selected intermediates or individual reaction steps, it is fundamentally incapable of being employed for calculating the entire mode of action of a sequential enzyme. Metal ion movement is an intrinsic part of catalysis and the model employed needs to be stable towards metal ion exchange as well as ionic movement relative to the amino acid residues that make up the bulk of the active site. Metal-ion centred template models are incapable of this; they simply fall apart *in silico* when the metal ions are removed or displaced from their optimal positions.

The only remedy to date is to employ a QM/MM method that is large enough to permit limited ionic movement. QM/MM approaches are based on a method gradient with the centre of the model being calculated using a higher-level, more accurate method and the larger periphery region with a lower-level (and thus faster) method (Fig. **(5)**). Due to technical difficulties, the overlap region is always accompanied by methodical errors; the results are unsystematic and can be quite inaccurate. Additional problems are encountered when one tries to move a metal ion (or just a simple point charge) across a method gradient. Most QM/MM software is simply not capable of this; chronic run-time errors result as the mathematics of the method change become extremely ill-defined.

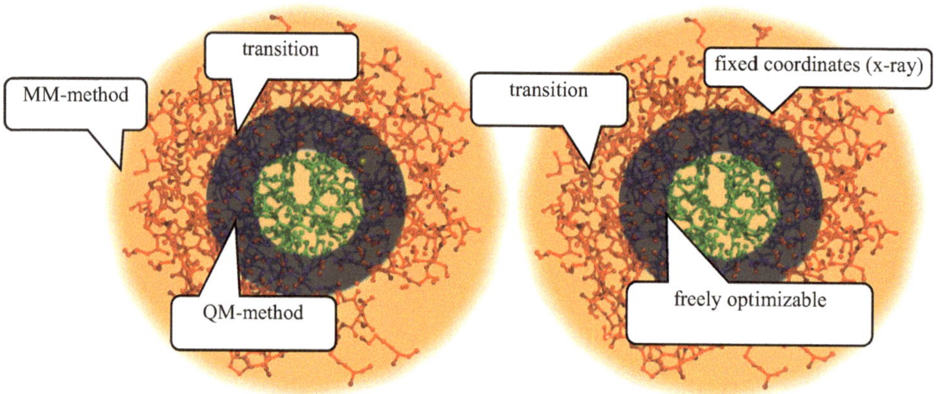

Figure 5: Left: Basic strategy of QM/MM models. Right: Structure gradient approach for designing models for sequential systems.

The author recommends using a different strategy that is based on a structure gradient [46]. In this approach, all known PDB structures of the enzyme are overlaid. As long as the mode of action is quasi-localized, the backbone residues will begin to overlap at some point in space moving out and away from the active site. At this point, a sphere containing the PDB coordinates of the active site is cut out and used to generate a Z-matrix. The positions of all heavy atoms located on the surface of this sphere are frozen in space. All open valences are completed with hydrogen atoms, thus creating a hard (fixed) outer shell. Moving from the fixed outside to the inside, one frees the dihedral angles, then the bond angles and, finally, the bond lengths to create a structure gradient that connects the frozen outside to a freely optimizable inside [46]. With a bit of input file manipulation, standard QM program packages are quite capable of performing calculations on these models.

An example of a structure gradient model for yeast enolase is illustrated in Fig. **(6)** and was optimized on a standard PC running under Linux and took on an average two days to complete. Solvent effects were partially taken into account using the COSMO method. Nonetheless, it has to be said that one needs use quasi-linear MARI-J convergence routines when trying to optimize systems of this size.

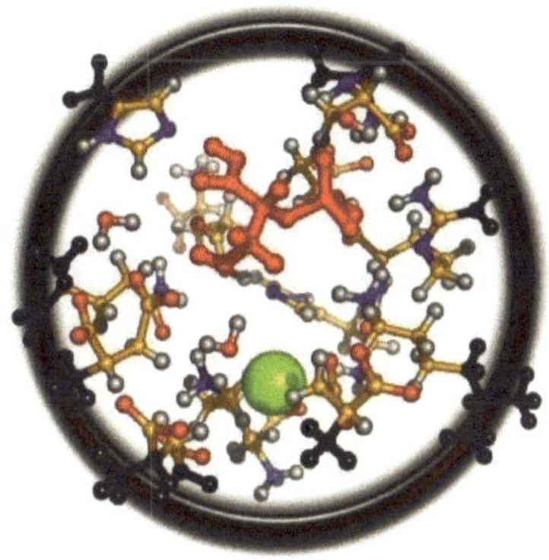

Figure 6: Structure gradient model for the [E-Mg^{2+}-S] intermediate of yeast enolase. The coordinates of the solid black atoms located on the outer sphere are fixed in space. Optimized at the TPSS(MARI-J; COSMO)/SV(P) level of theory using the Turbomole suite of programs.

Employing a structure gradient not only allows one to overcome the limitations of metal-ion centred models but also to employ a single computational method (DFT, for example) with approximately the same size/time advantages as QM/MM-methods but without the additional overlap error. The coordinated movement of metal ions, substrate, product as well as flexible side chain residues and specific solvation waters can now be explicitly studied. One can "put in", "take out" and even "move around" metal ions without the model falling apart.

ALLOSTERIC SYSTEMS

Some metaloenzymes exhibit an allosteric mode of action in which a metal ion binding/debinding event triggers structural changes in the *entire* enzyme. Indeed, it is difficult to speak of an active site as such. The metal ion or cofactor may bind to a regulatory site far away from where the chemical reaction of interest is taking place. Although the specific regions where the substrate is being transformed into product and where the regulatory site is can often be identified, it is impossible to consider these regions as being independent of the rest of the enzyme.

As a consequence, it is very difficult to perform mechanistic studies on allosteric systems – and even more difficult to develop computational models for studying their behaviour [47]. Even employing the most simple MM techniques, it is still usually beyond the capacity of modern computers to perform calculations on these enzymes which tend to be extremely large, often possessing several mutually interacting subunits [48]. At the moment, we are limited to small template or sequential models capable only of modelling selected chemical steps *that can be localized to a certain region* of the enzyme when studying such complex systems. Studies of this kind leave behind many unresolved questions concerning the influence of allosteric interactions on catalytic efficiency.

CONCLUSION

There has been considerable progress in understanding the underlying molecular modes of action that metalloenzymes employ in their catalytic circles in the last few years which has led to an increasing appreciation of the sheer complexity of nature's catalysts. In contrast to this, most quantum chemists are still using a rather limited picture of enzymatic catalysis when designing QM models to study these systems. Almost all of these models are template models, even though many enzymes exhibit a vastly more complex mechanistic behaviour. Nature basically uses a strategy that is fundamentally opposite to the theoretician's model building strategy. We rely on electrostatic effects to hold our models together whereas nature uses a peptide-template strategy. This can have drastic consequences on model validity. In addition, many enzymes exhibit a sequential activity that fundamentally differs from a template-type mode of action. Template models are simply not capable of correctly describing these enzymes. An alternative model-building strategy, based on a structure gradient is proposed to overcome the insufficiencies of template-based models.

ACKNOWLEDGEMENTS

Financial support by the DAAD (Deutscher Akademisher Austausch Dienst) and by the Universidad de los Andes (Bogotá, Colombia) is gratefully acknowledged.

CONFLICT OF INTEREST

The author confirms that this chapter contents have no conflict of interest.

ABBREVIATIONS

2-PGA	=	2-phosphoglycerate
B3LYP	=	a hybrid density functional
COSMO	=	Conductor-like screening model; a solvation model
DFT	=	Density Functional Theory
E	=	Enzyme
HF	=	Hartree-Fock method
MARI-J	=	Multipole Accelerated Resolution of Identity Approximation
MM	=	Molecular Modelling
MPW1K	=	a hybrid density functional
NMR	=	Nuclear Magnetic Resonance
PC	=	personal computer
PDB	=	Protein Data Bank
P	=	Product(s)
PEP	=	phosphoenol pyruvate
QM	=	Quantum Mechanics
QSAR	=	Quantative Structure-Activity Relationship
S	=	Substrate(s)
SV(P)	=	a standard split-valence basis set
SOD	=	Superoxide Dismutase

TPSS = a hybrid density functional

REFERENCES

[1] Breslow, R. Biomedical control of chemical selectivity. *Acc. Chem. Res.* **1980**, *13*, 170-177.

[2] a) Wolfenden, R. Degrees of difficulty of water-consuming reactions in the absence of enzimes. *Chem. Rev.* **2006**, *106*, 3379-3396. b) Wolfenden, R. Snider, M. J. The depth of chemical time and the power of enzimes as catalysts. *Acc. Chem. Res.* **2001**, *34*, 938-945.

[3] See, for example: a) Liang, J.; Fernández, J. M. Mechanochemistry: One bond at a time. *ACS Nano* **2009**, *3*, 1628-1645. b) Xu, W.; Kong, J. S.; Chen, P. Single-Molecule Kinetic Theory of Heterogeneous and Enzyme Catalysis. *J. Phys. Chem. C* **2009**, *113*, 2393-2404.

[4] See for example: a) Chang, Y. C.; Franch, W. R.; Oas, T. G. Probing the folding intermediate of Bacillus subtilis RNase P protein by NMR. *Biochemistry* **2010**, *49*, 9428-9437. b) Mangia, S.; Traaseth, N. J.; Veglia, G.; Garwood, M.; Michaeli, S. Probing slow protein dynamics by adiabatic r1ρ and r2ρ nmr experiments. *J. Am. Chem. Soc.* **2010**, *132*, 9979-9981.

[5] Alberty, R. A. Biochemical thermodynamics and rapid-equilibrium enzyme kinetics. *J. Phys. Chem. B.* **2010**, *114*, 17003-17012.

[6] Gao, J.; Ma, S.; Major, D. T.; Nam, K.; Pu, J.; Truhlar, D. G. Mechanisms and free energies of enzymatic reactions. *Chem. Rev.* **2006**, *106*, 3188-3209

[7] Hu, L. H.; Söderhjelm, P.; Ryde, U. On the convergence of QM/MM energies. *J. Chem. Thor. Comput.* **2011**, *7*, 761-777.

[8] Zhao, Y.; Truhlar, D. G. Density functionals with broad applicability in chemistry. *Acc. Chem. Res.* **2008**, *41*, 157-167.

[9] Lingwood, M.; Hammond, J. R.; Hrovat, D. A.; Mayer, J. M.; Borden, W. T. MPW1K performs much better than B3LYP in DFT calculations on reactions that proceed by proton-coupled electron transfer (PCET). *J. Chem. Theory Comput.* **2006**, *2*, 740-745.

[10] Schwabe, T.; Grimme, S. Theoretical thermodynamics for large molecules: walking the thin line between accuracy and computational cost. *Acc. Chem. Res.* **2008**, *41*, 569-579.

[11] Fischer, E. Einfluss der configuration auf die wirkung den. enzyme. *Ber. Dtsch. Chem. Ges.* **1894**, *27*, 2985-2993.

[12] Polanyi, M. *Z. Elektrochemie* **1921**, *27*, 143.

[13] Pauling, L. The nature of forces between large molecules of biological interest. *Nature*, **1948**, *161*, 707-709.

[14] Michaelis, L.; Menten, M. L. Die kinetik der invertinwirkung. *Biochem. Z.* **1913**, *49*, 333-369.

[15] a) Callahan, B. P. OMP Decarboxylase: An experimental test of electrostatic destabilization of the enzyme–substrate complex. *J. Am. Chem. Soc.* **2004**, *126*, 14698-14699. b) Jencks, W. P. Binding energy, specificity, and enzymic catalysis: the circe effect. *Adv. Enzymol.* **1975**, *43*, 219-410.

[16] Borman, S. Researchers are devising strategies to catch toxic drugs before they reach clinical trials. *Chem. & Eng. News.* **2004**, *82*, 35-39.

[17] Warshel, A.; Sharma, P. K.; Kato, M.; Xiang, Y.; Liu, H.; Olsson, M. H. M. Electrostatic basis for enzyme catalysis. *Chem. Rev.* **2006**, *106*, 3210-3235.

[18] Bruice, T. C.; Bruice, P. Y. Covalent intermediates and enzyme proficiency. *J. Am. Chem. Soc.* **2005**, *127*, 12478-12479 and literature contained therein.

[19] a) Zhang, X.; Houk, K. N. Why enzymes are proficient catalysts: beyond the pauling paradigm. *Acc. Chem. Res.* **2005**, *38*, 379-385. b) Garcia-Viloca, M.; Gao, J.; Karplus, M.; Truhlar, D. G. How Enzymes Work: Analysis by Modern Rate Theory and Computer Simulations. *Science,* **2004**, *303*, 186-195.

[20] a) Olsson, M. H. M.; Parson, W. W.; Warshel, A. Dynamical contributions to enzyme catalysis: Critical tests of a popular hypothesis. *Chem. Rev.* **2006**, *106*, 1737-1756. b) Hammes-Schiffer, S. Hydrogen tunneling and protein motion in enzyme reactions. *Acc. Chem. Res.* **2006**, *39*, 93-100. c) Benkovic, S. J.; Hammes-Schiffer, S. Enzyme motions inside and out. *Science* **2006**, *312*, 208-209.

[21] A typical example: Bains, J.; Kaufman, L.; Farnell, B.; Boulanger, M. J. A product analog bound form of 3-oxoadipate-enol- lactonase (pcad) reveals a multifunctional role for the divergent cap domain. *J. Mol. Biol.* **2011**, *406*, 649-658.

[22] Ragsdale, S. W. Metals and their scaffolds to promote difficult enzymatic reactions. *Chem. Rev.* **2006**, *106*, 3317-3337.

[23] Busch, D. H.; Stephenson, N. A. Molecular organization, portal to supramolecular chemistry: structural analysis of the factors associated with molecular organization in coordination and inclusion chemistry, including the coordination template effect. *Coord. Chem. Rev.* **1990**, *100*, 119-154.

[24] Johnson, F.; Giulivi, C. Superoxide dismutases and their impact upon human health. *Mol. Asp. Med.* **2005**, *26*, 340-352.

[25] A search in Science Direct (http://www.sciencedirect.com/science) yields > 60,000 publications linking oxidative damage to human diseases and pathologies.

[26] a) Barondeau, D. P.;Kassmann, C. J.; Bruns, C. K.; Tainer, J. A.; Getzoff, E. D. Nickel superoxide dismutase structure and mechanism. *Biochemistry* **2004**, *43*, 8038-8047. b) Wuerges, J.; Lee, J. W.; Yim, Y. I.; Yim, H. S.; Kang, S. O.; Carugo, K. D. Crystal structure of nickel-containing superoxide dismutase reveals another type of active site. *Proc. Natl. Acad. Sci. USA* **2004**, *101*, 8569-8574.

[27] a) Krause, M. E.; Glass, A. M.; Jackson, T. A.; Laurence, J. S. Novel tripeptide model of nickel superoxide dismutase. *Inorg. Chem.* **2010**, *49*, 362-364. b) Schmidt, M.; Zahn, S.; Carella, M.; Ohlenschläger, O.; Görlach, M.; Kothe, E.; Weston, J. Solution structure of a functional biomimetic and mechanistic implications for nickel superoxide dismutases. *ChemBioChem* **2008**, *9*, 2135-2146. c) Neupane, K. P.; Gearty, K.; Francis, A.; Shearer, J. Probing variable axial ligation in nickel superoxide dismutase utilizing metallopeptide-based models: insight into the superoxide disproportionation mechanism. *J. Am. Chem. Soc.* **2007**, *129*, 14605-14618. d) Shearer, J.; Long, L. M. A nickel superoxide dismutase maquette that reproduces the spectroscopic and functional properties of the metalloenzyme. *Inorg. Chem.* **2006**, *45*, 2358-2360.

[28] Tabares, L. C.; Gätjens, J.; Un, S. Understanding the influence of the protein environment on the mn(ii) centers in superoxide dismutases using high-field electron paramagnetic resonance. *Biochim. Biophys. Acta* **2010**, *1804*, 308-317.

[29] a) Borovik, A. S. Bioinspired Hydrogen Bond Motifs in Ligand Design: The role of noncovalent interactions in metal ion mediated activation of dioxygen. *Acc. Chem. Res.* **2005**, *38*, 54-61. b) Yamaguchi, S.; Kumagai, A.; Funahashi, Y.; Jitsukawa, K.; Masuda, H. An Accurately-Constructed Structural Model for an Active Site of Fe-Containing

Superoxide Dismutases (Fe-SODs). *Inorg. Chem.* **2003**, *42*, 7698-7700. c) Warzeska, S. T.; Micciche, F.; Mimmi, M. C.; Bouwman, E.; Kooijman, H.; Spek, A. L.; Reedijk, J. Tuning the coordination mode in mononuclear manganese complexes by changing the steric bulk of the carboxylates. *J. Chem. Soc. Dalton Trans.* **2001**, 3507-3512. d) Erre, L. S.; Micera, G.; Garribba, E.; Bényei, A. C. Molecular structure and spectral properties of bis(2,6-dimethoxybenzoato)(2,2′:6′,2″-terpyridine)manganese(II): a five-coordinate Mn(II) complex. *New J. Chem.* **2000**, *24*, 725-728.

[30] Li, J.; Fisher, C. L.; Konecny, R.; Bashford, D.; Noodleman, L. Density Functional and Electrostatic Calculations of Manganese Superoxide Dismutase Active Site Complexes in Protein Environments. *Inorg. Chem.* **1999**, *38*, 929-939.

[31] Miller, A. F. Redox tuning over almost 1 V in a structurally conserved active site: lessons from Fe-containing superoxide dismutase. *Acc. Chem. Res.* **2008**, *41*, 501-510.

[32] Duquerroy, S.; Camus C.; Janin, J. X-ray structure and catalytic mechanism of lobster enolase. *Biochemistry* **1995**, *34*, 12513-12523.

[33] Cowan, J. A. Structural and catalytic chemistry of magnesium-dependent enzymes. *BioMetals* **2002**, *15*, 225-235.

[34] Warburg, O.; Christian, W. Isolierung und Kristallisation des Garungsferments Enolase. *Biochem. Z.* **1941**, *310*, 384-421.

[35] Wang, S.; Scott, R. A.; Lebioda, L.; Zhou, Z. H.; Brewer, J. M. An x-ray absorption spectroscopy study of the interactions of Ni2+ with yeast enolase. *J. Inorg. Biochem.* **1995**, *58*, 209-221.

[36] a) Poyner, R. R.; Cleland, W. W.; Reed, G. H. Role of metal ions in catalysis by enolase: an ordered kinetic mechanism for a single substrate enzyme. *Biochemistry* **2001**, *40*, 8009-8017. b) Brewer, J. M. Specificity and mechanism of action of metal ions in yeast enolase. *FEBS Lett.* **1985**, *182*, 8-14.

[37] a) Stec, B.; Leiboda, L. Refined structure of yeast apo-enolase at 2.25 A resolution. *J. Mol. Biol.* **1990**, *211*, 235-248. b) Wedekind. J. E.; Reed, G. H.; Rayment, I.. Octahedral coordination at the high-affinity metal site in enolase: crystallographic analysis of the Mg^{II}-enzyme complex from yeast at 1.9 Å resolution. *Biochemistry* **1995**, *34*, 4325-4330. c) Leiboda, L.; Stec, B. Crystal structure of holoenolase refined at 1.9Å resolution: trigonal-bipyramidal geometry of the cation binding site. *J. Am. Chem. Soc.* **1989**, *111*, 8511-8513. d) Zhang, E.; Brewer, J. M.; Minor, W.; Carreira, L. A.; Lebioda, L. Mechanism of enolase: the crystal structure of asymmetric dimer enolase-2-phospho-D-glycerate/enolase-phosphoenolpyruvate at 2.0 A resolution. *Biochemistry* **1997**, *36*, 12526-12534. e) Leiboda, L.; Stec, B. Mechanism of enolase: the crystal structure of enolase-magnesium-2-phosphoglycerate/phosphoenolpyruvate complex at 2.2-.ANG. resolution. *Biochemistry* **1991**, *30*, 2817-2822. f) Lebioda, L.; Zhang, E.; Lewinski, K.; Brewer, J. M. Fluoride inhibition of yeast enolase: Crystal structure of the enolase–Mg2+–F−–Pi complex at 2.6 Å resolution. *Proteins Struc. Func. Gen.* **1993**, *16*, 219-225.g) Reed, G. H.; Poyner, R. R.; Larsen, T. M.; Wedekind, J. E.; Rayment, I. Structural and mechanistic studies of enolase. *Cur. Opin. Struct. Biol.* **1996**, *6*, 736-743. h) Larsen, T. M.; Wedekind, J. E.; Rayment, I.; Reed, G. H. A carboxylate oxygen of the substrate bridges the magnesium ions at the active site of enolase: structure of the yeast enzyme complexed with the equilibrium mixture of 2-phosphoglycerate and phosphoenolpyruvate at 1.8 Å resolution. *Biochemistry* **1996**, *35*, 4349-4358.

[38] Lebioda, L.; Stec, B.; Brewer, J. M. The structure of yeast enolase at 2.25-A resolution. An 8-fold beta + alpha-barrel with a novel beta beta alpha alpha (beta alpha)6 topology. *J. Biol. Chem.* **1989**, *264*, 3685-3693.

[39] Zhang, E.; Hatada, M.; Brewer, J. M.; Lebioda, L. Catalytic Metal Ion Binding in Enolase: The Crystal Structure of an Enolase-Mn2+-Phosphonoacetohydroxamate Complex at 2.4-.ANG. Resolution. *Biochemistry* **1994**, *33*, 6295-6300.

[40] Schreier, B.; Höcker, B. Engineering the Enolase Magnesium II Binding Site: Implications for Its Evolution. *Biochemistry* **2010**, *49*, 7582-7589.

[41] a) Poyner, R. R.; Larsen, T. M.; Wong, S. W.; Reed, G. H. Functional and structural changes due to a serine to alanine mutation in the active-site flap of enolase. *Arch. Biochem. Biophys.* **2002**, *401*, 155-163. b) Brewer, J. M.; Wampler, J. E. A differential scanning calorimetric study of the effects of metal ions, substrate/product, substrate analogues and chaotropic anions on the thermal denaturation of yeast enolase 1. *Int. J. Biol. Macromol.* **2001**, *28*, 213-218. c) Brewer, J. M.; Glover, C. V. C.; Holland, M. J.; Lebioda, L. Enzymatic function of loop movement in enolase: preparation and some properties of H159N, H159A, H195F and N207A enolases. *J. Prot. Chem.* **2003**, *22*, 353-361. d) Brewer, J. M.; Holland, M. J.; Lebioda, L. The H159A mutant of yeast enolase 1 has significant activity. *Biochem. Biophys. Res. Commun.* **2000**, *276*, 1199-1202. d) Vinarov, D. A.; Nowak, T. Role of His159 in yeast enolase catalysis. *Biochemistry* **1999**, *38*, 12138-12149. f) Brewer, J. M.; Glover, C. V. C.; Holland, M. J.; Lebioda, L. Significance of the enzymatic properties of yeast S39A enolase to the catalytic mechanism. *Biochim. Biophys. Acta* **1998**, *1383*, 351-355. g) Wedekind, J. E.; Poyner, R. R.; Reed, G. H.; Rayment, I. Chelation of serine 39 to mg2+latches a gate at the active site of enolase. *Biochemistry* **1994**, *33*, 9333-9342.

[42] Sims, P. A.; Menefee, A. L.; Larsen, T. M.; Mansoorabadi, S. O.; Reed, G. H. Structure and catalytic properties of an engineered heterodimer of enolase composed of one active and one inactive subunit. *J. Mol. Biol.* **2006**, *355*, 422-431.

[43] Qin, J.; Chai, G.; Brewer, J. M.; Lovelace, L. L.; Lebioda, L. Fluoride inhibition of enolase: crystal structure and thermodynamics. *Biochemistry* **2006**, *45*, 793-800.

[44] Anderson, V. E.; Cleland, W. W. Phosphonate analog substrates for enolase *Biochemistry* **1990**, *29*, 10498-10503.

[45] a) Liu, H.; Zhang, Y.; Yang, Y. How Is the Active Site of Enolase Organized To Catalyze Two Different Reaction Steps? *J. Am. Chem. Soc.* **2000**, *122*, 6560-6570. b) Alhambra, C.; Gao, J.; Corchado, J. C.; Villà, J.; Truhlar, D. G. Quantum mechanical dynamical effects in an enzyme-catalyzed proton transfer reaction *J. Am. Chem. Soc.* **1999**, *121*, 2253-2258. c) Hilal, S. H.; Brewer, J. M.; Lebioda, L.; Carreira, L. A.: Calculated effects of the chemical environment of 2-phospho-D-glycerate on the pK a of its carbon-2 and correlations with the proposed mechanism of action of enolase. *Biochem. Biophys. Res. Commun.* **1995**, *211*, 607-613.

[46] a) Kluge, S PhD Dissertation, "*Development of a Model for Inositol Monophosphatase*" Friedrich-Schiller-Universität, Jena, Germany, **2007**. b) Mollenhauer, D. Diplomarbeit (Master's thesis), "*Extending a Quantumchemical Model and First Investigations on the Mode of Action of Yeast Enolase*", Friedrich-Schiller-Universität, Jena, Germany, **2007**. c) Wagner, I. Diplomarbeit (Master's thesis), "*Development of a Quantumchemical Model for Sequential Enzymes on the Example of Yeast Enolase*", Friedrich-Schiller-Universität, Jena, Germany, **2006**.

[47] Hardy, J. A.; Wells, J. A. Searching for new allosteric sites in enzymes. *Cur. Opin. Struct. Biol.* **2004**, *14*, 706-715.

[48] Kern, D.; Zuiderweg, E. R. P. The role of dynamics in allosteric regulation. *Cur. Opin. Struct. Biol.* **2003**, *13*, 748-757.

CHAPTER 12

The Multi-Factor Coupled Protein Folding: Insights from Molecular Dynamics Simulations

Xiaomin Wu[2,3], Gang Yang[1,3,*] and Lijun Zhou[1]

[1]Resources and Environmental Department, Southwest University, Chongqing 400715, P.R. China; [2]Key Laboratory of Plant Resources and Biology, College of Life Sciences, Huaibei Normal University, Huaibei 235000, P.R. China and [3]Engineering Research Center of Forest Bio-preparation, Ministry of Education, Northeast Forestry University, Harbin 150040, P. R. China

Abstract: Deciphering the folding mechanism of proteins is significant to comprehend their physiological functions. In this chapter, several significant and yet common factors of protein folding have been discussed: 1) Space limitation (confinement and macromolecular crowding). Proteins are confined and crowded in cellular circumstances, which facilitates the folding and enhances the stability through the entropic reduction of the unfolded states. 2) Solvent effects. The various solvation models have been described. Water is more than the environment, and can also participate in the folding by mediating the collapse of protein chains and searching for the native topologies along the free energy landscapes. 3) Pressure, temperature and pH. The high hydrostatic pressure induces the volume decrease, destroys the non-covalent interactions and increases the roughness of free energy landscape, which generally drives the equilibrium toward the unfolded states. pH modulates protein structure and dynamics through protonation/deprotonation of sidechains and sometimes causes the misfolding. Temperature changes alter the conformational dynamics but not the folding pathway. 4) Structural modifications (mutation, truncation/insertion and protonation/deprotonation). The mutations of key residues significantly alter the folding by distorting the cooperative interactions, which can result in the misfolding or aggregation; nonetheless, the rational design by mutations can be beneficial to protein folding. The proper truncations do not show obvious influences on protein structure and dynamics, and the loop insertions may reduce the unfolding free energy barrier and facilitate the unfolding kinetics. Protonation of key residues affects significantly the folding/unfolding equilibrium by altering the non-covalent interactions.

Keywords: Confinement, conformational dynamics, cooperative interactions, denaturation and aggregation, folding/unfolding kinetics, free energy landscape,

***Corresponding author Gang Yang:** Engineering Research Center of Forest Bio-preparation, Ministry of Education, No. 26, Hexing Road, Xiangfang District, Harbin, Heilongjiang Province, P.R. China; Tel: 0086-451-82192223; Fax: 0086-451-82102082; E-mail: theobiochem@gmail.com

Subhash C. Basak, Guillermo Restrepo & José L. Villaveces (Eds)
Copyright © 2015 Bentham Science Publishers Ltd. Published by Elsevier Inc. All rights reserved.
10.1016/B978-1-68108-053-6.50012-0

h-bonding, hydrophobic core, macromolecular crowding, misfolding, molecular dynamics, mutation, non-covalent interactions, ph, pressure, protein folding, solvation models, solvent effects, structural modification, temperature.

INTRODUCTION

In the past few decades, remarkable progress has been made on structural genomics determining a large number of protein three-dimensional (3-D) structures. Nonetheless, it still represents a grand challenge to understand the folding mechanism and specific function of proteins [1, 2]. As compared to the DNA information transfer, the mapping from protein sequence to the native state is definitely more complicated [3-5], which has been achieved by the so-called "folding". Protein folding is envisioned as a communication process and closely associated with the physiological activity in living organisms, such as the allosteric regulation of cell signaling, reversible rotation of ATP-synthase and so on [6, 7]. Deciphering the folding code has been considered as a second translation of genetic message [8] and attracted focusing attention from around the world [9-12].

The early studies by Anfinsen *et al.* [13, 14] showed that the native structure of Ribonuclease A could be recovered completely by removing the denaturing chemicals or lowering the temperatures. As a matter of fact, protein folding takes place spontaneously in cellular circumstances, driven towards the native states by thermodynamics (enthalpy, entropy and free energy) and kinetics. In cellular circumstances, temperature, pressure, pH and ionic strength change within relatively small ranges, and the space allowed for protein folding is rather limited [15, 16]. Although proteins have a conformational diversity during the folding process and at the equilibrium state, only a few conformers will mediate their function [17, 18]. The adjustment of folding conditions is likely to affect the conformational distribution and induces the misfolding or aggregation [7, 13, 14, 19-23]. The confinement and macromolecular crowding limit the space for protein folding, which favor the folded states and increase their stability by narrowing the conformational distribution [24-29]. In presence of steric interferences, compressive stress or/and chemical denaturants, proteins will be aggregated to a certain degree [30-33].

Recently, it has been demonstrated by Tokuriki *et al.* [16] that the improper mutations to proteins decrease the configurational entropy and active-site flexibility, thus causing the latent promiscuous functions; on the other hand, the interferences of proteins by foreign cofactors such as antibodies are potential to shift the conformational equilibrium and enable the binding of unrelated ligands. Accordingly, the folding and assembly of proteins were determined not only by their own native topologies, but also by the linked intracellular cofactors. Either mutations or adverse cofactors can lead to the abnormality of protein functions. The mutations may significantly alter the folding kinetics and lead to the partially unfolded states, pathogenic oligomers or aggregations [34-36]. Our molecular dynamics (MD) simulations on the Trp-cage miniprotein indicated that only a few key residues show observable influences on its stability, dynamics and folding process [37]. On the other hand, Andersen and coworkers [38, 39] suggested that the replacement of glycines or other residues with those of long sidechains or the increase of proline content decreases the entropy of the unfolded states and thus improves protein stability [38]. The rational truncation of proteins does not show a dramatic effect on the conformations, and the β-hairpin in protein GB1 is a well-known example to support this viewpoint [40, 41]. In addition, the folding pathways of proteins can also be modulated by structural modifications, such as protonation/deprotonation [42], salt-bridge destruction [43-46] and so on.

From the above descriptions it can be seen that the folding of proteins is a rather complicated and multi-factor coupled process. In this chapter, we will give an abundant discussion of these factors, which are significant and yet common issues of protein folding. The outline was organized as follows: Section 2 for space limitation, including the confinement and macromolecular crowding; Section 3 for solvent effects, wherein the several commonly used solvation models were reviewed and commented as well as their application in the field of protein folding; Section 4 for pressure, temperature and pH; and Section 5 for structural modifications, including mutation, truncation/insertion and protonation/deprotonation. Concluding remark was then made in Section 6.

SPACE LIMITATION

It has long been recognized that the efficient and accurate folding of proteins towards their native states depends not only on their native topologies but also on the surrounding circumstances [15, 47-49]. In cells crowded with cytoskeletons, lipids and other macromolecules, the space allowed for protein folding is rather limited; in addition, the folding behavior is largely dependent on the surrounding media such as pH, ionic strength, temperature, pressure and so on [26, 29, 49-51]. Elucidating these effects is helpful to understand the complicated folding and function of proteins [51-57]. In this Section, we will discuss exclusively the space limitation (confinement and macromolecular crowding).

As shown in Fig. (**1**), the confinement and macromolecular crowding are mimicked by placing the protein in a spherical cage with the radius of R_c (Fig. **1**a) or among the crowded macromolecular particles with the radius of a_m (Fig. **1**b), and the folded and unfolded protein states are represented by solid spheres and polymer chains, respectively [26]. These simplified polymer physics-based models provide valuable insights into how the cell encapsulations influence the protein folding [58-65]. For example, molecular dynamics (MD) simulations of the off-lattice model performed on the β-hairpin detect that the folding rate increases in the spherical pore as compared with the bulk, and the confinement results in the enhanced folding stability due to the substantial entropic reduction of the unfolded state (see Fig. **2**a) [64]. The similar results have been derived for the all the β-sheet WW domain by Thirumalai *et al.* [65] (Fig. **2**b); besides, the authors observe that the confinement effects cause the increase of the folding rate.

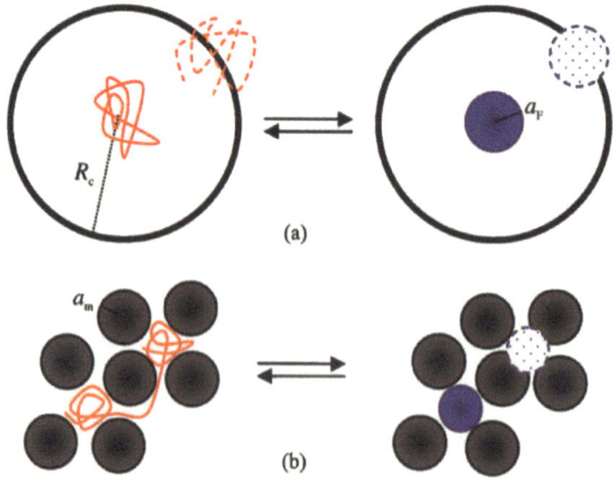

Figure 1: Illustration of protein folding (a) in the spherical cage (with radius R_c) and (b) in solution of macromolecular crowders (with radius a_m). [26] (Reproduced with permission from Elsevier) The unfolded protein is denoted by a red chain whereas the folded one by a blue sphere. In a spherical cage with radius R_c, a shell with thickness (a_F) is excluded at the center of the folded protein.

(a) (b)

Figure 2: Encapsulation of the β-hairpin in the spherical pore (a) and the WW protein with large spherical particles (b) according to the off-lattice models. [64, 65] (Reproduced with permission from PNAS) The radius of the spherical pore (R_s) is set to be larger than that of the bulk native conformation of the confined β-hairpin. ($R_s \geq 1.4\ R^b_{g,N}$).

In contrast to the confinement, the macromolecular crowding agents show relatively modest stabilization effects to proteins. A spherical cage is suitable to describe a spherical protein rather than other molecular shapes. For a helix-forming peptide, Ziv *et al.* [60] performed Langevin dynamics simulations to study how the cylindrical confinement exerts influences on its folding and formation. As in the above confinement, the α-helix structure has been identified to be entropically stabilized in the cylindrical cavity. This is in good agreement with experimental observations of protein folding in the artificially crowded conditions [66-74].

In order to understand the confinement- and solvent-mediated folding mechanisms, Pande and co-workers [56, 57] employed molecular dynamics (MD) simulations and all-atom models to study the folding of helical peptides and small proteins inside carbon nanotubes (CNTs) filled with explicit water molecules. As shown in Fig. (**3**), the hydration shell is tightly packed between the helical structure and the inner surfaces of CNTs. It suggests that the folding processes of helical peptides and small proteins are disfavored in the solvated CNTs. Both of the folded helical peptides and small proteins are not properly stabilized in the confined CNTs but substantially destabilized. It is contrary to the stabilization effects of the off-lattice models. The nonpolarity of the confined CNT surface increases the activity of the confined water molecules and causes a remarkable loss in the translational entropy, which should be responsible for the destabilization effects.

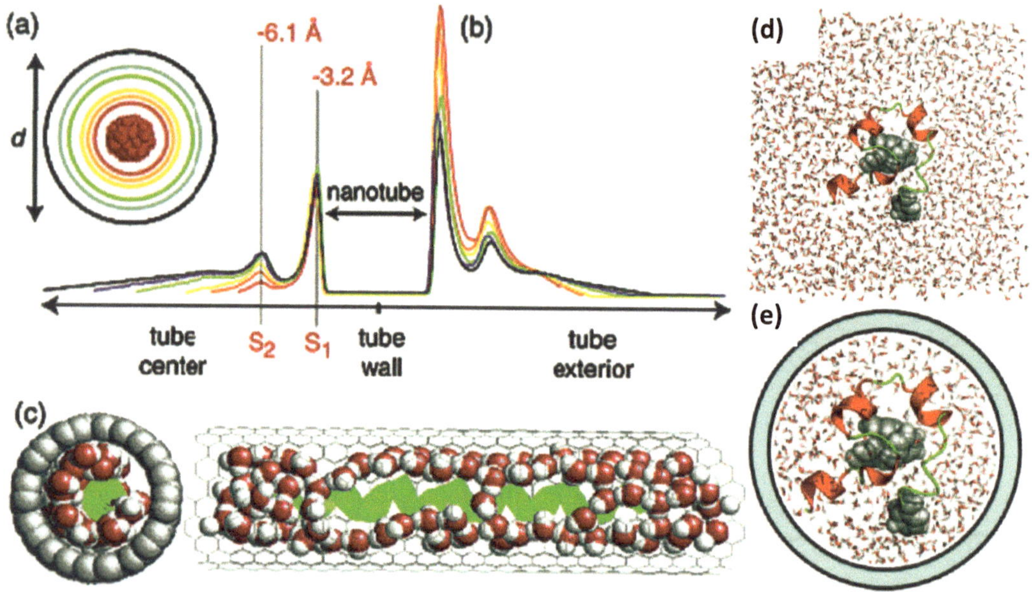

Figure 3: The diameters of carbon nanotubes (CNTs) relative to the all-atom van der Waals surfaces of the ideal α-helical peptide (a); Radial distribution of the water O atoms inside the confined solvated CNTs has been normalized to unity, with S1 and S2 representing the first and second hydration shells inside the CNTs (b); the ideal α-helix in the confined solvated CNTs is shown in green ribbon, water molecules are shown in white and red surfaces, and CNTs in gray (c); The villin headpiece is solvated in solution (d) and confined in the solvated CNTs (e). [56, 57] (Reproduced with permission from PNAS and ACS).

Using replica exchange molecular dynamics (REMD), Garicia *et al*. [75] study the folding/unfolding equilibrium of the Trp-cage miniprotein within the polar and nonpolar fullerene balls (Fig. **4**). It is found that the weak nonpolar confinement of the fullerene balls stabilizes the folded Trp-cage whereas the polar confinement destabilizes the folded state. The observations are in accord with those under the macromolecular crowding conditions showing that the stabilizing effects can be regulated by nonspecific interactions of protein chains with the confined surfaces or macromolecular crowding particles [75-77]. Thus, it can be inferred that there is a balance among the numerous competing effects that govern the protein folding behavior under the confined and crowded conditions. The confinement and macromolecular crowding give rise to the excluded volume effects on proteins, which alter their folding dynamics and conformations and further their function [78].

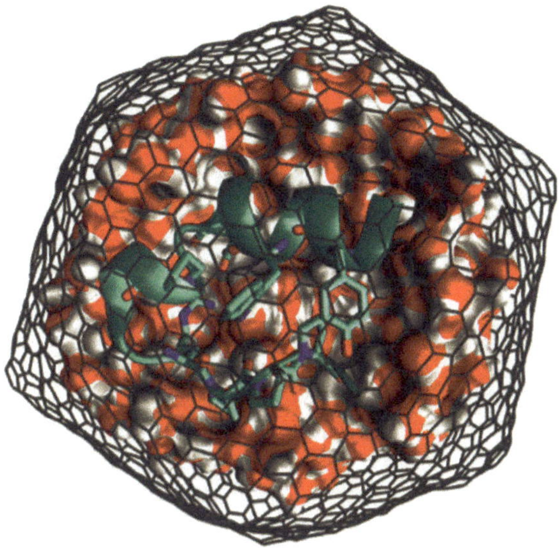

Figure 4: The Trp-cage miniprotein in the fullerene ball. [75] (Reproduced with permission from ACS) The Trp-cage and fullerene are shown in cyan and gray, respectively. The water molecules are represented by white and red surfaces.

It is known that amino acids in gas phase can not exist as the zwitterionic form ($NH_3^+CHRCOO^-$), because these zwitterions are not local minima on the potential energy surfaces (PES) [79]. The adsorption of amino acid isomers on the acidic sites of zeolites results in the protonated species ($NH_3^+CHRCOOH$) [80]. Recently, we have found that the amino acid zwitterions will be formed by interactions of the depronated form with the acidic sites of zeolites [81], as in the case of metal oxide surfaces [82, 83]. The zwitterionic and canonical isomers of glycine within ZSM-5 zeolite are given in Fig. **5**, and the energy barrier of proton transfer between these two isomers is calculated to be 4.5 kcal/mol, indicative of the facile transformations even at ambient conditions. The roles of negative charges and lattices of zeolites are evaluated during the stabilization of zwitterions. Their contributions amount to -8.6 and -6.3 kcal/mol, respectively. Accordingly, zeolite lattices play an important role in the zwitterionic formation and stabilization, albeit somewhat less than that of negative charges. Inspired by these results, we are now studying protein folding within confined zeolitic materials and clarifying the special confinement effects during the folding and stabilization of proteins.

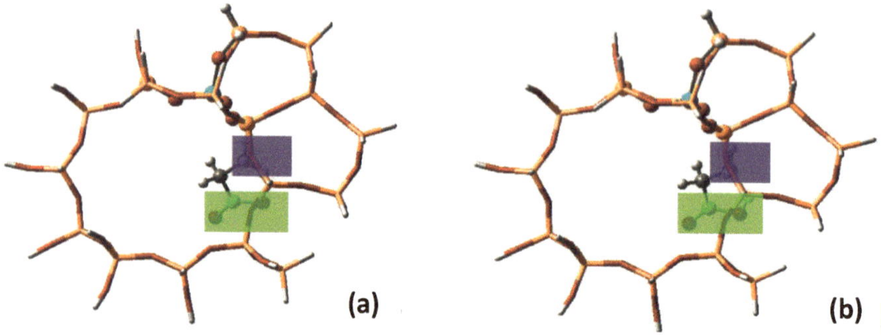

Figure 5: The zwitterionic (a) and canonical (b) isomers of glycine within the ZSM-5 zeolite [81] (Reproduced with permission from ACS). The amino and carboxyl groups of glycine are marked with blue and green backgrounds, respectively. In the zwitterionic ($NH_3^+CH_2COO^-$) isomer of glycine, the amino group is protonated whereas the carboxyl group deprotonated.

WATER SOLVENT

Protein folding usually takes place in aqueous solutions. Water is known to play an indispensable role in governing the structural stability, dynamics and function of proteins, such as the maintenance of structural integrity [84-86], mediation of protein folding [56, 87-91] and molecular recognition [92-97] and improvement of catalytic activity [98-103]. That is, water should not be envisioned as an inert environment, but instead as an integral and active structural component during the folding and function of proteins [97]. In particular, water molecules can be buried in the cavities of proteins and form direct interactions, such as H-bonding with the polar groups, electrostatic interactions with the charged amino-acid sidechains and hydrophobic interactions with the nonpolar atoms [103-108]. In computer simulations, the aqueous solutions can be efficiently described by the explicit solvation models, such as SPC, SPC/E, TIP3P, TIP4P, TIP4P/EW, TIP5P and so on. More explicit solvation models have been reported, but no "the best" has been found for protein simulations. The parameters of several common ones are listed in Table **1** [109-111].

Table 1: A comparison of potential parameters of several explicit solvation models commonly used in computer simulations. [a, b, c] [111] (Adapted with permission from AIP)

Model	d_{OH} (Å)	H-O-H	σ (Å)	(ε/k) (K)	q_H (e)	d_{OM} (Å)	d_{OL}(Å)
SPC	1.0	109.47	3.1656	78.20	0.41	0	---
SPC/E	1.0	109.47	3.1656	78.20	0.4238	0	---
TIP3P	0.9572	104.52	3.1506	76.54	0.417	0	---
TIP4P	0.9572	104.52	3.1540	78.02	0.52	0.15	---
TIP4P/EW	0.9572	104.52	3.1643	81.90	0.52422	0.1250	---
TIP5P	0.9572	104.52	3.1200	80.51	0.241	---	0.70

[a] d_{OH} is the distance between the O and H atoms of the water molecules, and H-O-H is the angle between two H-O bonds of the water molecules;
[b] The L-J site is located on the O atom with the parameters given as σ and ε/k; whereas the charge on the proton is denoted as q_H;
[c] All these models except TIP5P place the negative charge in a point M at a distance d_{OM} from the oxygen atom along the H-O-H bisector. For TIP5P, d_{OL} represents the distance between the oxygen and the site L to place the lone electron pairs.

The water solvent has also been considered by the implicit solvation models, based on the Generalized Born (GB) and Poisson-Boltzmann (PB) theories [112-114], dielectric screening functions [115], solvent-accessible surface area and etc [116]. In comparison with the explicit solvation models, the implicit solvation models save a lot of computational resources at the expense of accuracy. It is probably due to the absence of direct interactions between proteins and solvent molecules, which have been implicated to give rise to incorrect preferences for secondary structural motif [117] and structural distortion [114]. Albeit there are controversies over the degrees to which the implicit solvation models can mimic the aqueous environments [108, 118, 119], persistent efforts have been and are being made to improve the efficiency of the implicit solvation models. Indeed, the explicit solvation models are able to provide a detailed and accurate description of protein conformational sampling under physiological conditions, but a majority of computational resources have been spent on the simulations of the water molecules rather than the protein systems [97, 120]. It has shown that for small proteins the implicit and explicit solvation models obtain similar folding rates but different folding mechanisms [56, 108]. To tackle the conflicts between computational resources and model deficiencies, more efficient and yet accurate approaches have been introduced to study the contributions of water-mediated interactions and dynamics to protein structure and folding, such as the hybrid explicit/implicit solvation methods [120-122]. As shown in Fig. **(6)**, this approach incorporates both explicit and implicit solvation models. The inner solvent molecules that may directly interact with protein system are modeled with the explicit solvation model and the outer solvent molecules that are treated as the environment are modeled with the implicit solvation model. In this way, the solvent-mediated folding and function of proteins can be accurately described whereas the computational costs will not increase dramatically.

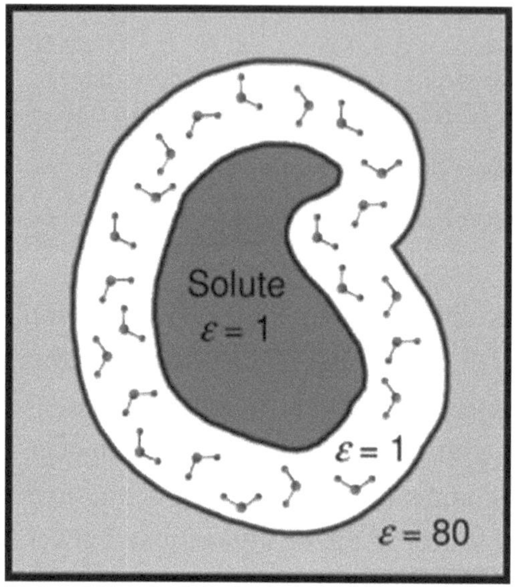

Figure 6: Representation of the hybrid explicit/implicit solvation models. [120] (Reproduced with permission from WILEY) The protein is immersed in a layer of explicit water molecules (explicit solvation model), and the gray outer region represents the environmental region treated by the implicit solvation model such as the bulk dielectric continuum of the GB theory.

The roles of water solvent during protein folding can be exactly probed by the solvation models discussed above. Using the replica exchange and explicit solvation model, Xu *et al.* [123] observe a high dehydration in the hydrophobic core during the Trp-cage folding. The polarity of the water molecules induces the nonpolar groups to be shielded from the water solvent and causes the collapse into a tightly packed hydrophobic core. Our recent explicit solvent MD simulations have also detected that the presence of water solvent induces the hydrophobic collapse and drives the correct structural rotations as the whole Trp-cage comparatively closes [37, 124]. The simulated studies of protein A three-helix bundle [125], protein G [126] and L [127] and other peptide models show the water solvent as a lubricant for the packing of the hydrophobic core. The hydrophobic effects, as proposed by

Kauzmann [128], are driven by the decrease of the unfavorable structural entropy caused by forming a large surface area of nonpolar groups within the water solvent. The water-induced hydrophobic interactions are one of the main driving forces during protein folding [37, 107, 124, 129]. In addition, García *et al.* [130] identify that one or two internal water molecules are completely buried in the interior of the Trp-cage miniprotein and H-bonding to its polar groups (See Fig. 7), and these buried H-bonds favorably stabilize the folded configurations. This structural feature has been highlighted by their previous simulations on the hydration of internal cavities in the wild-type staphylococcal nuclease and two mutants [131]. The absence of internal water molecules in V66K causes the structural rearrangement of loop elements. The role of water-mediated critical H-bonds has been confirmed by the NMR studies [132]. It indicates that the thermal folding of the hydrated lysozyme is closely related to the number of H-bonds in which the water molecules are closely involved, and in this way can the water solvent modulate the structural and dynamic properties.

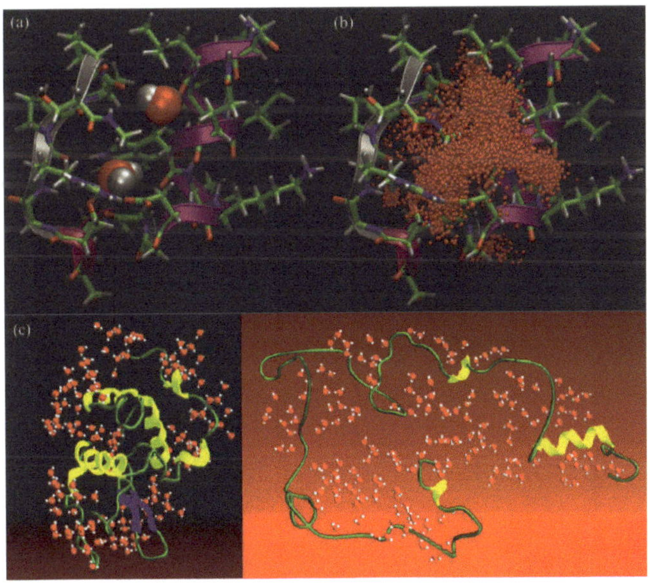

Figure 7: Ribbon representation of the folded Trp-cage with two buried water molecules in the interior (a); Position of the O atoms of the two closest water molecules (Trp-cage-water center of mass distance) (b); Snapshots of the folded (on the left, in black) and unfolded (on the right, in red) states obtained from the thermal folding of the hydrated lysozyme (c). [130, 132] (Reproduced with permission from Elsevier and ACS) In (a), two internal water molecules are shown in CPK.

Unlike the water molecules buried in the cavities of protein interiors, the surface water molecules may not form H-bonds with the large nonpolar surfaces of proteins but instead tend to form H-bonding interactions within themselves. Besides, the surface water molecules can form electrostatic interactions with the charged sidechains of proteins. The theoretical studies on the interactions from the buried and surface water molecules address that the water-mediated ionic interactions between the charged residue pairs contribute significantly to the stability and plasticity [56, 57, 75, 133-136] (Fig. **8**). Regardless of being in the interiors or surfaces of proteins, the water molecules participate in the folding through mediating the collapse of protein chains and the search for the native topologies along the free energy landscapes.

Recently, Pande *et al.* [56, 57] have proposed a distinct role of water solvent for the α-helical peptide confined in a single-wall carbon nanotube (CNT) surrounded with water molecules, wherein a significant hindrance to the helix formation has been observed. It is probably due to the differences in the solvent entropy caused by the confinement effects. We have also identified that the increase or decrease of simulation temperatures can alter the water-mediated H-bonding interactions and give rise to the different solvent effects on protein conformational dynamics and stability [137]. That is, the local environments and temperatures will affect the contribution of water solvent during the protein folding [138].

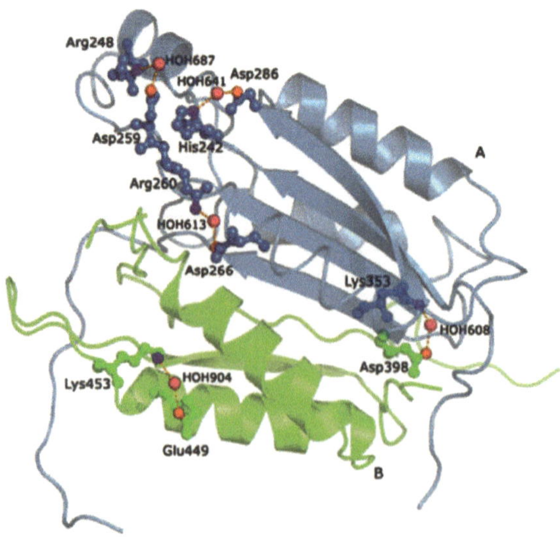

Figure 8: The water-mediated ionic interactions between the charged residue pairs in two subunits (A and B) of protein caspase-8. [134] (Reproduced with permission from IAS).

PRESSURE, TEMPERATURE AND pH

More than half of the earth's biosphere is occupied by oceans, and the deep-sea organisms are subjected to high hydrostatic pressure [139]. It has been accepted that the high hydrostatic pressure response is governed by the volume change of protein-solvent systems (ΔV) associated with the folding, unfolding and aggregations [140-145]. The decrease in volume induced by the high hydrostatic pressure is attributed to the exposure of nonpolar groups of proteins to the water solvent [141]. In contrast, Kauzmann and Hummer [142, 143] propose that the pressure-decreased volume is due to the penetration of water molecules into the protein interiors.

The application of the high hydrostatic pressure to protein systems generally results in the perturbation of the equilibrium toward the unfolded state and causes the volume reduction [140, 141, 145, 146]. The covalent bonds in proteins are not readily affected, and the non-covalent interactions are mainly responsible for the pressure-modulated denaturation [147]. The involved non-covalent interactions are classified as: 1) Electrostriction of the water solvent around the newly exposed charged and polar groups [148-150]; 2) Hydration of nonpolar groups in protein interiors [142, 143, 145, 151]; 3) Elimination of internal cavities upon unfolding [146, 152, 153] (Fig. **9**). The second factor (hydration of nonpolar groups in protein interiors) has been suggested to be the main driving force [154]. García *et al.* [141] use the off-lattice model to study the pressure-induced protein folding/unfolding kinetics, showing that the denaturation corresponds to the transfer of water molecules into the hydrophobic core. The high hydrostatic pressure slows down the formation and collapse kinetics of the hydrophobic pair, and as a result increases the roughness of the free-energy landscapes, which agrees well with the experimental studies on the staphylococcal nuclease and Trp repressor [155-158].

In addition to modulating the folding behavior, the high hydrostatic pressure induces the formation of a molten globule intermediate [154, 160]. As shown in Fig. **(10)**, the intermediate conformation of the Ras-binding domain of RalGDS is characterized by a local melting of the specific structural elements near the hydrophobic cavities while the overall folded structure has been finely retained

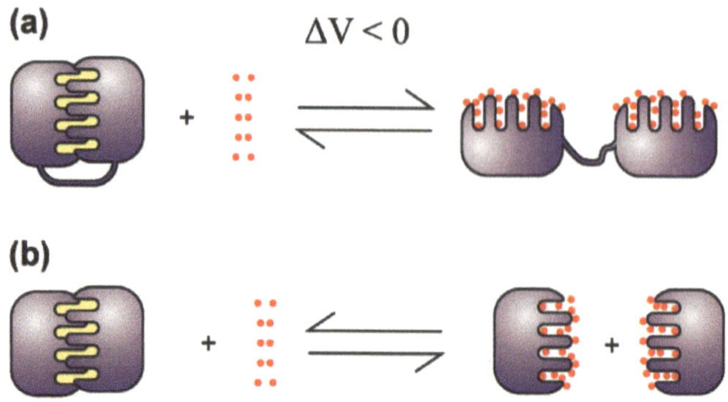

Figure 9: Effects of the high hydrostatic pressures on protein denaturation and dissociation. The pressure-induced denaturation by the combined effects of the breakage of intraprotein interactions and the elimination of internal cavities followed by water penetration (a); Dissociation of a protein dimer (b). [159] (Reproduced with permission from Elsevier) The water molecules are represented by small red circles and protein cavities in yellow.

[161]. The pressure-induced intermediates have also been observed in other protein systems such as hen lysozyme [162], dihydrofolate reductase [163] and the E2 DNA binding domain (E2-DBD) of human papillomavirus [164], and they are transient intermediates occurring during the folding process [165-168]. On the other hand, the high hydrostatic pressure can cause the dissociation of oligomeric proteins as well as reverse the aggregation and facilitate the refolding of proteins [169, 170]. Owing to the presence of high hydrostatic pressure, the P22 tailspike aggregation is dissociated and results in the formation of monomers and native folded trimers. The high hydrostatic pressure to combat aggregation has been corroborated by the studies of rhodanese [169], amyloidogenic protein transthyretin [171] and myoglobin [172]. The pressure-induced effects are explained by the stabilization of H-bonding interactions and destabilization of hydrophobic and electrostatic interactions [140]. To conclude, the high hydrostatic pressure as a potentially informative approach can be used to explore protein denaturation and refolding as well as to determine the significant folding/unfolding intermediates.

The folding thermodynamics and kinetics of proteins have been understood by the proposed free-energy landscape and funnel theories [47, 173-178]. Many studies have been done with the temperature-, pH-dependent and other protocols, in order to provide a description of the intrinsic folding/unfolding process [137, 176-180].

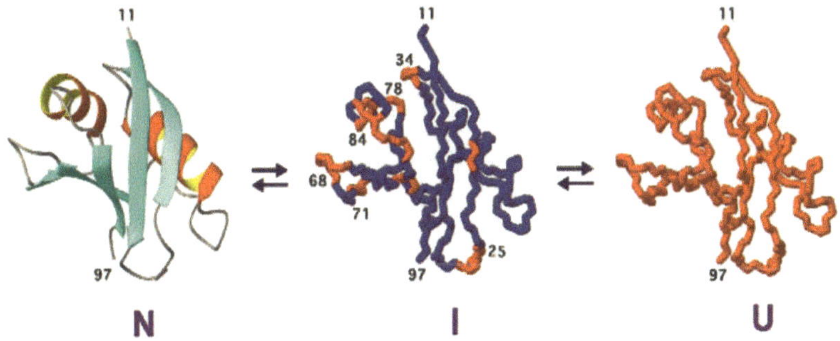

Figure 10: Illustration of three conformations occurring during the pressured-induced denaturation of RalGDS-RBD. [161] (Reproduced with permission from NPG) The NMR native state (N) is displayed in ribbon and the unfolded state (U) in red. The intermediate state (I) is structurally similar to the native state (in violet) but with local melting regions (in red).

In particular, the pH-dependent conformational dynamics is involved in a variety of biological processes, such as protein folding/misfolding and aggregation, ATP synthesis, enzyme catalysis and so on [181-187]. The role of pH played in modulating the protein structure and dynamics has been primarily attributed to the effects of protonation/deprotonation of the sidechains of the comprising residues. In the human prion protein (PrP) there is an "uneven" distribution of positive and negative charges at low pH values, as compared to the neutral pH values, see their respective electrostatic potential surfaces in Fig. (**11**). This is due to the pH-induced protonation effects on the two key histidine residues [184]. However, these two histidine residues will be deprotonated at elevated pH values [188]. Using molecular dynamics (MD) simulations, Barone *et al.* [185] identify that the protonation of the histidine residues results in a significant loss of the α-helix content and the conversion of monomeric human prion protein (PrP) to the pathogenic isoform under mild acidic conditions. The acidic conditions trigger the PrP transitions to the misfolded isoform [186, 187]. Daggett and Barone *et al.* [184, 185, 189] observe that the conformational transitions and folding instability are more apparent under acidic than neutral pH conditions.

The pressure-induced denaturation and aggregation are often associated with the temperature change [171, 190]. As a matter of fact, temperature has often been used as an external perturbation in protein folding. The thermal and cold denaturation is a common phenomenon found in globular proteins, and the unfolding pathways and associated denatured states have fundamental implications for comprehending the folding behavior [191, 192]. The folding/unfolding dynamics during the cold- and heat-denaturation processes are quite distinct. According to the conformational dynamics and thermal stability of the αβ-tubulin dimer under the various temperature conditions, we found that the αβ-tubulin dimer is frozen and its structural dynamics is rather slow at the extremely low temperatures. On the contrary, the elevation of temperatures induces the protein unfolding and destroys the native folded structures, probably due to the severe damage of non-covalent interactions and decrease of water polarity at high temperatures [137].

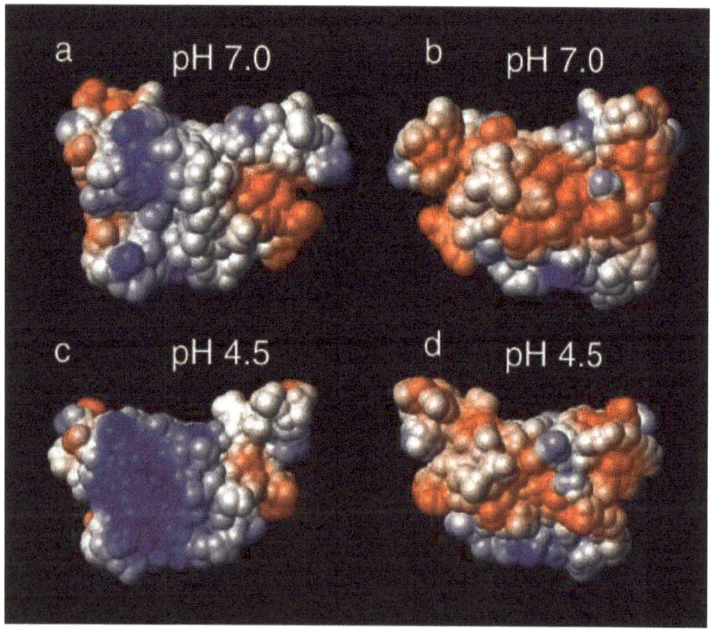

Figure 11: Surface distribution of electrostatic potential for human PrP globular domain under different pH conditions. [184] (Reproduced with permission from HIGHWIRE) Two-side views of Prion protein at neutral (a and b) and low (c and d) pH values. The positively and negatively charged surfaces are represented in blue and red, respectively.

Although the temperature changes lead to the protein unfolding and denaturation, the folding landscapes and transition pathways may not be distorted [193]. As a result, the thermal unfolding has been widely used providing indirect and yet valuable evidence for the protein folding mechanisms [194-198]. By comparing the multiple unfolding processes of chymotrypsin inhibitor 2 (CL2) at seven different temperatures (298 to 498 K), Daggett *et al.* [193] find that the increase of simulation temperatures enables the efficient conformational sampling by traversing high energy barriers and accelerating the folding towards the global lowest-energy structure. It has been further validated by the temperature-induced unfolding of cold-adapted type III antifreeze protein (AFPIII) from the Antarctic eelpout *Lycodichthys dearborni* (AFPIII) [194], see Fig. (**12**). As the elevation of temperatures only affects the protein folding rate without changing the folding pathway, replica exchange MD (REMD), wherein multiple replicas of identical systems are simulated in parallel at different temperatures, have been developed [199]. We also perform the temperature-controlled molecular dynamics (TCMD) simulations to probe the protein folding mechanisms [124, 200]. In TCMD, the MD simulations are initiated from a high temperature and then the simulation temperatures gradually decrease until room temperature (300 K). The characteristic conformations of the covered temperatures can be considered as intermediates along the folding pathway, and thus with the TCMD simulations, the clear folding landscapes have been provided for the tubulin active peptide ($Pep^{1\text{-}28}$) [200] and Trp-cage miniprotein [124]. That is, the temperature-dependent MD studies have been envisioned as a good means for elucidating the protein folding mechanisms.

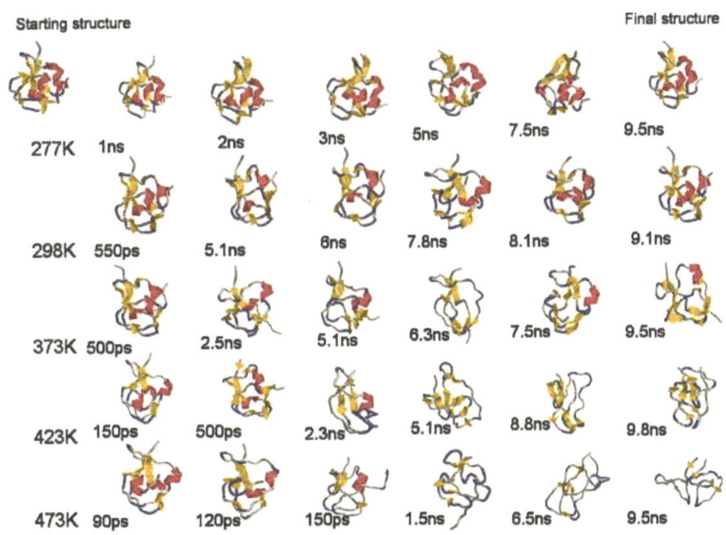

Figure 12: Snapshots of the representative conformations along the unfolding trajectories at various temperatures. [194] (Reproduced with permission from Elsevier).

STRUCTURAL MODIFICATIONS

In this section, we will discuss how structural modifications (mutation, truncation/insertion, protonation/deprotonation) affect the structure, dynamics and folding proteins, which have been the subject of many recent experimental and computational studies [201-211].

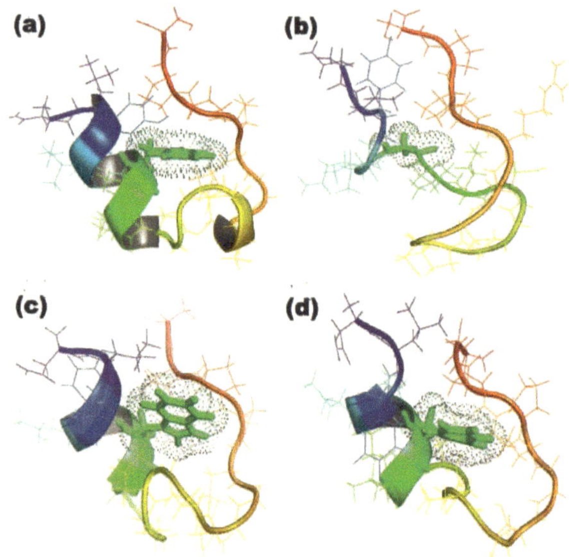

Figure 13: The representative structures of the Trp-cage miniprotein: (a) Wild-type; (b) W6G mutation; (c) L7G mutation; (d) P19G mutation. (Reproduced with permission from Elsevier).

The single-site mutations at the key positions of proteins exhibit significant effects on the structure, dynamics and folding, as revealed by our recent MD simulations on the Trp-cage miniprotein [37]. The mutations inside the hydrophobic core of the Trp-cage seriously distort the cooperative interactions among key residues, especially in the cases of W6G, L7G and P19G mutations (Fig. **13**); instead, much slighter influences have been caused by other mutations. The key residues are usually involved in a number of non-covalent interactions (*e.g.* electrostatic, H-bonding, hydrophobic, *etc.*). The role of some residues has also been reported by other groups. Barua *et al.* [201] find that the W6H mutation destroys the buried long-range H-bonding interactions between the Trp6-εNH and Arg16 carbonyl-O atoms, resulting in the partial exposure to water solvent. In the wild type, this buried H-bond contributes significantly to the stability and folding by burial of the Trp6 indole ring [37, 45, 202]. Perczel *et al.* [45, 46] carry out the D9S, D9N and R16A mutations and find that the damage of the D9/R16 salt bridge disrupts the folding cooperativity by destabilizing the central 3_{10}-helix and H-bonding network. The effects of the charged-residue mutations have been studied by Wang *et al.* [203]. The E46A and R3E mutations of the cold shock protein from themophile *Bacillus caldolyticus* (Bc-Csp) obviously decrease the folding kinetics whereas the L66E mutation results in only slight decrease. However, neither of these two mutations causes serious alteration to the transition states and folding pathway.

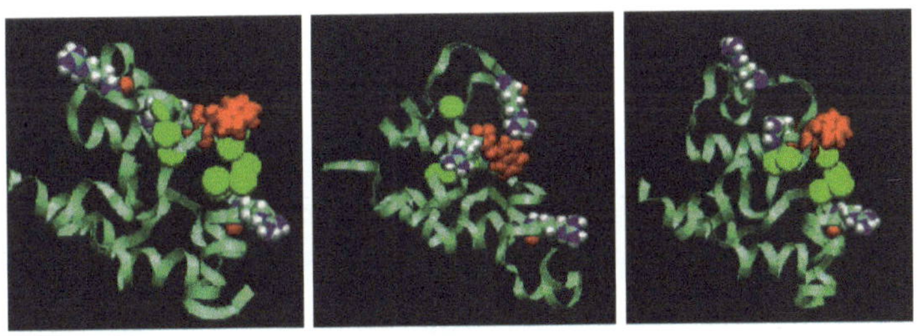

Wild-type

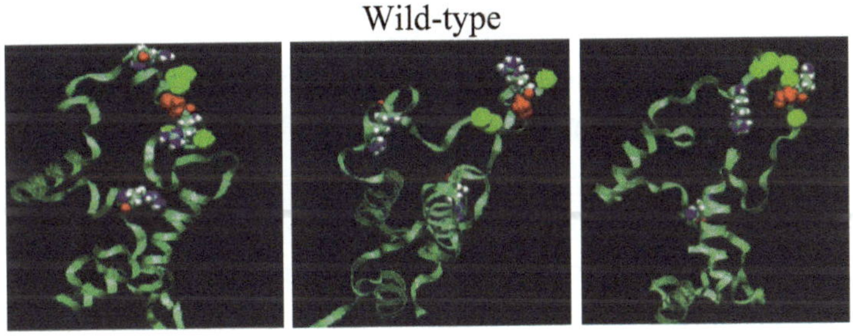

Mutant

Figure 14: Representative structures of the wild-type lysozyme and W62G mutant during the 100 to 1000 ns MD simulations. [204] (Reproduced with permission from PNAS) The Trp62/Gly62 residue is shown as red CPK model, and the three nearby basic residues (Arg73, Lys97, Arg112) as van der Waals space-fills. The green balls represent the residues that have local contacts with the Trp62/Gly62 residue.

Some mutations in proteins can even lead to the misfolding or aggregation. Zhou *et al.* [204] find that the W62G mutation of the hen egg-white lysozyme causes the misfolding, and the misfolded structure is much less stable than the wild type, see Fig. **(14).** Albeit located on the protein surface instead of in the hydrophobic core, Trp62 is closely involved in the long-range hydrophobic interactions and acts as a bridge between two neighboring basic residues, thus constructing a nucleation site for the folding process. However, the above interactions will be damaged or absent in the mutated W62G structure [205]. The experimental mutagenesis studies [206-208] support the simulated results. The incorrect folding behavior of the lysozyme have been identified by other single-site mutations; *e.g.*, A9G, W111G and W123G [205].

On the contrary, some mutations can be beneficial to the protein folding. Andersen *et al.* [202] computationally design the N9D and A16R mutations to the 20-residue TC5b and reveal that the insertion of the D9/R16 salt bridge results in the obvious stabilization effects to the hydrophobic core and thus accelerates the folding rate. The W6F simulations performed by Chowdhury *et al.* [209] indicate that the wild-type Trp-cage and W6F mutant experience similar folding pathways, but the mutant corresponds to the much more rapid folding rate. The encapsulation of the Trp6 indole ring inside the cage is the rate-limiting step for the folding of the Trp-cage miniprotein. Owing to the W6F mutation, the asymmetric entropic cost is reduced for the Trp6 sidechain, and the free energy barrier of packing the hydrophobic core decreases accordingly. Based on the understanding of the Trp-cage folding mechanisms, Gai *et al.* [210] construct a P12W mutant with the presence of Trp-Trp interactions. Thus, the ruggedness of the folding funnel of the wild type can be minimized, which then causes the acceleration of the folding process.

As a matter of fact, the Trp-cage and β-hairpin, which have been widely used as models to study protein folding, are truncated from the protein exendin-4 (Ex-4) [201, 211-215] and the B1 domain of streptococcal protein G (protein GB1) [126, 216-219], respectively. It shows that the isolated Trp-cage and β-hairpin can fold towards their native states that resemble the corresponding fragments within the proteins. The structural truncation can be used to understand the length dependence of protein folding and specific interactions during the folding process. As indicated by our MD simulations on the truncated peptides of the αβ-tubulin dimer with different lengths ($Pep^{1\text{-}50}$, $Pep^{1\text{-}40}$, $Pep^{15\text{-}39}$, $Pep^{1\text{-}25}$, $Pep^{15\text{-}25}$ and $Pep^{19\text{-}25}$), the truncated peptides may undergo distinct structural transitions and folding pathways. The native conformation of the active-site peptide ($Ap^{15\text{-}25}$) can be finely retained in the case of $Pep^{1\text{-}50}$ instead of the others, because the residues of the active-site peptide are closely involved in the long-range interactions with other residues (*e.g.*, β-Arg48) [41]. In the case of the Trp-cage miniprotein (TC5b), the truncation of the N-terminal helix-capping residue reduces the folding efficiency by almost 40%; likewise, the truncation at the C-terminal Ser20 decreases the folding degree albeit slightly (3%) [212]. The length dependence of protein folding has been carefully studied by assessing a series of truncations to the BRCT domain [220]. It is found that this protein can tolerate a truncation of up to eight amino acids, and the further truncation will result in drastic folding defects. On the other hand, the loop regions are tolerant of flexible residue insertion without obviously affecting the native conformations. If the loop regions have been fully shaped in the mechanical unfolding transition state as in the folded state, the increase of its length will show close destabilization effects to the native and transition states (Fig. **15**b). Thus, the free energy barrier for the mechanical unfolding will not be affected. If the loop is partially deformed in the transition state, the increase of the loop length will affect the free energy barrier of the mechanical unfolding (Fig. **15**c). As shown in Fig. **15**a, the flexible linker sequences are inserted between residues 39 and 40 of the loop region of protein GB1. The loop insertions reduce the free energy barriers and accelerate the unfolding kinetics [221], which is as a result of the loss of configurational entropy upon closing an unstructured flexible loop during the unfolding process.

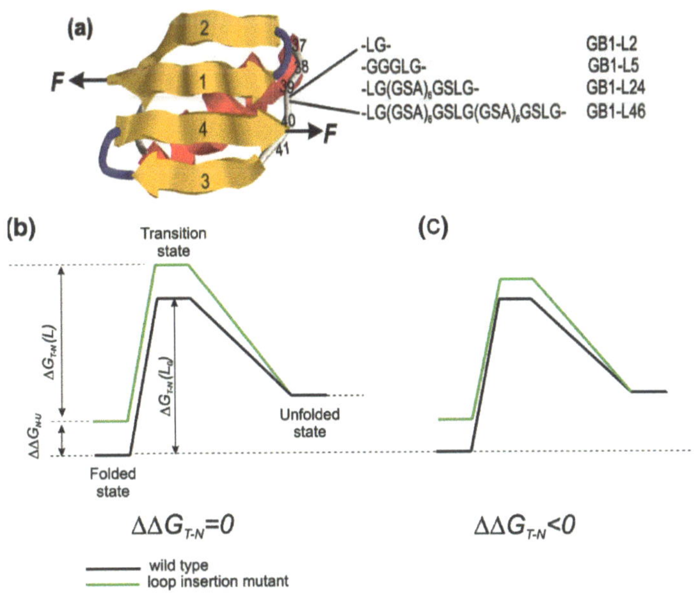

Figure 15: The loop insertions in protein GB1 (a) as well as the free energy diagrams accounting for the two different effects of loop insertions on the mechanical unfolding (b and c). [221] (Reproduced with permission from Elsevier)

As described in Section 4, pH modulates the protein folding by protonation/deprotonation of the sidechains. Here will discuss the effects of protonation/deprotonation on the specific residues. García *et al.* [42] utilize the unbiased replica exchange molecular dynamics (REMD) simulations to study the protonation/deprotonation effects on the Trp-cage folding, stability and kinetics within a wide range of temperatures. The results suggest that the protonation of Asp9 generates multiple H-bonds with the backbone atoms in an unfolded configuration with a significant fraction, which is however not present in the charged state, see Fig. (**16**). The drastic decrease of the folded structure ensemble in the protonated state is caused by absence of the Asp9/Arg16 salt bridge.

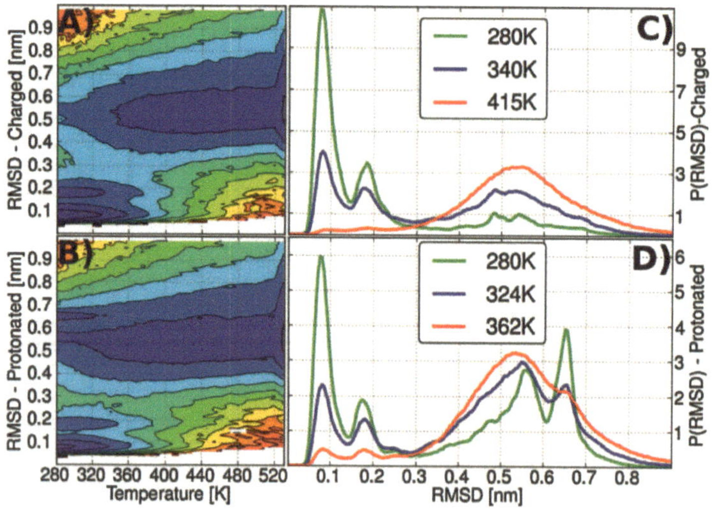

Figure 16: The Cα RMSD *vs.* temperature density maps for the charged (A) and protonated (B) Asp9 in the Trp-cage miniprotein as well as the corresponding histograms of the population of states as a function of their RMSD values (C and D). [42] (Reproduced with permission from RSC) At low temperatures the charged state is mainly in the folded state, while the protonated state occupies both the folded (RMSD < 2.2 Å) and unfolded states (RMSD > 2.2 Å).

CONCLUDING REMARKS

Protein folding is a rather complex and multi-factor coupled process. Many proteins fold on the millisecond timescale that is so "long" and beyond the capability of current computer modeling. On the other hand, these millisecond events are "transient" for the experimental observations so that it is intractable to capture the exact and comprehensive information, although fruitful results have been derived by using various techniques such as atomic force microscopy (AFM), nuclear magnetic resonance (NMR) and circular dichroism (CD).

For the studies of protein folding, computer modeling is superior to experiments in that it can selectively investigate one factor, capture the detailed conformational dynamics and observe the whole folding process. The roles of numerous factors have been clarified in this way, such as space limitation (confinement and macromolecular crowding), solvent effects, pressure, temperature, pH and structural modifications (mutation, truncation/insertion and protonation/deprotonation). These have been discussed elaborately in this chapter. There are other potential factors that may play a role, for example, topological frustrations [223-224], chaperonins [225, 226], tethering induced biophysical determinants [227] and external fields (electric and magnetic) [228]. The clarification of these factors is definitely beneficial to comprehend the protein folding process in realistic conditions such as cellular organelles, where the above factors are synergistic resulting in the special functions and activity [15, 16, 229-231]. Computer modeling is expected to continuously play a significant role in this area.

ACKNOWLEDGEMENTS

We are grateful for the financial supports from the National Natural Science Foundation (No. 20903019), Anhui Provincial Natural Science Foundation (No. 1208085QC58), Natural Science Foundation from Educational Commission of Anhui Province (No. KJ2012B163) and the Talented Funds of Huaibei Normal University (No. 600584).

CONFLICT OF INTEREST

The authors confirm that this chapter contents have no conflict of interest.

REFERENCES

[1] Kennedy, D.; Norman, C. What don't we know? *Science*, **2005**, 309, 75-75.

[2] Buchner, G.S.; Murphy, R.D.; Buchete, N.; Kubelka, J. Dynamics of protein folding: Probing the kinetic network of folding-unfolding transitions with experiment and theory. *Biochim. Biophys. Acta*, **2011**, 1814, 1001-1020.

[3] Crescenzi, P.; Goldman, D.; Papadimitriou, C.; Piccolboni, A.; Yannakakis, M. On the complexity of protein folding. *J. Comput. Biol.*, **1998**, 5, 1-36.

[4] Bowman, G.R.; Voelz, V.A.; Pande, V.S. Taming the complexity of protein folding. *Curr. Opin. Struct. Biol.*, **2011**, 21, 4-11.

[5] Haspel, N.; Tsai, C.J.; Wolfson, H.; Nussinov, R. Reducing the computational complexity of protein folding *via* fragment folding and assembly. *Protein Sci.*, **2003**, 12, 1177-1187.

[6] Vinson, V. J. Proteins in motion. *Science*, **2009**, 324, 197.

[7] Brooks III, C.L.; Gruebele, M.; Qnuchic, J.N.; Wolynes, P.G. Chemical physics of protein folding. *Proc. Natl. Acad. Sci. USA*, **1998**, 95, 11037-11038.

[8] Goldberg, M.E. The second translation of the genetic message: protein folding and assembly. *Trends. Biochem. Sci.*, **1985**, 10, 388-391.

[9] Dill, K.A.; Ozkan, S.B.; Shell, M.S.; Weikl, T.R. The protein folding problem. *Annu. Rev. Biophys.*, **2008**. 37, 289-316.

[10] Udgaonkar, J.B. Multiple routes and structural heterogeneity in protein folding. *Annu. Rev. Biophys.*, **2008**, 37, 489-510

[11] Mńoz, V. Conformational dynamics and ensembles in protein folding. *Annu. Rev. Biophys. Biomol. Struct.*, **2007**, 36, 395-412.

[12] Lindorff-Larsen, K.; Piana, S.; Dror, R.O.; Shaw, D.E. How fast-folding proteins fold. *Science*, **2011**, 334, 517-520.

[13] Anfinsen, C.B.; Redfield, R.R.; Choate, W.L.; Page, J.; Carroll, W.R. Studies on the gross structure, cross-linkages, and terminal sequences in ribonuclease. *J. Biol. Chem.*, **1954**, 207, 201-210.

[14] Anfinsen, C.B.; Haber, E.; Sela, M.; White, F.H. The kinetics of formation of native ribonuclease during oxidation of the reduced polypeptide chain. *Proc. Natl. Acad. Sci. USA*, **1961**, 47, 1309-1314.

[15] Yon, J.M.; Betton, J.M. Protein folding *in vitro* and in the cellular environment. *Biol. Cell*, **1911**, 71, 17-23.

[16] Tokuriki, N.; Tawfik, D.S. Protein dynamism and evolvability. *Science*, **2009**, 324, 203-207.

[17] James, L.C.; Tawfik, D.S. Conformational diversity and protein evolution--a 60-year-old hypothesis revisited. *Trends Biochem. Sci.*, **2003**, 28, 361-368.

[18] Meier, S.; Özbek, S. A biological cosmos of parallel universes: does protein structural plasticity facilitate evolution? *Bioessays*, **2007**, 29, 1095-1104.

[19] Onuchic, J.N.; Wolynes, P.G. Theory of protein folding. *Curr. Opin. Struct. Biol.*, **2004**, 14, 70-75.

[20] Du, R.; Pande, V.S.; Grosberg, A.Y.; Tanaka, T.; Shakhnovich, E; On the role of conformational geometry in protein folding. *J. Chem. Phys.*, **1999**, 111, 10375-10380.

[21] Radford, S.E. Protein folding: progress made and promises ahead. *Trends Biochem. Sci.*, **2000**, 25, 611-618.

[22] Clementi, C.; Nymeyer, H.; Onuchic, J.N. Topological and energetic factors: what determines the structural details of the transition state ensemble and en-route intermediates for protein folding? An investigation for small globular proteins. *J. Mol. Biol.*, **2000**, 298, 937-953.

[23] Abkevich, V.I.; Gutin, A.M.; Shakhnovich, E.I. Improved design of stable and fast-folding model proteins. *Fold Des.*, **1996**, 1, 221-230.

[24] Ellis, R.J.; Minton, A.P. Protein aggregation in crowded environments. *Biol. Chem.*, **2006**, 387, 485-497.

[25] Pincus, D.L.; Thirumalai, D. Crowding effects on the mechanical stability and unfolding pathways of ubiquitin. *J. Phys. Chem. B.*, **2009**, 113, 359-368.

[26] Zhou, H.X. Protein folding in confined and crowded environments. *Arch. Biochem. Biophys.*, **2008**, 469, 76-82.

[27] Zhou, H.X.; Rivas, G.; Minton, A.P. Macromolecular crowding and confinement: biochemical, biophysical, and potential physiological consequences. *Annu. Rev. Biophys.*, **2008**, 37, 375-397.

[28] Tsao, D.; Minton, A.P.; Dokholyan, N.V. A didactic model of macromolecular crowding effects on protein folding. *PLoS ONE*, **2010**, 8, e11936.

[29] Zimmerman, S.B.; Minton, A.P. Macromolecular crowding: biochemical, biophysical, and physiological consequences. *Annu. Rev. Biophy. Biomol. Struct.*, **1993**, 22, 27-65.

[30] Kim, P.; Abkarian, M.; Stone, H.A. Hierarchical folding of elastic membranes under biaxial compressive stress. *Nat. Mater.*, **2011**, 10, 952-957.

[31] Paoli, B.; Pellarin, R.; Caflisch, A. Slow folding of cross-linked alpha-helical peptides due to steric hindrance. *J. Phys. Chem. B.*, **2010**, 114, 2023-2027.

[32] Gao, M. She, Z.S. Zhou, R. Key residues that play a critical role in urea-induced lysozyme unfolding. *J. Phys. Chem. B.*, **2010**, 114, 15687-15693.

[33] Wafer, L.N.; Streiche, W.W.; Makhatadze, G.I. Thermodynamics of the Trp-cage miniprotein unfolding in urea. *Proteins*, **2010**, 78, 1376-1381.

[34] Hills Jr., R.D.; Brooks III, C.L. Insights from coarse-grained Gō models for protein folding and dynamics. *Int. J. Mol. Sci.*, **2009**, 10, 889-905.

[35] Dyson, H.J.; Wright, P.E. Intrinsically unstructured proteins and their functions. *Nat. Rev. Mol. Cell Biol.*, **2005**, 6, 197-208.

[36] Thirumalai, D.; Klimov, D.K.; Dima, R.I. Emerging ideas on the molecular basis of protein and peptide aggregation. *Curr. Opin. Struct. Biol.*, **2003**, 13, 146-159.

[37] Wu, X.M.; Yang, G.; Zu, Y.G.; Yang, Z.W.; Fu, Y.J.; Zhou, L.J.; Yuan, X.H. The Trp-cage miniprotein with single-site mutations: studies of stability and dynamics using molecular dynamics. *Comput. Theor. Chem.*, **2011**, 973, 1-8.

[38] Williams, D.V.; Barua, B.; Andersen, N.H. Hyperstable miniproteins: additive effects of D- and L-Ala mutations. *Org. Biomol. Chem.*, **2008**, 6, 4287-4289.

[39] Zondlo, N.J. Aromatic-proline interactions: Electronically tunable CH/π interactions. *Acc. Chem. Res.*, **2013**, 46, 1039-1049.

[40] Vainshtein, I.; Atrazhev, A.; Eom, S.H.; Elliott, J.F.; Wishart, D.S.; Malcolm B.A. Peptide rescue of an N-terminal truncation of the Stoffel fragment of taq DNA polymerase. *Protein Sci.*, **1996**, 5, 1785-1792.

[41] Yang, G.; Wu, X.M.; Zu, Y.G.; Yang, Z.W.; Fu, Y.J.; Zhou, L.J. Molecular dynamic simulations on the folding and conformational insights of the truncated peptides. *J. Theo. Comput. Chem.*, **2009**, 8, 317-331.

[42] Jimenez-Cruz, C.A.; Makhatadze, G.I.; García, A.E. Protonation/deprotonation effects on the stability of the Trp-cage miniprotein. *Phys. Chem. Chem. Phys.*, **2011**, 13, 17056-17063.

[43] Williams, D.V.; Byrne, A.; Stewart, J.; Andersen, N.H. Optimal salt bridge for Trp-cage stabilization. *Biochemistry*, **2011**, 50, 1143-1152.

[44] Streicher, W.W.; Makhatadze, G.I. Unfolding thermodynamics of Trp-cage, a 20 residue miniprotein, studied by differential scanning calorimetry and circular dichroism spectroscopy. *Biochemistry*, **2007**, 46, 2876-2880.

[45] Rovó, P.; Farkas, V.; Hegyi, O.; Szolomájer-Csikós, O.; Tóth, G.K.; Perczel, A. Cooperativity network of Trp-cage miniproteins: probing salt-bridges. *J. Pept. Sci.*, **2011**, 17, 610-619.

[46] Hudáky, P.; Stráner, P.; Farkas, V.; Váradi, G.; Tóth, G.; Perczel, A. Cooperation between a salt bridge and the hydrophobic core triggers fold stabilization in a Trp-cage miniprotein. *Biochemistry*, **2008**, 47, 1007-1016.

[47] Anfinsen, C.B. Principles that govern folding of protein chains. *Science*, **1973**, 181, 223-230.

[48] England, J.L.; Haran, G. Role of solvation effects in protein denaturation: From thermodynamics to single molecules and back. *Annu. Rev. Phys. Chem.*, **2011**, 62, 257-277.

[49] Cheung, M.S.; Thirumalai, D. Effects of crowding and confinement on the structures of the transition state ensemble in proteins. *J. Phys. Chem. B.*, **2007**, 111, 8250-8257.

[50] Mittal, J.; Best, R.B. Thermodynamics and kinetics of protein folding under confinement. *Proc. Natl. Acad. Sci. USA,* **2008**, 105, 20233-20238.

[51] Samiotakis, A.; Cheung, M.S. Folding dynamics of Trp-cage in the presence of chemical interference and macromolecular crowding. I. *J. Chem. Phys.*, **2011**, 135, 175101-175116

[52] Fersht, A.R. From the first protein structures to our current knowledge of protein folding: delights and skepticisms. *Nat. Rev. Mol. Cell Biol.*, **2008**, 9, 650-654.

[53] Benton, L.A.; Smith, A.E.; Young, G.B.; Pielak, G.J. Unexpected effects of macromolecular crowding on protein stability. *Biochemistry*, **2012**, 51, 9773-9775.

[54] Gierasch, L.M.; Ignatova, Z.; Krishnan, B.; Hong, J.; Wang, Q.; Sahoo, H.; Budyak, I. The influence of the cellular environment on protein folding and stability. *FASEB J.*, **2008**, 22, 528-531.

[55] Zhou, H.X. Helix formation inside a nanotube: Possible influence of backbone-water hydrogen bonding by the confining surface through modulation of water activity. *J. Chem. Phys.*, **2007**, 127, 245101-245104.

[56] Lucent, D.; Vishal, V.; Pande, V.S. Protein folding under confinement: A role for solvent. *Proc. Natl. Acad. Sci. USA*, **2007**, 104, 10430-10434.

[57] Sorin, E.J.; Pande, V.S. Nanotube confinement denatures protein helices. *J. Am. Chem. Soc.*, **2006**, 128, 6316-6317.

[58] Jewett, A.I.; Baumketner, A.; Shea, J.E. Accelerated folding in the weak hydrophobic environment of a chaperonin cavity: Creation of an alternate fast folding pathway. *Proc. Natl. Acad. Sci. USA*, **2004**, 101, 13192-13197.

[59] Ojeda, P.; García, A.E.; Londono, A.; Chen, N.Y. Monte Carlo simulations of proteins in cages: influence of confinement on the stability of intermediate states. *Biophys. J.*, **2009**, 96, 1076-1082.

[60] Ziv, G.; Haran, G.; Thirumalai, D. Ribosome exit tunnel can entropically stabilize α-helices. *Proc. Natl. Acad. Sci. USA*, **2005**, 102, 18956-18961.

[61] Friedel, M.; Sheeler, D.J.; Shea, J.E. Effects of confinement and crowding on the thermodynamics and kinetics of folding of a minimalist β-barrel protein. *J. Chem. Phys.*, **2003**, 118, 8106-8113.

[62] Takagi, F.; Koga, N.; Takada, S. How protein thermodynamics and folding mechanisms are altered by the chaperonin cage: Molecular simulations. *Proc. Natl. Acad. Sci. USA*, **2003**, 100, 11367-11372.

[63] Cheung, M.S.; Thirumalai, D. Nanopore-protein interactions dramatically alter stability and yield of the native state in restricted spaces. *J. Mol. Biol.*, **2006**, 357, 632-643.

[64] Klimov, D.K.; Newfield, D.; Thirumalai, D. Simulations of β-hairpin folding confined to spherical pores using distributed computing. *Proc. Natl. Acad. Sci. USA*, **2002**, 99, 8019-8024.

[65] Cheung, M.S.; Klimov, D.; Thirumalai, D. Molecular crowding enhances native state stability and refolding rates of globular proteins. *Proc. Natl. Acad. Sci. USA*, **2005**, 102, 4753-4758.

[66] Ravindra, R.; Zhao, S.; Gies, H.; Winter, R. Protein encapsulation in mesoporous silicate: The effects of confinement on protein stability, hydration, and volumetric properties. *J. Am. Chem. Soc.*, **2004**, 126, 12224-12225.

[67] Campanini, B.; Bologna, S.; Cannone, F.; Chirico, G.; Mozzarelli, A.; Bettati, S. Unfolding of green fluorescent protein mut2 in wet nanoporous silica gels. *Protein Sci.*, **2005**, 14, 1125-1133.

[68] Wheeler, K.E.; Nocek, J.M.; Hoffman, B.M. NMR spectroscopy can characterize proteins encapsulated in a sol-gel matrix. *J. Am. Chem. Soc.*, **2006**, 128, 14782-14783.

[69] Bolis, D.; Politou, A.S.; Kelly, G.; Pastore, A.; Temussi, P. Protein stability in nanocages: a novel approach for influencing protein stability by molecular confinement. *J. Mol. Biol.*, **2004**, 336, 203-212.

[70] Babu, C.R.; Hilser, V.J.; Wand, A.J. Direct access to the cooperative substructures of proteins and the protein ensemble *via* cold denaturation. *Nat. Struct. Mol. Biol.*, **2004**, 11, 352-357.

[71] Peterson, R.W.; Anbalagan, K.; Tommos, C.; Wand, A.J. Forced folding and structural analysis of metastable proteins. *J. Am. Chem. Soc.*, **2004**, 126, 9498-9499.

[72] Shi, Z.S.; Peterson, R.W.; Wand, A.J. New reverse micelle surfactant systems optimized for high-resolution NMR spectroscopy of encapsulated proteins. *Langmuir*, **2005**, 21, 10632-10637.

[73] Simorellis, A.K.; Flynn, P.F. Fast local backbone dynamics of encapsulated ubiquitin. *J. Am. Chem. Soc.*, **2006**, 128, 9580-9581.

[74] Mukherjee, S.; Chowdhury, P.; Gai, F. Tuning the cooperativity of the helix-coil transition by aqueous reverse micelles. *J. Phys. Chem. B*, **2006**, 110, 11615-11619.

[75] Tian, J.; García, A.E. Simulation studies of protein folding/unfolding equilibrium under polar and nonpolar confinement. *J. Am. Chem. Soc.*, **2011**, 133, 15157-15164.

[76] Miklos, A.C.; Li, C.; Sharaf, N.G.; Pielak, G.J. Protein crowding tunes protein stability. *Biochemistry*, **2010**, 49, 6984-6991.

[77] Schlesinger, A.; Wang, Y.; Tadeo, X.; Millet, O.; Pielak, G. Macromolecular crowding fails to fold a globular protein in Cells. *J. Am. Chem. Soc.* **2011**, 133, 8082-8085.

[78] Homouz, D.; Perham, M.; Samiotakis, A.; Cheung, M.S.; Wittung-Stafshede, P. Crowded, cell-like environment induces shape changes in aspherical protein. *Proc. Natl. Acad. Sci. USA*, **2008**, 105, 11754-11759.

[79] Gutowski, M.; Skurski, P.; Simons, J. Dipole-bound anions of glycine based on the zwitterion and neutral structures. *J. Am. Chem. Soc.*, **2000**, 122, 10159-10162.

[80] Boekfa, B.; Pantu, P.; Limtrakul, J. Interactions of amino acids with H-ZSM-5 zeolite: An embedded ONIOM study. *J. Mol. Struct.*, **2008**, 889, 81-88.

[81] Yang, G.; Zhou, L.J.; Liu, C.B. Glycine canonical and zwitterionic isomers within zeolites. *J. Phys. Chem. B*, **2009**, 113, 10399-10402.

[82] Rimola, A.; Corno, M.; Zicovich-Wilson, C.M.; Ugliengo, P. *Ab initio* modeling of protein/biomaterial interactions: glycine adsorption at hydroxyapatite surfaces. *J. Am. Chem. Soc.*, **2008**, 130, 16181-16183.

[83] Costa, D.; Garrain, P.A.; Diawara, B.; Marcus, P. Biomolecule-biomaterial interaction: A DFT-D study of glycine adsorption and self-assembly on hydroxylated Cr_2O_3 surfaces. *Langmuir*, **2011** 27, 2747-2760.

[84] Papoian, G.A.; Ulander, J.; Eastwood, M.P.; Luthey-Schulten, Z.; Wolynes, P.G. Water in protein structure prediction . *Proc. Natl. Acad. Sci. USA,* **2004**, 101, 3352-3357.

[85] De Simone, A.; Dodson, G.G.; Verma, C.S.; Zagari, A.; Fraternali, F. Prion and water: Tight and dynamical hydration sites have a key role in structural stability. *Proc. Natl. Acad. Sci. USA,* **2005**, 102, 7535-7540.

[86] Mukherjee, S.; Chowdhury, P.; Gai, F. The effect of dehydration on the aggregation kinetics of two amyloid peptides. *J. Phys. Chem. B,* **2009**, 113, 531-535.

[87] Cruz, L.; Urbanc, B.; Borreguero, J.M.; Lazo, N.D.; Teplow, D.B.; Stanley, H. E. Solvent and mutation effects on the nucleation of amyloid β-protein folding. *Proc. Natl. Acad. Sci. USA,* **2005**, 102, 18258-18263.

[88] Cheung, M.S.; García, A.E.; Onuchic, J.N. Protein folding mediated by solvation: Water expulsion and formation of the hydrophobic core occur after the structural collapse. *Proc. Natl. Acad. Sci. USA,* **2002**, 99, 685-690.

[89] Kim, S.J.; Born, B.; Havenith, M.; Gruebele, M. Real-time detection of protein-water dynamics upon protein folding by terahertz absorption spectroscopy. *Angew. Chem. Int. Ed.,* **2008**, 47, 6486-6489.

[90] Kimura, T.; Maeda, A.; Nishiguchi, S.; Ishimori, K.; Morishima, I.; Konno, T.; Goto, Y.; Takahashi, S. Dehydration of main-chain amides in the final folding step of single-chain monellin revealed by time-resolved infrared spectroscopy. *Proc. Natl. Acad. Sci. USA,* **2008**, 105, 13391-13396.

[91] Daidone, I.; Ulmschneider, M.B.; Di Nola, A.; Amadei, A.; Smith, J.C. Dehydration-driven solvent exposure of hydrophobic surfaces as a driving force in peptide folding. *Proc. Natl. Acad. Sci. USA,* **2007**, 104, 15230-15235.

[92] Zhou, R.; Huang, X.; Margulius, C.J.; Berne, B.J. Hydrophobic collapse in multi-domain protein folding. *Science*, **2004**, 305, 1605-1609.

[93] Billeter, M.; Guntert, P.; Luginbuhl, P.; Wuthrich, K. A molecular code dictates sequence-specific DNA recognition by homeodomains. *Cell*, **1996**, 85, 1057-1065.

[94] Zhong, D.; Pal, S.K.; Zewail, A.H. Femtosecond studies of protein–DNA binding and dynamics: Histone I. *ChemPhysChem.*, **2001**, 2, 219-227.

[95] Jayaram, B.; Jain, T. The role of water in protein-DNA recognition. *Annu. Rev. Biophys. Biomol. Struct.*, **2004**, 33, 343-361.

[96] Zhong, D.; Douhal, A.; Zewail, A.H. Femtosecond studies of protein-ligand hydrophobic binding and dynamics: Human serum albumin. *Proc. Natl. Acad. Sci. USA*, **2000**, 97, 14056-14061.

[97] Levy, Y.; Onuchic, J.N. Water mediation in protein folding and molecular recognition. *Annu. Rev. Biophys. Biomol. Struct.*, **2006**, 35, 389-415.

[98] Garczarek, F.; Gerwert, K. Functional waters in intraprotein proton transfer monitored by FTIR difference spectroscopy. *Nature*, **2006**, 439, 109-112.

[99] Lin, J.; Balabin, I.A.; Beratan, D.N. The nature of aqueous tunneling pathways between electron-transfer proteins *Science*, **2005**, 310, 1311-1313.

[100] Herbst, D.; Peper, S.; Niemeyer, B. Enzyme catalysis in organic solvents: influence of water content, solvent composition and temperature on Candida rugosa lipase catalyzed transesterification. *J. Biotechnol.*, **2012**, 162, 398-403.

[101] Shrimpton, P.; Allemann, R.K. Role of water in the catalytic cycle of E. coli dihydrofolate reductase. *Protein Sci.,* **2002**, 11, 1442-1451.

[102] Bon, C.; Dianoux, A.J.; Ferrand, M.; Lehmann, M.S. A model for water motion in crystals of lysozyme based on an incoherent quasielastic neutron-scattering study. *Biophys. J.*, **2002**, 83, 1578-1588.

[103] Pocker, Y. Water in enzyme reactions: biophysical aspects of hydration-dehydration processes. *Cell. Mol. Life Sci.,* **2000**, 57, 1008-1017.

[104] Edsall, J.T.; McKenzie, H.A. Water and proteins. I. The significance and structure of water; its interaction with electrolytes and non-electrolytes. *Adv. Biophys.*, **1978**, 10, 137-203.

[105] Edsall, J.T.; McKenzie, H.A. Water and proteins. II. The location and dynamics of water in protein systems and its relation to their stability and properties. *Adv. Biophys.*, **1983**, 16, 53-183.

[106] Zhang, L.; Yang, Y.; Kao, Y.T.; Wang, L.; Zhong, D. Protein hydration dynamics and molecular mechanism of coupled water-protein fluctuations. *J. Am. Chem. Soc.*, **2009**, 131, 10677-10691.

[107] Gao, M.; Zhu, H.Q.; Yao, X.Q.; She, Z.S. Water dynamics clue to key residues in protein folding. *Biochem. Biophys. Res. Commun.*, **2010**, 392, 95-99.

[108] Rhee, Y.M.; Sorin, E.J.; Jayachandran, G.; Lindahl, E.; Pande, V.S. Simulations of the role of water in the protein folding mechanism. *Proc. Natl. Acad. Sci. USA,* **2004**, 101, 6456-6461.

[109] Vega, C.; Abascal, J.L.F.; Sanz, E.; MacDowell, L.G.; McBride, C. Can simple models describe the phase diagram of water? *J. Phys.: Condens. Matter*, **2005**, 17, S3283.

[110] King, G.; Warshel, A. A surface constrained all-atom solvent model for effective simulations of polar solutions. *J. Chem. Phys.*, **1989**, 91, 3647-3661.

[111] Vega, C.; Sanz, E.; Abascal, F. Melting temperature of common water models. *J. Chem. Phys.*, **2005**, 122, 114507-114515.

[112] Still, W.C.; Tempczyk, A.; Hawley, R.C.; Hendrickson, T. Semianalytical treatment of solvation for molecular mechanics and dynamics. *J. Am. Chem. Soc.*, **1990**, 112, 6127-6129.

[113] Onufriev, A.; Bashford D.; Case, D.A. Effective Born radii in the generalized Born approximation: The importance of being perfect. *J. Comp. Chem.,* **2002**, 23, 1297-1304.

[114] Calimet, N.; Schaefer, M.; Simonson, T. Protein molecular dynamics with the generalized Born/ACE solvent model. *Proteins.*, **2001**, 45, 144-158.

[115] Lazaridis, T.; Karplus, M. Effective energy function for protein in solution. *Proteins.*, **1999**, 35, 133-152.

[116] Eisenberg, D.; McLachlan, A.D. Solvation energy in protein folding and binding. *Nature,* **1986**, 319, 199-203.

[117] Mezei, M.; Fleming, P.J.; Srinivasan, R.; Rose, G.D. The solvation free energy of the peptide backbone is conformation-dependent. *Proteins.*, **2004**, 55, 502-507.

[118] Nymeyer, H.; García, A.E. Simulation of the folding equilibrium of alpha-helical peptides: a comparison of the generalized Born approximation with explicit solvent. *Proc. Natl. Acad. Sci. USA,* **2003**, 100, 13934-13939.

[119] Zhou, R.H.; Berne, B.J. Can a continuum solvent model reproduce the free energy landscape of a beta-hairpin folding in water? *Proc. Natl. Acad. Sci. USA,* **2002**, 99, 12777-12782.

[120] Lee, M.S.; Salsbury, F.R.Jr.; Olson, M.A. An efficient hybrid explicit/implicit solvent method for biomolecular simulations. *J. Comput. Chem.*, **2004**, 25, 1967-1978.

[121] Neumann, M.; Steinhauser, O. The influence of boundary conditions used in machine simulations on the structure of polar systems. *Mol. Phys.*, **1980**, 39, 437-454.

[122] Brooks III, C.L.; Brünger, A.; Karplus, M. Active site dynamics in protein molecules: a stochastic boundary molecular-dynamics approach. *Biopolymers*, **1985**, 24, 843-865.

[123] Xu, W.X.; Mu, Y.G. *Ab initio* folding simulation of Trp-cage by replica exchange with hybrid Hamiltonian. *Biophys. Chem.*, **2008**, 137, 116-125.

[124] Wu, X.M.; Yang, G.; Zu, Y.G.; Yang, Z.W.; Fu, Y.J.; Zhou, L.J.; Yuan, X.H. Molecular dynamics characterizations of the Trp-cage folding mechanisms: in the absence and presence of water solvents. *Mol. Simulat.*, **2012**, 38, 161-171.

[125] García, A.E.; Onuchic, J.N. Folding a protein in a computer: an atomic description of the folding/unfolding of protein. A. *Proc. Natl. Acad. Sci. USA*, **2003**, 24, 13898-13903.

[126] Sheinerman, F.B.; Brooks III, C.L. Calculations on folding of segment B1 of Streptococcal protein G. *J. Mol. Biol.*, **1998**, 278, 439-456.

[127] Karanicolas, J.; Brooks III, C.L. The structural basis for biphasic kinetics in the folding of the WWdomain from a formin-binding protein: lessons for protein design? *Proc. Natl. Acad. Sci. USA*, **2003**, 100, 3954-3959.

[128] Kauzmann, W. Some factors in the interpretation of protein denaturation. *Adv. Protein Chem.,* **1959**, 14, 1-59.

[129] Dill, K.A. Dominant forces in protein folding. *Biochemistry*, **1990**, 29, 7133-7155.

[130] Paschek, D.; Nymeyer, H.; García, A.E. Replica exchange simulation of reversible folding/unfolding of the Trp-cage miniprotein in explicit solvent: On the structure and possible role of internal water. *J. Struct. Biol.*, **2007**, 157, 524-533.

[131] Damjanović, A.; García-Moreno, B.; Lattman, E.E.; García, A.E. Molecular dynamics study of water penetration in staphylococcal nuclease. *Proteins.*, **2005**, 60, 433-449.

[132] Mallamace, F.; Corsaro, C.; Mallamace, D.; Baglioni, P.; Stanley, H.E.; Chen, S.H. A possible role of water in the protein folding process. *J. Phys. Chem. B*, **2005**, 60, 450-463.

[133] Teyra, J.; Pisabarro, M.T. Characterization of interfacial solvent in protein complexes and contribution of wet spots to the interface description. *Proteins.*, **2007**, 67, 1087-1095.

[134] Sabarinathan, R.; Aishwarya, K.; Sarani, R.; Vaishnavi, M.K.; Sekar, K. Water-mediated ionic interactions in protein structures. *J. Biosci.*, **2011**, 36, 253-263.

[135] Mehl, A.F.; Demeler, B.; Zraikat, A. A water mediated electrostatic interaction gives thermal stability to the 'tail' region of the GrpE protein from E. coli. *Protein J.*, **2007**, 26, 239-245.

[136] Bogin, O.; Levin, I.; Hacham, Y.; Tel-Or, S.; Peretz, M.; Frolow, F.; Burstein, Y. Structural basis for the enhanced thermal stability of alcohol dehydrogenase mutants from the mesophilic bacterium Clostridium beijerinckii: contribution of salt bridging. *Protein Sci.*, **2002**, 11, 2561-2574.

[137] Wu, X.M.; Yang, G.; Zu Y.G.; Yang, Z.W.; Zhou, L.J. Conformational dynamics and thermal stabilities of the αβ-tubulin dimer: A molecular dynamics simulation study. *In Silico Biol.*, **2009**, 9, 271-284.

[138] Sessions, R.B.; Thomas, G.L.; Parker, M.J. Water as a conformational editor in protein folding. *J. Mol. Biol.*, **2004**, 343, 1125-1133.

[139] Gross, M.; Jaenicke, R. Proteins under pressure. *Eur. J. Biochem.*, **1994**, 221, 617-630.

[140] García, A.F.; Heindl, P.; Voigt, H.; Büttner, M.; Butz, P.; Tauber, N.; Tauscher, B.; Pfaff, E. Dual nature of the infectious prion protein revealed by high pressure. *J. Biol. Chem.*, **2005**, 280, 9842-9847.

[141] Hillson, N.; Onuchic, J.N.; García, A.E. Pressure-induced protein-folding/unfolding kinetics. *Proc. Natl. Acad. Sci. USA*, **1999**, 100, 14848-14853.

[142] Kauzmann, W. Thermodynamics of unfolding. *Nature*, **1987**, 325, 763-764.

[143] Hummer, G.; Garde, S.; García, A.E.; Paulaitis, M.E.; Pratt, L.R. The pressure dependence of hydrophobic interactions is consistent with the observed pressure denaturation of proteins. *Proc. Natl. Acad. Sci. USA*, **1998**, 95, 1552-1555.

[144] Royer, C.A. Revisiting volume change in pressure-induced protein unfolding. *Biochim. Biophys. Acta*, **2002**, 1595, 201-209.

[145] Woenckhaus, J.; Köhling, R.; Thiyagarajan, P.; Littrell, K.C.; Seifert, S.; Royer, C.A.; Winter, R. Pressure-jump small-angle X-ray scattering detected kinetics of staphylococcal nuclease folding. *Biophys. J.*, **2001**, 80, 1518-1523.

[146] Frye, K.J.; Royer, C.A. Probing the contribution of internal cavities to the volume change of protein unfolding under pressure. *Protein Sci.*, **1998**, 7, 2217-2222.

[147] Balny, C. Pressure effects on weak interactions in biological systems. *J. Phys.: Condens. Matter*, **2004**, 16, S1245-S1253.

[148] Mozhaev, V.V.; Heremans, K.; Frank, J.; Masson, P.; Balny, C. High pressure effects on protein structure and function. *Proteins.*, **1996**, 24, 81-91.

[149] Kauzmann, W. Some factors in the interpretation of protein denaturation. *Adv. Protein Chem.*, **1959**, 14, 1-67.

[150] Neuman, R.C.; Kauzman, W.; Zipp, A. Pressure dependence of weak acid ionizations in aqueous buffers. *J. Phys. Chem.*, **1973**, 77, 2687-2691.

[151] Perrett, S.; Zhou, J.M. Expanding the pressure technique: insights into protein folding from combined use of pressure and chemical denaturants. *Biochim. Biophys. Acta*, **2002**, 1595, 210-233.

[152] Klapper, M.H. On the nature of the protein interior. *Biochim. Biophys. Acta*, **1971**, 229, 557-566.

[153] Roche, J.; Caro, J.A.; Norberto, D.R.; Barthe, P.; Roumestand, C.; Schlessman, J.L.; Garcia, A.E.; García-Moreno, B.E.; Royer, C.A. Cavities determine the pressure unfolding of proteins. *Proc. Natl. Acad. Sci. USA*, **2012**, 109, 6945-6950.

[154] Silva, J.L.; Oliveira, A.C.; Gomes, A.M.; Lima, L.M.; Mohana-Borges, R.; Pacheco, A.B.; Foguel, D. Pressure induces folding intermediates that are crucial for protein-DNA recognition and virus assembly. *Biochim. Biophys. Acta.*, **2002**, 1595, 250-265.

[155] Vidugiris, G.J.A.; Markley, J.L.; Royer, C.A. Evidence for a molten globule-like transition state in protein folding from determination of activation volumes. *Biochemistry*, **1995**, 34, 4909-4912.

[156] Panick, G.; Malessa, R.; Winter, R.; Rapp, G.; Frye, K.J.; Royer, C.A. Structural characterization of the pressure-denatured state and unfolding/refolding kinetics of staphylococcal nuclease by synchrotron small-angle X-ray scattering and Fourier-transform infrared spectroscopy. *J. Mol. Biol.*, **1998**, 275, 389-402.

[157] Desai, G.; Panick, G.; Zein, M.; Winter, R.; Royer, C.A. Pressure-jump studies of the folding/unfolding of *trp* repressor. *J. Mol. Biol.*, **1999**, 288, 461-475.

[158] Frye, K.J.; Royer, C.A. The kinetic basis for the stabilization of staphylococcal nuclease by xylose. *Protein Sci.*, **1997**, 6, 789-793.

[159] Mantulin, W.W.; Pownall, H.J. Reversible folding reactions of human apolioprotein A-I: pressure and guanidinium chloride effects. *Biochim. Biophys. Acta*, **1985**, 836, 215-221.

[160] Chapeaurouge, A.; Johansson, J.S.; Ferreira, S.T. Folding intermediates of a model three-helix bundle protein. *J. Biol. Chem.,* **2001**, 276, 14861-14866.

[161] Inoue, K.; Yamada, H.; Akasaka1, K.; Herrmann, C.; Kremer, W.; Maurer, T.; Döker, R.; Kalbitzer, H.R. Pressure-induced local unfolding of the Ras binding domain of RalGDS. *Nat. Struct. Biol.*, **2000**, 7, 547-550.

[162] Nash, D.P.; Jonas, J. Structure of pressure-assisted cold denatured lysozyme and comparison with lysozyme folding intermediates. *Biochemistry*, **1997**, 36, 14375-14383.

[163] Kitahara, R.; Sareth, S.; Yamada, H.; Ohmae, E.; Gekko, K.; Akasaka, K. High pressure NMR reveals active-site hinge motion of folate-bound *Escherichia coli* dihydrofolate reductase. *Biochemistry*, **2000**, 39, 12789-12795.

[164] Lima, L.M.; Foguel, D.; Silva, J.L. DNA tightens the dimeric DNA binding domain of human papillomavirus E2 protein without changes in volume. *Proc. Natl. Acad. Sci. U. S. A*, **2000**, 97, 14289-14294.

[165] Sasahara, K.; Nitta, K. Pressure-induced unfolding of lysozyme in aqueous guanidinium chloride solution, *Protein Sci.*, **1999**, 8, 1469-1474.

[166] Silva, J.L.; Weber, G. Pressure stability of proteins. *Annu. Rev. Phys. Chem.*, **1993**, 44, 89-113.

[167] Shoemaker, B.A.; Wolynes, P.G. Exploring structures in protein folding funnels with free energy functionals: the denatured ensemble. *J. Mol. Biol.*, **1999**, 287, 657-674.

[168] Onuchic, J.N.; Nymeyer, H.; García, A.E.; Chahine, J.; Socci, N.D. The energy landscape theory of protein folding: insights into folding mechanisms and scenarios. *Adv. Protein Chem.*, **2000**, 53, 87-152.

[169] Gorovits, B.M.; Horowitz, P.M. High hydrostatic pressure can reverse aggregation of protein folding intermediates and facilitate acquisition of native structure. *Biochemistry*, **1998,** 37, 6132-6135.

[170] Foguel, D.; Robinson, C.R.; de Sousa Jr., P.C.; Silva, J.L.; Robinson, A.S. Hydrostatic pressure rescues native protein form aggregates. *Biotechnol. Bioeng.*, **1999**, 63, 552-558.

[171] Ferrao-Gonzales, A.D.; Souto, S.O.; Silva, J.L.; Foguel, D. The preaggregated state of an amyloidogenic protein: Hydrostatic pressure converts native transthyretin into the amyloidogenic state. *Proc. Natl. Acad. Sci. USA*, **2000**, 97, 6445-6450.

[172] Smeller, L.; Rubens, P.; Heremans, K. Pressure effect on the temperature-induced unfolding and tendency to aggregate of myoglobin. *Biochemistry*, **1999**, 38, 3816-3820.

[173] Kundrot, C.E.; Richards, F.M. Crystal structure of hen egg-white lysozyme at a hydrostatic pressure of 1000 atmospheres. *J. Mol. Biol.*, **1987**, 193, 157-170.

[174] Kundrot, C.E.; Richards, F.M. Effect of hydrostatic pressure on the solvent in crystals of hen egg-white lysozyme. *J. Mol. Biol.*, **1988**, 200, 401-410.

[175] Anfinsen, C.B.; Scheraga, H.A. Experimental and theoretical aspects of protein folding. *Adv. Protein Chem.* **1975**, 29, 205-300.

[176] Makhatadze, G.I.; Privalov, P.L. Energetics of protein structure. *Adv. Protein Chem.*, **1995**, 47, 308-425.

[177] Eftink, M.R. Ionescu, R. Thermodynamics of protein folding: questions pertinent to testing the validity of the two state model. *Biophys. Chem.*, **1997**, 64, 175-197.
[178] Taulier, N.; Chalikian, T.V. Compressibility of protein transitions. *Biochim. Biophys. Acta*, **2002**, 1595, 48-70.
[179] Freire, E. Thermodynamics of partly folded intermediates in proteins. *Annu. Rev. Biophys. Biomol. Struct.,* **1995**, 24, 141-165.
[180] Robertson, A.D.; Murphy, K.P. Protein structure and the energetics of protein stability. *Chem. Rev.*, **1997**, 97, 1251-1267.
[181] Tanford, C. Protein denaturation. Parts A and B. *Adv. Protein Chem.*, **1968**, 23, 121-282.
[182] Tanford, C. Protein denaturation. Part C. *Adv. Protein Chem.*, **1970**, 24, 1-95.
[183] Khandogin, J.; Chen, J.; Brooks III, C.L. Exploring atomistic details of pH-dependent peptide folding. *Proc. Natl. Acad. Sci. USA*, **2006**, 103, 18546-18550.
[184] Calzolai, L.; Zahn, R. Influence of pH on NMR structure and stability of the human prion protein globular domain. *J. Biol. Chem.*, **2003**, 278, 35592-35596.
[185] Langella, E.; Improta, R.; Barone, V. Checking the pH-induced conformational transition of prion protein by molecular dynamics simulations: effect of protonation of histidine residues. *Biophys. J.*, **2004**, 87, 3623-3632.
[186] Zanusso, G; Farinazzo, A.; Fiorini, M.; Gelati, M.; Castagna, A.; Righetti, P.G.; Rizzuto, N.; Monaco, S. pH-dependent prion protein conformation in classical Creutzfeldt-Jakob disease. *J. Biol. Chem.*, **2001**, 276, 40377-40380.
[187] Alonso, D.O.V.; DeArmond, S.J.; Cohen, F.E.; Daggett, V. Mapping the early steps in the pH-induced conformational conversion of the prion protein. *Proc. Natl. Acad. Sci. USA*, **2001**, 98, 2985-2989.
[188] Sato, S.; Raleigh, D.P. pH-dependent stability and folding kinetics of a protein with an unusual α–β topology: The C-terminal domain of the ribosomal protein L9. *J. Mol. Biol.*, **2002**, 318, 571-582.
[189] Talley, K.; Alexov, E. On the pH-optimum of activity and stability of proteins. *Proteins*, **2010**, 78, 2699-2706.
[190] Paschek, D.; Gnanakaran, S.; Garcia, A.E. Simulations of the pressure and temperature unfolding of an α-helical peptide. *Proc. Natl. Acad. Sci. USA*, **2005**, 102, 6765-6770.
[191] Brian Dyer, R.; Maness, S.J; Franzen, S.; Matthew Fesinmeyer, R.; Olsen, K.A.; Andersen, N.H. Hairpin folding dynamics: the cold-denatured state is predisposed for rapid refolding. *Biochemistry,* **2005**, 44, 10406-10415.
[192] Shao, Q.; Shi, J.Y.; Zhu, W.L. Molecular dynamics simulation indicating cold denaturation of β-hairpins. *J. Chem. Phys.,* **2013**, 138, 085102-085112.
[193] Day, R.; Bennion, B.J.; Ham, S.; Daggett, V. Increasing temperature accelerates protein unfolding without changing the pathway of unfolding. *J. Mol. Biol.*, **2002**, 322, 189-203.
[194] Kundu, S.; Roy, D. Temperature-induced unfolding pathway of a type III antifreeze protein: Insight from molecular dynamics simulation. *J. Mol. Graph. Model.*, **2008**, 27, 88-94
[195] Li, A.; Daggett, V. Characterization of the transition state of protein unfolding by use of molecular dynamics: chymotrypsin inhibitor 2. *Proc. Natl Acad. Sci. USA*, **1994**, 91, 10430-10434.
[196] Li, A.; Daggett, V. Identification and characterization of the unfolding transition state of chymotrypsin inhibitor 2 by molecular dynamics simulations. *J. Mol. Biol.*, **1996**, 257, 412-429.

[197] Kazmirski, S.L.; Li, A.; Daggett, V. Analysis methods for comparison of multiple molecular dynamics trajectories: applications to protein unfolding pathways and denatured ensembles. *J. Mol. Biol.*, **1999**, 290, 283-304.

[198] Hałabis, A.; Żmudzińska, W.; Liwo, A.; Ołdziej, S. Conformational dynamics of the Trp-cage miniprotein at its folding temperature. *J. Phys. Chem. B*, **2012**, 116, 6898-6907.

[199] Sugita, Y.; Okamoto, Y. Replica-exchange molecular dynamics method for protein folding. *Chem. Phys. Lett.*, **1999**, 314, 141-151.

[200] Wu, X.M.; Zu, Y.G.; Yang, Z.W.; Fu, Y.J.; Zhou, L.J.; Yang, G. Temperature-controlled molecular dynamics studies on the folding mechanism of the tubulin active peptides. *Acta Phys. Chim. Sin.*, **2009**, 25, 773-782.

[201] Barua, B.; Andersen, N.H. Determinants of miniprotein stability: can anything replace a buried H-bonded Trp sidechain. *Lett. Pept. Sci.*, **2002**, 8, 221-226.

[202] Neidigh, J.; Fesinmeyer, R.; Andersen, N. Designing a 20-residue protein. *Nat. Struct. Biol.*, **2002**, 9, 425-430.

[203] Su, J.G.; Chen, W.Z.; Wang, C.X. Role of electrostatic interactions for the stability and folding behavior of cold shock protein. *Proteins*, **2010**, 78, 2157-2169.

[204] Zhou, R.; Eleftheriou, M.; Royyuru, A.K.; Berne, B.J. Destruction of long-range interactions by a single mutation in lysozyme. *Proc. Natl. Acad. Sci. USA*, **2007**, 104, 5824-5829.

[205] Klein-Seetharaman, J.; Oikawa, M.; Grimshaw, S.B.; Wirmer, J.; Duchardt, E.; Ueda, T.; Imoto, T.; Smith, L.J.; Dobson, C.M.; Schwalbe, H. Long-range interactions within a nonnative protein. *Science*, **2002**, 295, 1719-1722.

[206] Kagan, B.L.; Dobson, C.M. Amyloidosis and protein folding. *Science*, **2005**, 307, 42-43.

[207] Mishima, T.; Ohkuri, T.; Monji, A.; Imoto, T.; Ueda, T. Amyloid formation in denatured single-mutant lysozymes where residual structures are modulated. *Protein Sci.*, **2006**, 15, 2448-2452.

[208] Dumoulin, M.; Last, A.; Desmyter, A.; Decanniere, K.; Canet, D.; Spencer, A.; Archer, D.; Muyldermans, S.; Wyns, L.; Matagne, A. A camelid antibody fragment inhibits the formation of amyloid fibrils by human lysozyme. *Nature*, **2003**, 424, 783-788.

[209] Chowdhury, S.; Lee, M.C.; Duan, Y. Characterizing the rate-limiting step of Trp-cage folding by all-atom molecular dynamics simulations. *J. Phys. Chem. B,* **2004**, 108, 13855-13865.

[210] Bunagan, M.R.; Yang, X.; Saven, J.G.; Gai, F. Ultrafast folding of a computationally designed Trp-cage mutant: Trp2-cage. *J. Phys. Chem. B*, **2006**, 110, 3759-3763.

[211] Neidigh, J.; Fesinmeyer, R.; Prickett, K.; Andersen, N. Exendin-4 and glucagon-like-peptide-1: NMR structural comparisons in the solution andmicelle-associated states. *Biochemistry*, **2001**, 40, 13188-13200.

[212] Barua, B.; Lin, J.C.; Williams, V.D.; Kummler, P.; Neidigh, J.W.; Andersen, N.H. The Trp-cage: optimizing the stability of a globular miniprotein. *Protein Eng. Des. Sel.*, **2008**, 21, 171-185.

[213] Wu, X.M.; Yang G.; Zu, Y.G.; Fu, Y.J.; Zhou, L.J. Molecular dynamics studies on the β-hairpin folding with the pressence of the sodium ion. *Comput. Biol. Chem.*, **2012**, 38, 1-9.

[214] Paschek, D.; Day, R.; García, A.E. Influence of water-protein hydrogen bonding on the Trp-cage miniprotein. A comparison between the TIP3P and TIP4P-Ew water models. *J. Chem. Phys.*, **2011**, 13, 19840-19847.

[215] Culik, R.M.; Serrano, A.L.; Bunagan, M.R.; Gai, F. Achieving secondary structural resolution in kinetic measurements of protein folding: A case study of the folding mechanism of Trp-cage. *Angew. Chem. Int. Ed.*, **2011**, 50, 10884-10887.

[216] Gronenborn, A.M.; Filpula, D.R.; Essig, N.Z.; Achari, A.; Whitlow, M.; Clore, G.M. A novel, highly stable fold of the immunoglobulin binding domain of streptococcal protein G. *Science*, **1991**, 253, 657-661.

[217] Choi, E.J.; Mayo, S.L. Generation and analysis of proline mutants in protein G. *Protein Eng. Des. Sel.*, **2006**, 19, 285-289.

[218] Juraszek, J.; Bolhuis, P.G. Effects of a mutation on the folding mechanism of a β-hairpin. *J. Phys. Chem. B,* **2009,** 113, 16184-16196.

[219] Wu, X.M.; Yang G.; Zhou, L.J. Identifying the intermediates during the folding/unfolding of protein GB1 with MD simulations. *Theor. Chem. Acc.*, **2012**, 131, 1299-1303.

[220] Williams, R.S.; Chasman, D.I.; Hau, D.D.; Hui, B.; Lau, A.Y.; Glover, J.N. Detection of protein folding defects caused by BRCA1-BRCT truncation and missense mutations. *J. Biol. Chem.*, **2003**, 278, 53007-53016.

[221] Li, H.; Wang, H.C.; Cao, Y.; Sharma, D.; Wang, M. Configurational entropy modulates the mechanical stability of protein GB1. *J. Mol. Biol.*, **2008**, 379, 871-880.

[222] Lee, J.Y.; Duan, L.; Iverson, T.M.; Dima, R.I. Exploring the role of topological frustration in action refolding with molecular simulations. *J. Phys. Chem. B*, **2012**, 116, 1677-1686.

[223] Norcross, T.S.; Yeates, T.O. A framework for describing topological frustration in models of protein folding. *J. Mol. Biol.*, **2006**, 362, 605-621.

[224] Gosavi, S.; Chavez, L.L.; Jennings, P.A.; Onuchic, J.N. Topological frustration and the folding of interleukin-1 beta. *J. Mol. Biol.*, **2006**, 357, 986-996.

[225] Kmiecik, S.; Kolinski, A. Simulation of chaperonin effect on protein folding: A shift from nucleation-condensation to framework mechanism. J. Am. Chem. Soc., 2011, 133, 10283-10289.

[226] Bechtluft, P.; Kedrov, A.; Slotboom, D.J.; Nouwen, N.; Tans, S.J.; Driessen, A.J. Tight hydrophobic contacts with the SecB chaperone prevent folding of substrate proteins. *Biochemistry*, **2010**, 49, 2380-2388.

[227] Arviv, O.; Levy, Y. Folding of multidomain proteins: Biophysical consequences of tethering even in apparently independent folding. *Proteins*, **2012**, 80, 2780-2798.

[228] Solomentsev, G. Y.; English, N. J.; Mooney, D. A. Effects of external electromagnetic fields on the conformational sampling of a short alanine peptide. J. Comput. Chem., 2012, 33, 917-923.

[229] Wang, Y.; Chu, X.; Suo, Z.; Wang, E.; Wang, J. Multidomain protein solves the folding problem by multifunnel combined landscape: theoretical investigation of a Y-family DNA polymerase. J. Am. Chem. Soc., 2012, 134, 13755-13764.

[230] Braselmann, E.; Clark, P.L. Autotransporters: The cellular environment reshapes a folding mechanism to promote protein transport. *J. Phys. Chem. Lett.*, **2012**, 3, 1063-1071.

[231] Renault, M.; Bos, M.P.; Tommassen, J.; Baldus, M. Solid-state NMR on a large multidomain integral membrane protein: the outer membrane protein assembly factor BamA. *J. Am. Chem. Soc.*, **2011**, 133, 4175-4177.

CHAPTER 13

Generalized Topologies: Hypergraphs, Chemical Reactions, and Biological Evolution

Christoph Flamm[1], Bärbel M. R. Stadler[2] and Peter F. Stadler[1–6,*]

[1]Institute for Theoretical Chemistry, University of Vienna, Währingerstraße 17, A-1090 Vienna, Austria; [2]Max Planck Institute for Mathematics in the Sciences, Inselstraße 22, D-04103 Leipzig, Germany; [3]Bioinformatics Group, Department of Computer Science, and Interdisciplinary Center of Bioinformatics, University of Leipzig, Härtelstraße 16-18, D-04107 Leipzig, Germany; [4]RNomics Group, Fraunhofer IZI, Perlickstraße 1, D-04103 Leipzig, Germany; [5]Center for non-coding RNA in Technology and Health, University of Copenhagen, Grønnegårdsvej 3, DK-1870 Frederiksberg, Denmark and [6]The Santa Fe Institute, 1399 Hyde Park Rd., Santa Fe, New Mexico, USA

Abstract: In the analysis of complex networks, the description of evolutionary processes, or investigations into dynamics on fitness or energy landscapes notions such as similarity, neighborhood, connectedness, or continuity of change appear in a natural way. These concepts are of an inherently topological nature. Nevertheless, the connection to the mathematical discipline of point set topology is rarely made in the literature, presumably because in most applications there is no natural object corresponding to an open or closed set. The link to textbook topology thus cannot be made in a straightforward manner. Many of the deep results of point set topology still remain valid, however, when open sets are abandoned and generalizations of the closure operator are used as the foundation of the mathematical theory. Here we survey some applications of such generalized point set topologies to chemistry and biology, providing an overview of the underlying mathematical structures.

Keywords: Barrier tree, chemical reaction, closure, continuity, energy landscape, evolution, fitness landscape, folding, graph, graph grammar, hypergraph, neighborhood, network, RNA structure, similarity, topological structures, topology.

***Corresponding author Peter F. Stadler:** Härtelstraße 16-18, D-04107 Leipzig, Germany; Tel: + 49 341 97 16690; Fax: + 49 341 97 16709; E-mail: peter.stadler@bioinf.uni-leipzig.de

Subhash C. Basak, Guillermo Restrepo & José L. Villaveces (Eds)
Copyright © 2015 Bentham Science Publishers Ltd. Published by Elsevier Inc. All rights reserved.
10.1016/B978-1-68108-053-6.50013-2

INTRODUCTION

The mathematical theories of chemistry that students are consistently exposed to throughout their education are thermodynamics and quantum mechanics, both of which are formalized in a continuum setting of vector spaces and differential operators. Similarly, we learn to think of molecules and their internal motions as trajectories in an appropriate phase space with dynamical laws given either by quantum mechanics, or its Newtonian approximation in the realm of macromolecular systems such as proteins. At the same time, molecules have natural representations as graphs, endowing them with a rich, albeit discrete mathematical structure, see *e.g.* [1, 2] for conceptual historical discussions of the discrete aspects of mathematical chemistry.

Investigations into general principles underlying chemical reaction networks, molecular phylogenetics, evolutionary changes, or the folding of biopolymers naturally encounter notions such as *similarity*, *proximity*, *connectedness*, or *continuity of change*. To a mathematician, all of these concepts are intrinsically topological in nature. All these systems are very far away from the "continuum" picture that familiar constructions such as ε -balls seem to be of little use. Even worse, *the* starting point of textbook topology, the *open set*, does not have a natural counterpart in the paradigmatic examples that we will consider throughout this contribution.

Eduard Čech's treatise of Topological Spaces [3], however, showed that the classical theory of Point Set Topology can be constructed in a different way, starting from Kuratowski's axioms of closure functions or an equivalent notion of neighborhoods. As we shall see below, these can be seen as formalizations of a vague notion of accessibility that lies at the heart of many of the questions that we are interested in. Before proceeding to the mathematical framework, we briefly introduce a few of our model systems. We note, finally, that generalized topological spaces have applications in various applied domains of computer science, including digital image processing, information representation, the semantics of modal logic, or handwriting recognition [4-11]. In mathematical chemistry, such ideas have been used *e.g.* in [12] to construct a topological space of chemical elements.

Chemical Reaction Networks

We start from a fixed set X of molecular types or chemical species. A chemical reaction ρ is a transformation rule of the form

$$\sum_{x \in X} s^{-}_{\rho,x}\, x \rightarrow \sum_{x \in X} s^{+}_{\rho,x}\, x \tag{1}$$

where $E^{-}_{\rho} := \{x \in X | s^{-}_{\rho,x} > 0\}$ are the educts and $E^{+}_{\rho} := \{x \in X | s^{+}_{\rho,x} > 0\}$ are the products of reaction ρ. Thus we can interpret reaction ρ as a hyperedge $\mathscr{E}_{\rho} = (E^{-}_{\rho}, E^{+}_{\rho})$ in a directed hypergraph $\mathfrak{h}$. It may be convenient, even, to interpret E^{+} and E^{-} as multi-sets to incorporate the stoichiometric coefficients $s^{\pm}_{\rho,x}$ as multiplicities [13]. The coefficients $s^{\pm}_{\rho,x}$ are connected to the *stoichiometric matrix* **S** *via* $s_{\rho,x} = s^{+}_{\rho,x} - s^{-}_{\rho,x}$. Fig. **1** shows a graphical description of a small chemical reaction network.

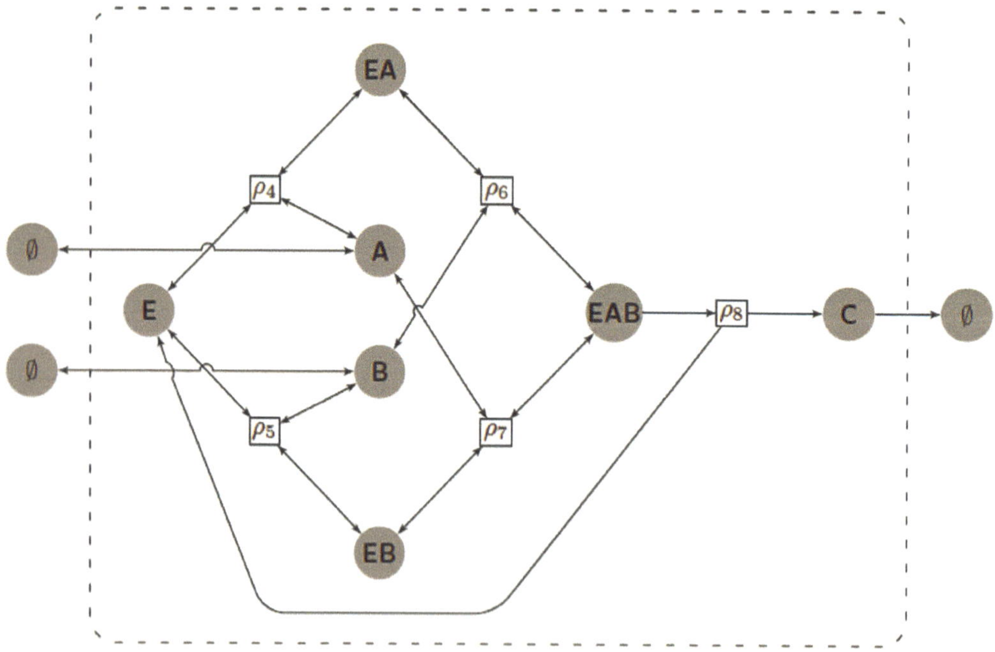

Figure 1: Hypergraph representation of the reaction network of elementary reaction steps for the overall enzyme catalyzed reaction $\boldsymbol{A} + \boldsymbol{B} \overset{E}{\rightarrow} \boldsymbol{C}$ with unordered substrate binding. Reaction nodes are drawn as squares and species nodes as gray circles. (Note for clarity the in- and outflow reaction nodes ($\boldsymbol{\rho_1} - \boldsymbol{\rho_3}$) have been omitted. The gray nodes outside the dashed oval are dummy nodes for the in- and outflow reactions).

Within chemical networks pathways, *i.e.* sequences of reactions/hyperedges that connect chemical input with output molecules, and catalytic cycles are of particular interest. Chemical organizations, that is, closed and self-maintaining subsets of $\mathfrak{h}$ [14, 15], furthermore, are closely related to the limit set of the corresponding reaction kinetics [16]. The most useful of these structural features are related to algebraic invariants that can be expressed in terms of **S**, see [17] for a recent review.

The topological description of chemical networks centers around describing what can be produced instantaneously from set $A \subseteq X$ of molecular species, *i.e.*,

$$p(A) = \bigcup_{\rho : E_\rho^-} \{x \in X \mid s_{\rho,x} > 0\} \tag{2}$$

A species x is *maintainable* in A if no reaction ρ with $E_\rho^- \subseteq A$ inevitably leads to the depletion of x. The set of maintainable species in A will be denoted by $m(A)$. A set A is closed if $p(A) = A$ and self-maintaining if $m(A) = A$. So-called *flow systems* make the additional assumption that every species x can flow out of the system. Both p and m are set-valued set functions, *i.e.* they are of the form $p : \mathfrak{P}(X) \to \mathfrak{P}(X)$, where $\mathfrak{P}(X)$ denotes the power-set of X. One can show that in flow systems both p and m are isotonic (see below), and hence impose a generalized topological structure on X [18].

Evolutionary Computation and Genetic Evolution

A (combinatorial) optimization problem is usually specified in terms of a set X of configurations and a cost function $f : X \to \mathbf{R}$, where $\mathbf{R}$ is an ordered set, or, in the case of multi-objective optimization [19], a partially ordered set [20]. A large class of heuristic algorithms, including Simulated Annealing, Genetic Algorithms, Evolutionary Strategies, or Genetic Programming, attempt to find optimal solutions by moving through the set X and evaluating the cost function at different points $x \in X$. This search procedure imposes an implicit mathematical structure on the set X that determines how points or, more generally, subsets are mutually accessible. In a more biologically inspired setting, this *search space* is uniquely determined by the genetic operators, *i.e.* by mutation, recombination, duplication, deletion, or rearrangement of gene order.

A natural way of abstracting the action of these operators is to determine for each "population" $A \subseteq X$, the set $c(A)$ of configurations that can be reached from A by applying a single one of these operator. This is most easily visualized for mutations: Each parent x may give rise to a set $c(\{x\})$ of possible offsprings (mutants). In this case, (X, c) defines a (possibly directed) graph. The situation becomes more complicated, however, when recombination (crossover) is considered [22]. The analogue of the adjacency relation of the graph is the recombination set $\mathscr{R}(x,y)$, which is defined as the set of all (possible) recombinants of two parents x and y. Recombination sets are usually required to satisfy (1) $\{x,y\} \in \mathscr{R}(x,y)$, and (2) $\mathscr{R}(x,y) = \mathscr{R}(y,x)$. Often (3) $\mathscr{R}(x,x) = \{x\}$ is assumed, which is, however, not satisfied by models of unequal crossover [23, 24]. Functions $\mathscr{R}: X \times X \to \mathfrak{P}(X)$ satisfying these three axioms were also considered under different names, *e.g. transit functions* [25] and as *P-structures* [26, 27]. We note that recombination can be seen as a ternary relation $(x, v, y) \in \mathscr{R}$ if and only if $v \in \mathscr{R}(x,y)$ closely connected with betweenness relations [28].

Similar to the chemical networks, we may base a formal treatment of topological structures on a set-valued set-function that encodes reachability at the level of sets by setting

$$c(A) = \bigcup_{x,y \in A} \mathscr{R}(x, y) \tag{3}$$

The topological structure defined by c also brings with it a concept of connectedness [29, 30] (Fig. **2**), and hence allows a construction of level sets or "basins" for the cost function $f: X \to \mathbf{R}$ as connected components (w.r.t. c) of the sets $X_h = \{x | f(x) \leq h\}$. The basin have a hierarchical structure: if A and B are connected components of $X_{h'}$ and $X_{h''}$, $h' < h''$, resp., then either $A \subseteq B$ or $A \cap B = \emptyset$. This gives rise to a tree structure representing the landscape [31-37], called the barrier tree. Local minima of the cost function correspond to the leaves of the tree, interior nodes are saddle points that define the fitness barriers between local minima. This notion of a local minimum is clearly a topological concept, as it can be defined in terms of the behavior of the cost function in the neighborhood around a point: x is a local minimum if there is a neighborhood N of x such that $f(y) \geq f(x)$ for all $x \in N$. The concept of a saddle point is much less clear. A number of similar but not-equivalent constructions is discussed in detail in [35]. A connection to combinatorial vector fields is drawn in [38].

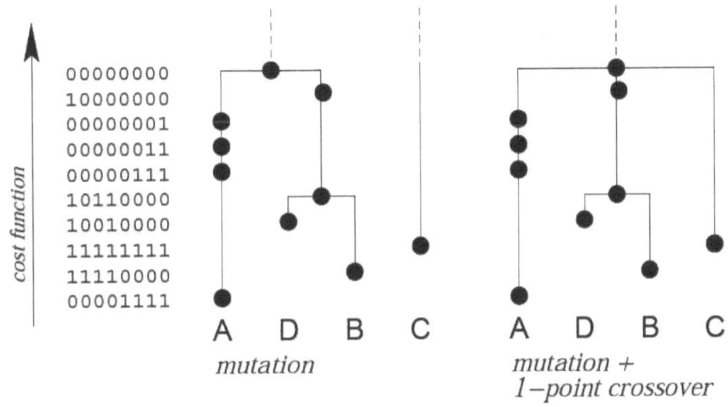

Figure 2: Barrier tree of a simple landscape with mutation (left) and mutation plus 1-point-crossover. Figure adapted from [21].

The Genotype-Phenotype Map

Two distinct and largely independent processes drive biological evolution and lie at the heart of Charles Darwin's theory: the generation of variation and selection of the variants according to their fitness. With the advent of molecular genetics it has become clear that variation is produced (primarily) at the molecular level by mutation, recombination, and other rearrangements of the genomic sequence information. Selection, in contrast, acts on the macroscopic living organism. The genotype-phenotype map, which relates genetic information to organismal appearance, properties, and behavior, thus plays a central role in modern theories of evolution [39].

RNA secondary structures have played a major role for the understanding of genotype-phenotype maps in general since they can be readily explored computationally [40]. In this setting, the primary sequence of the RNA is the genotype, while its secondary structure takes on the role of phenotype. At the phenotypic level, evolutionary processes are governed by the accessibility of phenotypic variants. The biophysical properties of RNA molecules allow at least a qualitative understanding of the likely structural effects of simple point mutations and thus the formulation of phenotypic rules of accessibility [41, 42], see Fig. **3**.

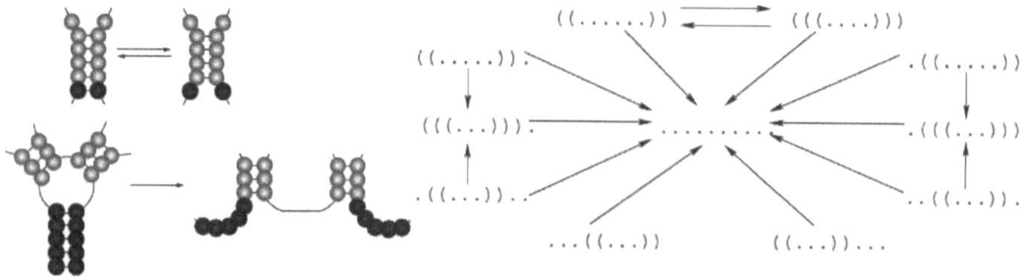

Figure 3: Fontana-Schuster topology of RNA phenotype space. L.H.S.: Frequent transitions between RNA secondary structures in response to point mutations are the opening or closing of singe base pairs and the opening of constrained stems. The latter transition is not reversible since most sequences are not pre-disposed to closing a constrained stem in response to a point mutation. R.H.S.: Space of secondary structures from GC sequences of length 10. Arrows indicate accessibility defined as frequent transitions in the sense of the two rules on the R.H.S. Figure adapted from [43, 44].

An evolutionary trajectory can be regarded as a function *f* from the time axis into phenotype space, where $f(t)$ represents *e.g.* the dominating phenotype in a population at time *t*. Computer simulations of the RNA model reveal a pattern of periods of stasis with intermittent bursts of adaptive evolution. With few exceptions, consecutive phenotypes arise from each other through one of the easily accessible structural changes of Fig. (**3**). These transitions are continuous in the usual topological sense [45, 46].

A phenotype is usually described by a set of "characters", that is, properties that can be used to differentiate between different types of organisms. The particular state of a given character (*e.g.* the presence of 5 fingers) is interpreted as an evolutionary adaptation caused by natural selection. This explanation requires the assumption that the character state can be produced by mutation without significantly affecting the functionality and/or structure of the rest of the body, *i.e.* "quasi-independence" [47]. As argued in [48], this is a statement about the structure of accessible sets in phenotype space, namely the requirement that the phenotype space can be represented as a product of a generalized topological space corresponding to the individual characters, Fig. (**4**). Since characters are meaningfully defined only in a local or regional subset of the entire phenotype space, the product structure also can only be a local approximation. These ideas have led to the development of a theory of "Approximative Graph Products" and research into algorithms for local approximate factorization of graphs [49, 50].

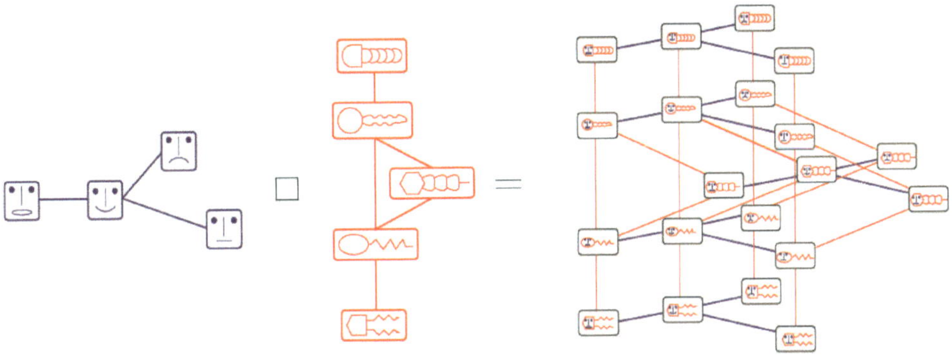

Figure 4: Phenotype space as a product of two characters ("body shape" and "faces"). For illustrative purpose the Cartesian graph product rather than the strong graph product, which corresponds to the standard product of topological spaces, is shown. The latter is obtained by adding the two diagonal edges to each quadrangle. Figure adapted from [51].

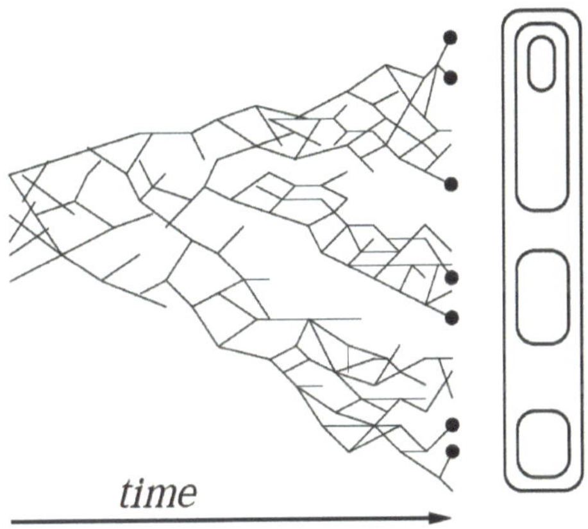

Figure 5: Sketch of the DAG $G(V,E)$. Fat dots indicate the vertices in the extant sample $\boldsymbol{X} \subseteq V$. A possible clustering that conforms to the phylogenetic relationships is indicated. For details we refer to [52].

The "Directed Acyclic Graph of Everything"

Instead of looking at the potential of evolutionary processes it is also of interest to consider only the actual history of life. This idea was recently explored in [52] using a graph $G(V,A)$ whose vertex set V consists of all individuals that ever lived. A directed edge $x \to y$ indicates that x contributed genetic information to y. Pedigree graphs are special cases of this construction, where the in-degree is limited to 2, namely an arc from the father and an arc from the mother, see *e.g.* [53, 54]. This model also easily incorporates horizontal gene transfer or the formation of hybrid species, and hence forms also a basis to study arbitrary complex phylogenetic networks [55]. Given a subset of extant observable individuals $X \subset V$ several notions of connectedness of subsets of X can be defined. The main result of [52] is that these lead to natural collections of clusters that can be interpreted *e.g.* as phylogenies, Fig. (**5**). The directed acyclic graph (DAG) $G(V, A)$ might also be an attractive starting point to study more general forms of phylogenetic networks.

GENERALIZED TOPOLOGY

Abstract Closure Functions

Instead of open sets, most approaches to generalized topology start from an abstract *closure function* $c : \mathfrak{P}(X) \rightarrow \mathfrak{P}(X)$ that encapsulates a notion of reachability or accessibility from a given subset A. The natural conjugate $i : \mathfrak{P}(X) \rightarrow \mathfrak{P}(X)$ defined by $i(A) = X \setminus c(X \setminus A)$ identifies the interior of A as the part of A that is not accessible from the complement, *i.e.* the outside of A. Closure and, equivalently, interior give rise to a concept of neighborhood of points: N is a neighborhood of x, if its interior contains x. The collection of neighborhoods can be viewed as a function $\mathcal{N} : X\mathfrak{P}(\mathfrak{P}(X))$ so that $N \in \mathcal{N}(x)$ iff $x \in i(N), i.e.$ iff $x \notin c(X \setminus N)$. Closure, interior, and neighborhood are equivalent in specifying the generalized topology [56]. For later reference we note that the notion of neighborhoods naturally extends to sets: $N \in \mathcal{N}(A)$ iff $A \subseteq i(N), i.e. N$ is a neighborhood of all $x \in A$.

The notion of continuity lies at the heart of topological theory. In the most abstract setting it comes in two flavors. Let (X, c_X) and (Y, c_Y) be two sets, each endowed with its closure function. A function $f: X \rightarrow Y$ is

closure preserving if for all $A \in \mathfrak{P}(X)$, $f(c_x(A)) \subseteq c_y(f(A))$ holds;

continuous if for all $B \in \mathcal{P}(Y), c_x(f^{-1}(B)) \subseteq f^{-1}(c_y(B))$ holds.

One says that $f: X \rightarrow Y$ is *continuous in* x if $B \in \mathcal{N}(f(x))$ implies $f^{-1}(B) \in \mathcal{N}(x)$. It can be shown that $f: X \rightarrow Y$ is continuous if and only if it is continuous in each $x \in X$ [57, Thm.3.1.].

Obviously, the identity $\iota : (X, c) \to (X, c) : x \longmapsto x$ is both closure-preserving and continuous. Furthermore, the concatenation $h = g(f)$ of the closure-preserving (continuous) functions $f: X \to Y$ and $g: Y \to Z$ is again closure-preserving (continuous).

Note that at this point we have made no assumption at all on the properties of c. Almost all approaches to extend the framework of topology assume at least that the closure functions are isotonic [56, 58-61]. The importance of isotony is emphasized by several equivalent conditions [59, Lem.10] listed in Table **1** below. A (not necessarily non-empty) collection $\mathfrak{F} \subseteq \mathfrak{P}(X)$ is a *stack* if $F \in \mathfrak{F}$ and $F \subseteq G$ implies $G \in \mathfrak{F}$. The closure function c is isotonic if and only if the neighborhood system $\mathcal{N}(x)$ is a stack for all $x \in X$. In isotonic spaces, continuity and closure preservation are equivalent.

Kuratowski's Axioms

Kuratowski's axioms for the closure function of a topological space [62] may be seen as specializations of the very general closure functions that we have considered so far. It is interesting to note that each of them can be formulated equivalently for closure, interior, and neighborhoods. Different combinations of these axioms, summarized in Table **2** define generalized topological structures that have been studied to various degrees in the literature.

Connectedness and Separation

Topological connectedness is closely related to separation. The basic idea is to investigate under which conditions closure or neighborhoods of distinct points or sets do or do not intersect. We say that A and B are *separated* if they have disjoint neighborhoods, *i.e.* $\mathcal{N}(A) \cap \mathcal{N}(B) = \emptyset$. Two sets are *semi-separated* if there are neighborhoods $N' \in \mathcal{N}(A)$ and $N'' \in \mathcal{N}(B)$ such that $A \cap N'' = N' \cap B = \emptyset$. Consider a continuous function $v: (X, c) \to [0,1]$, where $[0,1]$ denotes the unit interval endowed with the usual topology of the real numbers. It is called an *Urysohn function* separating A and B if $v(A) \subseteq \{0\}$ and $v(B) \subseteq \{1\}$. If such a function exists, A and B are called Urysohn-separated. B is completely within $A, B \Subset A$, if B and $X \setminus A$ are Urysohn-separated. Any pair of sets A, B that is Urysohn-separated is also separated, and therefore semi-separated.

Table 1: Kuratowski's axioms
The properties below are meant to hold for all $\boldsymbol{A}, \boldsymbol{B} \in \mathfrak{P}(\boldsymbol{X})$ and all $\boldsymbol{x} \in \boldsymbol{X}$, respectively.

	Closure	Interior	Neighborhood
K0'	$\exists A: x \notin c(A)$	$\exists A: x \in i(A)$	$\mathrm{N}(x) \neq \emptyset$
K0	$c(\emptyset) = \emptyset$	$i(X) = X$	$X \in \mathrm{N}(x)$
K1 isotonic monotone	$A \subseteq B \Rightarrow c(A) \subseteq c(B)$ $c(A \cap B) \subseteq c(A) \cap c(B)$ $c(A) \cup c(B) \subseteq c(A \cup B)$	$A \subseteq B \Rightarrow i(A) \subseteq i(B)$ $i(A) \cup i(B) \subseteq i(A \cup B)$ $i(A \cap B) \subseteq i(A) \cap i(B)$	$N \in \mathrm{N}(x)$ and $N \subseteq N'$ $\Longrightarrow$ $N' \in \mathrm{N}(x)$
KA	$c(X) = X$	$i(\emptyset) = \emptyset$	$\emptyset \notin \mathrm{N}(x)$
KB	$A \cup B = X \Rightarrow$ $c(A) \cup c(B) = X$	$A \cap B = \emptyset \Rightarrow$ $i(A) \cap i(B) = \emptyset$	$N', N'' \in \mathrm{N}(x) \Rightarrow$ $N' \cap N'' \neq \emptyset$
K2 expansive	$A \subseteq c(A)$	$i(A) \subseteq A$	$N \in \mathrm{N}(x) \Rightarrow x \in N$
K3 sub-linear	$c(A \cup B) \subseteq c(A) \cup c(B)$	$i(A) \cap i(B) \subseteq i(A \cup B)$	$N', N'' \in \mathrm{N}(x) \Rightarrow$ $N' \cap N'' \in \mathrm{N}(x)$
K4 idempotent	$c(c(A)) = c(A)$	$i(i(A)) = i(A)$	$N \in \mathrm{N}(x) \Leftrightarrow$ $i(N) \in \mathrm{N}(x)$
K5 additive	$\bigcup_{i \in I} c(A_i) = c\left(\bigcup_{i \in I} A_i\right)$	$\bigcap_{i \in I} i(A_i) = i\left(\bigcap_{i \in I} A_i\right)$	$\mathrm{N}(x) = \emptyset$ or $\exists N(x)$: $N \in \mathrm{N}(x)$ $\Leftrightarrow \mathrm{N}(x) \subseteq N$

A large number of subtly different *separation axioms* have been considered in the literature, of which here we just list a few to give the flavor. In the following, we consider conditions for all distinct points x, y and all disjoint non-empty subsets A, B in (X, c).

(th0) there is $z \in X$ and $N \in \mathscr{N}(z)$ such that for all

$N' \subset N, N' \in \mathscr{N}(z)$ hold $x \in N'$ and $y \notin N'$ or *vice versa.*

(th1) $\mathscr{N}(x) = \mathscr{N}(y)$ implies $x = y$.

(T0) there is $N \in \mathscr{N}(x)$ such that $y \notin N$ or there is $N' \in \mathscr{N}(y)$ such that $x \notin N'$.

(T0') $y \notin c(\{x\})$ *or* $x \notin c(\{y\})$.

(T1) there is $N \in \mathscr{N}(x)$ and $N' \in \mathscr{N}(y)$ such that $x \notin N'$ and $y \notin N$.

(T1') $c(x) \subseteq \{x\}$.

(T2) there is $N \in \mathcal{N}(x)$ and $N' \in \mathcal{N}(y)$ such that $N \cap N' = \emptyset$.

(T2½) there is $N \in \mathcal{N}(x)$ and $N' \in \mathcal{N}(y)$ such that $c(N) \cap c(N') = \emptyset$.

(T2U) Any two distinct *x* and *y* are Urysohn-separated.

(R) If $x \notin c(A)$ then there are disjoint neighborhoods of *x* and *A*.

(R') For every neighborhood *N* of *x* there is also a neighborhood $U \in \mathcal{N}(x)$ such that $c(U) \subseteq N$.

(CR) For every neighborhood *N* of *x* there is also a neighborhood $U \in \mathcal{N}(x)$ such that $U \Subset N$.

(QN) If $c(A) \cap c(B) = \emptyset$ then *A* and *B* are separated.

(N) If $c(A) \cap c(B) = \emptyset$ then *c*(*A*) and *c*(*B*) are separated.

(UN) If $c(A) \cap c(B) = \emptyset$ then *A* and *B* are Urysohn-separated.

(CN) Any two semi-separated sets are separated.

Several symmetry conditions are close associated with separation

(R0) If *x* is contained in every neighborhood of *y* then *y* is contained in every neighborhood of *x*

(R0') $x \in c(\{y\})$ implies $y \in c(\{x\})$.

(S) If $x \in N$ for all $N \in \mathcal{N}(y)$ then $\mathcal{N}(y) = \mathcal{N}(y)$.

(S') If $c(A) \cap c(\{x\}) \neq \emptyset$ then $x \in c(A)$.

(RE) If $N \cap N' \neq \emptyset$ for all $N \in \mathcal{N}(x)$ and $N' \in \mathcal{N}(x)$, then $\mathcal{N}(y) = \mathcal{N}(y)$.

Table 2: Axioms for various types of closure space

Defining axioms are marked by •, further properties that are implied are marked by ○

Axiom	$c(\emptyset) = \emptyset$	$A \subseteq B \Rightarrow c(A) \subseteq c(B)$ **Isotonic**	$A \subseteq c(A)$ **Enlarging**	$c(A \cup B) \subseteq c(A) \cup c(B)$ **Sub-Linear**	$c(c(A)) = c(A)$ **Idempotent**	$c(\cup_i A_i) = \cup_i c(A_i)$ **Additive**	**Refs.**
	(K0)	**(K1)**	**(K2)**	**(K3)**	**(K4)**	**(K5)**	
Extended Topology	•	•					[59]
Brissaud	•		•				[60]
Neighborhood space	•	•	•				[63]
Closure space	(•)	•	•		•		[64]
Smyth space	•	•		•			[5]
Binary relation	•	○		○		•	[65, 66]
Pretopology	•	•	•	•			[3]
Topology	•	•	•	•	•		
Alexandroff space	•	○	•	○		•	
Alexandroff topology	•	○	•	○	•	•	[67]

The interesting point about these axioms is that there are elaborate chains of implications among them in (ordinary) topological spaces. Some of these carry over to pretopologies, neighborhood spaces, or even isotonic spaces satisfying only (K0) and (K1), see Fig. (**6**) and its caption. Separation axioms have close connections to generalized uniform structures [51] and the existence of metrics that allow the definition of ε -balls as basis for neighborhoods of points, see [3].

The separation axiom (T0) implies both *thinness* conditions (th0) and (th1), which do not seem to have been studied in detail in the context of topology. In particular (th1) plays an important role in the theory of graph products [68].

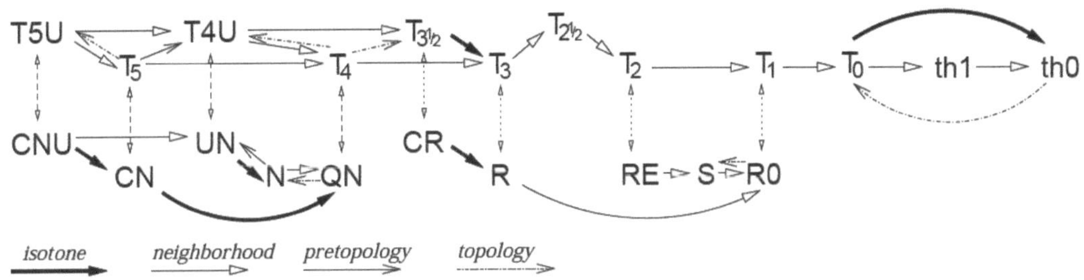

Figure 6: Relationships among separation and symmetry axioms. Different arrow types indicate implications that hold in all isotonic spaces, in all neighborhood spaces, in all pretopological spaces, or only in topological spaces. The (T)-separation axioms are defined by a normality condition and (T1), dashed double-arrow, or by a regularity/symmetry condition and (T0), dotted double-arrow, respectively. Most of the proofs for the implications are non-trivial. They can be found in the book [3], in the supplemental material to [44], or in [69].

There have been several attempts to use connectedness as the primitive notion in topological theory [29, 30, 70]. Here, we employ the more usual approach to defined connectedness as a derived property in closure spaces: A set Z is *connected* in (X, c) if it is not a disjoint union of semi-separated pairs of non-empty sets $A, Z \setminus A$. In isotonic spaces, connected sets are characterized by the Hausdorff-Lennes condition:

$$[c(A) \cap (Z \setminus A)] \cup [c(Z \setminus A) \cap A] \neq \emptyset \tag{4}$$

In neighborhood spaces, we arrive at a more familiar definition: A set is connected if and only if it is not the disjoint union of two non-empty closed (open) sets [71, Thm.5.2].

The collection of connected sets satisfies the following four properties in isotonic spaces [44, 71]:

(c1) If Z consists of a single point, then Z is connected.

(c2) If Y and Z are connected and $Y \cap Z \neq \emptyset$ then $Y \cup Z$ is connected

(c3) If Z is connected and $Z \subseteq c(Z)$, then $c(Z)$ is also connected.

(c4) Let I be an arbitrary index set and $x \in X$. Suppose Z_ι is connected and $x \in Z_\iota$ for $\iota \in I$. Then $W := \bigcup_{\iota \in I} Z_\iota$ is connected.

Consider a set A and a point $x \in A$ and let $A[x]$ be the union of all connected subsets of A that contain x. By $(c4)$, $A[x]$ is itself connected. We call it the connected component of A containing x. Note that the well-definedness of $A[x]$ is required *e.g.* for the construction of the barrier trees mentioned in the Introduction.

The relationship of connected components and semi-separation becomes counter-intuitive in non-additive spaces: Suppose A has a finite number $k > 1$ of connected components and let Q be such a component. Then Q and $A \backslash Q$ are not necessarily semi-separated unless (X, c) is a pretopology [21]. Without the benefit of additivity, furthermore, it becomes natural to investigate alternative definitions of connectedness: In [18], for instance, a less stringent definition of connectivity is introduced that is in particular suitable for chemical reaction networks: A and B are *productively separated* if for all $Z \subseteq A \cup B$ holds: (1) $c(Z \cap A) \cap B = \emptyset$ and $c(Z \cap B) \cap A = \emptyset$, and (2) $c(Z) = c(Z \cap A) \cup c(Z \cap B)$. A set is Z *productively connected* if it cannot be written as the disjoint union of two non-empty productively separated subsets.

Path-connectedness is a widely used notion of connectedness that in general is stronger than topological connectedness. From the topological point of view, a *path* is a *continuous* function $p : [0,1] \rightarrow X$ whose endpoints are $p(0)$ and $p(1)$. A set A is *path-connected* if for any two points $x, y \in A$, there is a path p with $p(0) = x$ and $p(1) = y$.

A useful lemma [21] characterizes connectedness in 2-point sets: Let (X, c) be a neighborhood space and $\{x, y\} \subseteq X$ a 2-point subset. Then the following three statements are equivalent: (1) $y \in c(\{x\})$ or $x \in c(\{y\})$; (2) $\{x, y\}$ is path connected; (3) $\{x, y\}$ is connected. We will return to this point later when considering finite generalized topologies.

Product Spaces

Let (X_1, c_1) and (X_2, c_2) be two isotonic spaces. Then the product space $(X_1 \times X_2, c_1 \times c_2)$ is defined by means of the neighborhood system

$\mathcal{N}(x_1, x_2)$, where

$$N \in \mathcal{N}(x_1, x_2) \Leftrightarrow \tag{5}$$

$$\exists N_1 \in \mathcal{N}_1(x_1) \text{ and } N_2 \in \mathcal{N}_2(x_2) \text{ such that } N_1 \times N_2 \subseteq N$$

For sets of the form $A_1 \times A_2$ this translates to $c(A_1 \times A_2) = c_1(A_1) \times c_2(A_2)$ in the product space, see [16, Thm.8.1]. If (X_1, c_1) and (X_2, c_2) satisfy (K2), (K3), or (K4), respectively, then so does their product. We remark that the construction of products can be extended to infinite families of spaces, even uncountable ones. The projections $\pi_i\colon \prod_j (X_j, c_j) \to (X_i, c_i)\colon x = (\dots, x_i, \dots) \mapsto x_i$ are continuous.

The inductive product $(X_1, c_1) \square (X_2, c_2)$ has the neighborhoods $N \in \mathcal{N}^{\square}(x_1, x_2)$ iff there is $N_1 \in \mathcal{N}_1(x_1)$ and $N_2 \in \mathcal{N}_2(x_2)$ such that $N_1 \times \{x_2\} \cup \{x_2\} \times N_2 \subseteq N$. This product is discussed briefly in [3].

Finite Generalized Topologies

For practical applications in computational biology and computational chemistry, finite spaces are of particular interest. Of course, in this case (K3) implies (K5). The by far best studied finite structures are finite pretopologies: these are the simple directed graphs. More precisely, if (X, c) is a finite pretopological space, the associated graph has vertex set X and (x, y) is a directed edge iff $y \in c(\{x\})$. Conversely, given a graph this recipe defines a finite pretopology on X [46]. Graphs that correspond to topological spaces are considered *e.g.* in [43].

Several of the topological concepts outlined above have independently been developed in graph theory. Continuous function, for instance, coincide with graph homomorphisms [72], *i.e.* functions $f\colon (X, E) \to (Y, F)$ such that $(x, y) \in E$ implies $(f(x), f(y)) \in F$ or $f(x) = f(y)$. The usual graph-theoretical definition of connectedness is also the same as pre-topological connectedness: a two-element subset $\{x, y\}$ is connected if and only if $(x, y) \in E$ or $(y, x) \in E$. Furthermore, connectedness and path-connectedness are the same [21]. The (R0) symmetry axiom, furthermore, characterizes undirected graphs. On the other hand, it appears to be unknown if strong connectedness in digraphs has a straightforward interpretation as a topological property.

The strong product of graphs coincides with the product of generalized topological spaces, while the inductive product is the same as the Cartesian graph product. For graphs, products have been studied in particular w.r.t. to the conditions under which a graph has a unique prime factorization. We refer to the book [73] for an extensive discussion of this topic. Interestingly, thinness conditions play an important role in this context. Local and approximative product structures have been explored recently [49, 50] motivated by the interpretation of characters as local factor spaces of phenotype space [48].

As mentioned in the introduction, topological spaces are typically studied in terms of their closed sets, *i.e.* the collection $\mathscr{C} = \{A | c(A) = A\}$. If (K1) to (K4) are satisfied, we have (I1) $X \in \mathscr{C}$, (I2) arbitrary intersections of closed sets are closed, and (I3) the union of two closed sets is closed. In lattice theory more general so-called *intersection structures* are considered that fulfill only (I2) see *e.g.* [74], in graph theory the same structures appear as convexities [75]. Since neither idempotency nor additivity of the closure function readily applies to the examples in the introduction we will not consider them further in this contribution.

Not much is known about finite neighborhood spaces. They correspond to the $\mathfrak{N}$-*systems* introduced in [76]. Consider a pair $(X, \mathfrak{N})$ consisting of a nonempty finite set X and a function $\mathfrak{N} : X \to \mathfrak{P}(\mathfrak{P}(X))$ that associates to each $x \in X$ a collection $\mathfrak{N}(x) = \{N^1(x), N^2(x), \ldots, N^{d(x)}\}$ of $d(x)$ subsets of X with the following properties:

(N0) $\mathfrak{N}(x) \neq \emptyset$.

(N1) $N^i(x) \subseteq N^j(x)$ implies $i = j$.

(N2) $x \in N^i(x)$, for $1 \leq i \leq d(x)$.

$\mathfrak{N}$ -systems are by construction exactly the finite neighborhood spaces, when the $N^i(x)$ are interpreted as minimal neighborhoods, i.e. $N \in \mathscr{N}(X)$ if and only if there is $N' \subseteq N$ with $N' \subseteq \mathfrak{N}(x)$. We may also interpret them as a special type of directed hypergraphs with hyperedges of the form $(\{x\}, N^i(x))$. Axiom (N1) ensures that it is simple.

The characterization of connectedness in 2-point sets [21] suggests to associate a graph $\vec{\Gamma}(X, c)$ with vertex set X with a given finite neighborhood space (X, c) such that (x, y) is a (directed) edge if $y \in c(\{x\})$. If c is additive, then $\vec{\Gamma}$ is exactly the graph representation of the pretopological space (X, c). In finite neighborhood spaces every path consists of a finite number of connected 2-point sets consisting of consecutive points. Thus a subset of X is path-connected if and only if the corresponding induced subgraph of $\vec{\Gamma}$ is path connected [21]. Connectedness is a strictly weaker property since a connected set with three points does not necessarily contain connected pairs. Finite neighborhood spaces admit a unique prime factorization w.r.t. the usual topological product under certain conditions on $\vec{\Gamma}$ [76].

Finite isotonic spaces, finally, have not been considered as combinatorial objects to our knowledge.

DIRECTED HYPERGRAPHS AS TOPOLOGICAL STRUCTURES

It may come as a surprise that hypergraphs have remained virtually unstudied from a topological perspective. While the interpretation of graphs as finite pretopological spaces is quite natural, we will see below that there does not seem to be a unique canonical translation for hypergraphs.

A directed hypergraph H consists of a vertex set X and a set $\mathcal{E}$ of directed hyperedges, each of which is a pair $E = (E^-, E^+)$ with $E^-, E^+ \subseteq X$. A hypergraph is simple if no edge is properly contained in another one, *i.e.* if $E^- \subseteq F^-$ and $E^+ \subseteq F^+$ implies $E = F$. One way of defining undirected hypergraphs as a special case of directed ones is to require $E^+ = E^-$ for all $E \in \mathcal{E}$.

Given a directed hypergraph H, it appears natural to consider the closure function

$$c(A) := \bigcup_{\mathcal{E}:E^- \subseteq A} E^+ \tag{6}$$

or

$$c(A) := A \cup \bigcup_{\mathcal{E}:E^- \subseteq A} E^+ \tag{7}$$

depending on whether we want to insist that c is enlarging or not. This approach indeed has been taken in [18] to describe chemical reaction networks. These constructions, however, become trivial for undirected hypergraphs since the condition $E^+ = E^-$ for all hyperedges ensures that $c(A) \subseteq A$.

As a possible remedy one might identify undirected hypergraphs with directed ones that satisfy the following symmetry condition:

(SH) For every directed hyperedge $\mathcal{E} = (E^-, E^+)$ every pair (A,B) of nonempty sets with $A \cup B = E^- \cup E^+$ is also a hyperedge.

In particular, in this case $(\{x\}, E)$ with $E := E^- \cup E^+$ and any $x \in E$ is a hyperedge. Thus, for a given hyperedge $\mathcal{E}$, if $E^- \subseteq A$ then there is $x \in E \cap A$ and hence $(\{x\}, E)$ also contributes to $c(A)$. Thus eq. (6) can be rephrased as

$$c(A) = \bigcup_{\mathcal{E}: E \cap A \neq \emptyset} E \tag{8}$$

This expression also makes perfect sense for undirected hypergraphs. It is equally unsatisfactory, however, since this closure function is obviously additive and hence describes a (symmetric) graph with adjacency relation $x \sim y$ iff there is a hyperedge $\mathcal{E}$ such that $x, y \in E$.

An alternative approach is to consider hyperedges as minimal neighborhoods in the sense of the $\mathfrak{N}$-systems introduced in the previous section. In finite pretopologies, *i.e.* directed graphs, the intersection of all $N \in \mathcal{N}(x)$ is again a neighborhood of x. It coincides with the "in-neighborhood" of x, given by

$$N(x) = \{y \in X | x \in c(y)\} \tag{9}$$

For directed hypergraphs, one would analogously postulate that hyper-edges that "produce" x are the minimal neighborhoods of x, *i.e.*

$$N \in \mathcal{N}(x) \Leftrightarrow \exists \mathcal{E} : x \in E^+ \text{ and } E^- \cup E^+ \subseteq N \tag{10}$$

It is straightforward to compute the corresponding closure function

$$c(A) = \{x | \exists \mathcal{E} : x \in E^+ \text{ implies } (E^- \cup E^+) \cap A\} \neq \emptyset \tag{11}$$

An undesirable feature remains, however. If we specialize this construction to directed graphs, *i.e.*, so that E^- and E^+ consist of two distinct points for each hyperedge, then $N \in \mathcal{N}(x)$ iff there is an arc (y, x) such that $x, y \in N$. The associated closure function is $c(A) = A \cup \{z | N_{in}(z) \subseteq A\}$. This is in general not additive.

A possibly fruitful alternative is to start from the notation of hypergraph homomorphisms $(X, \mathcal{E}) \to (Y, \mathcal{F})$, *i.e.* maps $\varphi: X \to X'$ so that for every $(E^-, E^+) \in \mathcal{E}$ there is $(F^-, F^+) \in \mathcal{F}$ so that $\varphi(E^-) \subseteq F^-$ and $\varphi(E^+) \subseteq F^+$. One may then ask if there exist isotonic closure functions so that the hypergraph homomorphisms are exactly the continuous functions as in the case of graphs. To our knowledge, this question is open.

We have to conclude that at present there is no satisfactory interpretation of hypergraphs as topological objects.

DYNAMICAL ASPECTS

The notion of *topological dynamical systems*, see *e.g.* [77] does not seem to have been explored in a systematic way for generalized topological spaces. A natural starting point would be flows of the form $\omega: T \times X \to X$ where T is a suitable (topological) semigroup such as the non-negative integers and $(\iota, x) \longmapsto x$, where ι denotes the neutral element of T. Orbits, for instance, are well-defined in this setting: $\Omega(x) = \{tx | t \in T\} \subseteq X$, where tx denotes the action of the semigroup element t on the point x. Similarly, trajectories are simply functions of the form $\xi: T \to X$ such that $t \longmapsto \xi(t) = tx_0$ for a fixed "initial condition" x_0. In [46], for instance, conditions are explored under which trajectories phenotypic evolution are continuous. It turns out that this matches well with intuition developed earlier in [41, 42] in the context of the evolution of RNA secondary structures.

Combinatorial vector fields offer a more specialized alternative formalization. Let M be a simplicial complex constructed over the set K of simplices, see *e.g.* [78]. We write $\sigma < \tau$ if the simplex σ lies in the boundary of the simplex τ. A *combinatorial vector field* on M [79] is a map $\eta: K \to K \cup \{\emptyset\}$ such that

1. If $\eta(\sigma) \neq \emptyset$ then $dim\,\eta(\sigma) = dim(\sigma) + 1$, and $\sigma < \eta(\sigma)$.

2. If $\eta(\sigma) = \tau \neq \emptyset$, then $\eta(\tau) = \emptyset$.

3. For all $\sigma \in K$, $|\eta^{-1}(\sigma)| \leq 1$.

We remark that combinatorial vector fields can in fact be defined on the much more general CW-complexes [79].

Combinatorial vector fields come with a natural notion of η-paths as the analog of trajectories, namely a finite sequence of simplices $\gamma = (\sigma_0, \tau_0, \sigma_1, \tau_1, \ldots \sigma_{n-1}, \tau_{n-1}, \sigma_n)$ such that $\eta(\sigma_i) = \tau_i$ for $0 \leq i < n$ and $\sigma_{i+1} < \tau_i$. A rest point is a simplex σ such that $\eta(\sigma) = \emptyset$ and $\eta^{-1}(\sigma) = \emptyset$. The rest points and the closed η-paths together form the so-called chain-recurrent set $\mathscr{R}$ of η. These play the role of attractors. All other trajectories lead towards $\mathscr{R}$.

A *Lyapunov function* of the combinatorial vector field η is a function $F : M \rightarrow \mathbb{R}$ such that

1. if $\sigma \notin \mathscr{R}$ and $\tau > \sigma$ then

   (a) $F(\sigma) < F(\tau)$ if $\tau \neq \eta(\sigma)$

   (b) $F(\sigma) \geq F(\tau)$ if $\tau = \eta(\sigma)$

2. if $\sigma \in \mathscr{R}$ and $\tau > \sigma$ then

   (a) $F(\sigma) = F(\tau)$ if $\sigma \sim \tau$

   (b) $F(\sigma) < F(\tau)$ if $\sigma \nsim \tau$

A combinatorial version of Conley's theorem ensures that there is a Lyapunov function for every combinatorial vector field [79]. Combinatorial vector fields thus form the discrete analog for gradient vector fields.

In [38] this observation was used to consider the collection of combinatorial vector fields on a graph G (with vertex set X) for which a given function $F : X$

$\rightarrow \mathbb{R}$ is a Lyapunov function. In this context, the function F is interpreted as energy function. The combinatorial vector field, on the other hand, can be interpreted as a partial orientation P on G so that $\{x,y\}$ is directed from x to y iff $\eta(x) = \{x,y\}$. Conversely, a system of directed edges P with the properties

1. $(x, y) \in P$ implies $\{x, y\} \in E$ (consistency with G)
2. $(x, y) \in P$ and $(x, z) \in P$ implies $y = z$ (uniqueness)
3. $(x, y) \in P$ implies $(y, x) \notin P$ (antisymmetry)

corresponds to a combinatorial vector field on G. A trajectory in η is then given by a sequence of vertices $(x_i), i = 1, \ldots, k$ so that $(x_i, x_{i+1}) \in P$.

These combinatorial vector fields provide a convenient description of the system of adaptive (downhill) walks on the energy landscape (G,F) and admits an alternative approach towards characterizing basins, barriers, and their hierarchical structure. In particular, they highlight the complications in the analysis of landscapes arising from degeneracies in the energy function. Given F, denote by G^F the subgraph of G with edges $\{x, y\} \in E(G)$ if $F(x) = F(y)$. The connected components $G^F(x)$, the so called shelves of the landscape, provide the main complication for practical computations, Fig. (7). While the vector fields for which F is a Lyapunov function necessarily point downward between shelves, it is complicated to handle the possible orientations with a single shelf since edges may be oriented in both directions depending on the particular choice of η.

A direct connection to the topology of the landscape (G,F) is obtained by means of the following construction:

Definition. A point y is reachable from x in (G,F) if there is a combinatorial vector field η for which F is a Lyapunov function and that admits a trajectory from x to y. Let $C(x)$ be the set of vertices reachable from x.

As shown in [38], C can be extended to an additive closure function that satisfies all of Kuratowski's axioms and hence defines a finite topological space on the vertex set X. This topology is intimately related to the barrier structure of the landscape by means of the notion of a valley: A *valley* in (G,F) is a maximal connected closed set with respect to the reachability topology. Equivalently a valley W is a maximal connected subset of X so that no vertex $y \notin W$ is reachable from any $x \in W$.

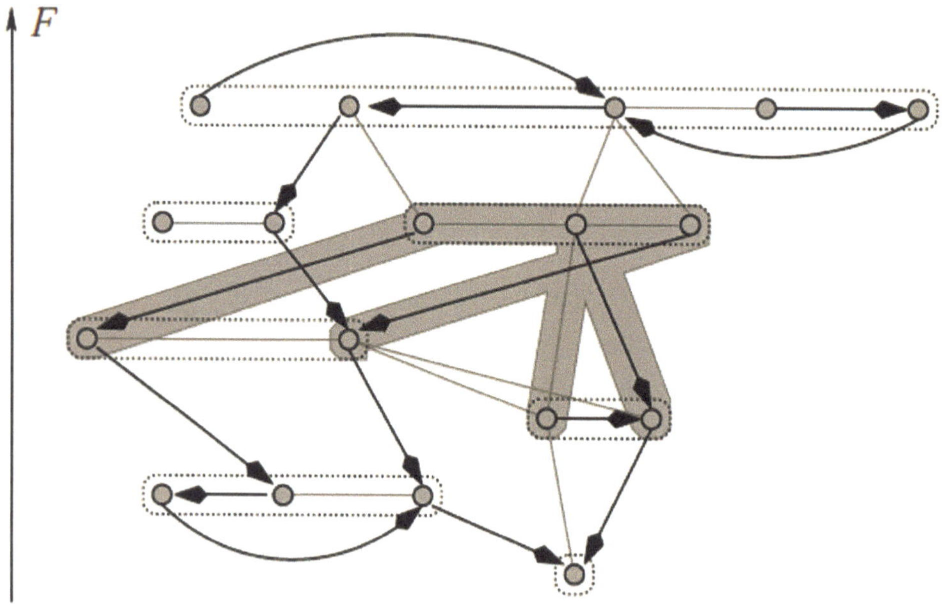

Figure 7: Shelves, *i.e.* maximal connected sets on which the energy function F is constant, are indicated by dotted boxes. Edges of G that are oriented by the combinatorial vector field η are shown as arrows, the remaining edges of G are shown in gray. Edges between shelves are either not oriented or point downward in any combinatorial vector field for which F is a Lyapunov function. Both orientations are possible for edges within shelves. Figure adapted from [38].

CONCLUDING REMARKS: COARSE GRAINING

Accessibility is of particular interest in the context of "constructive systems", *i.e.* models comprising a set of combinatorial objects and set of rules with which they can be combined. This was pioneered *e.g.* in the work of Fontana and Buss [14] using λ-calculus as a modeling platform for abstract chemistry. The state space X in such models is usually not finite. In simulations of (evolutionary) processes on such models, families of objects emerge that share many regularities and differ essentially in size only [14]. In chemistry, polymers are the most obvious example. Naturally, the question arises whether there are coarse-grained representations.

Recently the network of organic chemistry (NOC) [80, 81] has been constructed from all organic reactions reported in the chemical literature since 1779. Despite the apparent complexity of this network, which captures all synthetic knowledge, NOC shows an unexpected but well-defined topological structure [82]. In particular NOC possesses a densely wired core region of about 300 synthetically important building blocks and industrial compounds which are mutually connected by short synthetic pathways. The core region is embedded in a large and sparse periphery constituting approximately 78% of chemistry. Compounds in the periphery can be reached on average in 3-7 synthetic steps from the core region. Core region and periphery are surrounded by small isolated islands constituting about 18% of NOC. Besides the interesting topological properties of NOC, detailed investigations of its fine structure resulted in the discovery and experimental validation of one-pot reactions with direct relevance to experimental chemistry [83]. A one-pot reaction lumps together individual reactions into sequences, which can be performed under the same conditions in one pot. One-pot reactions can be seen as a form of coarse-graining of the chemical space.

Coarse-graining is an important issue also in the analysis of *e.g.* fitness landscapes. A rather simple example is the use of barrier trees, which represent a partition of the state space X. Quotient spaces $X/\sim$, obtained by identifying $\sim$ equivalent points of X, thus appear to be a natural formal structure to consider in this context. To our knowledge, this avenue has not been explored systematically for any of the model systems discussed above.

Recently, we have started to explore, from an algebraic perspective, whether compositions of transformation rules can be employed to capture family structures [84]. It will be interesting to see if the intuitive connection of rule compositions and quotient spaces can be given a precise topological meaning.

ACKNOWLEDGEMENTS

This work was supported in part by the Volkswagen Stiftung and the *Deutsche Forschungsgemeinschaft* within the EUROCORES Programme EUROGIGA (project GReGAS) of the European Science Foundation.

CONFLICT OF INTEREST

The authors confirm that this chapter contents have no conflict of interest.

LIST OF ABBREVIATIONS

DAG = directed acyclic graph

NOC = network of organic chemistry

RNA = ribo-nucleic acid

REFERENCES

[1] Balaban, A. T. Reflections about mathematical chemistry. *Found. Chem.* **2005**, *7*, 289–306.
[2] Restrepo, G.; Villaveces, J. L. Mathematical thinking in chemistry. *HYLE.* **2012**, *18*, 3–22.
[3] Čech, E. *Topological Spaces*; Wiley, London, 1966.
[4] Eckhardt, U.; Latecki, L. Digital topology. Technical Report 89, *Hamburger Beitr. z. Angew. Math. A*, **1994**.
[5] Smyth M. B. Semi-metric, closure spaces and digital topology. *Theor. Computer Sci.* **1995**, *151*, 257–276.
[6] Pfaltz, J. Closure lattices. *Discrete Math.* **1996**, *154*, 217–236.
[7] Galton, A. **2000**: 'Continuous motion in discrete space', in: A. G. Cohn; F. Giunchiglia; B. Selman (eds), *Principles of Knowledge Representation and Reasoning: Proceedings of the Seventh International Conference (KR2000)*, Morgan Kaufmann Publishers, San Francisco, CA, pp. 26–37.
[8] Marchand-Maillet, S.; Sharaiha, Y. M. Discrete convexity, straightness, and the 16-neighborhood. *Comput. Vis. Image Und.*, **1997**, *66*, 316–329.
[9] Largeron, C.; Bonnevay, S. A pretopological approach for structural analysis. *Inform. Sciences*, **2002**, *144*, 169–185.
[10] LeBourgeois, A. W. M.; Bouayad, M. Structure relation between classes for supervised learning using pretopology, in: *Fifth International Conference Document Analysis and Recognition*, **1999**, page 33.
[11] Pekalska, E.; Duin, R. P. W. *The Dissimilarity Representation for Pattern Recognition: Foundations and Applications*. World Scientific, Singapore, 2005.
[12] Restrepo, G.; Llanos, E. J.; Mesa, H. Topological space of the chemical elements and its properties. *J. Math. Chem.* **2006**, *39*, 401-416.
[13] Zeigarnik, V. **2000**: On hypercycles and hypercircuits in hypergraphs, in: P. Hansen; P. W. Fowler; M. Zheng (eds), *Discrete Mathematical Chemistry*, volume 51 of *DIMACS series in discrete mathematics and theoretical computer science*, American Mathematical Society, Providence, RI, pp. 377–383.
[14] Fontana, W.; Buss, L. W. What would be conserved if 'the tape were played twice'? *Proc. Natl. Acad. Sci. USA*, **1994**, *91*, 757–761.

[15] Dittrich, P.; Speroni di Fenizio, P. Chemical organization theory. *Bull. Math. Biol.* **2007**, *69*, 1199–1231.

[16] Stadler, P.; Dittrich, P. On the relation between organizations and limit sets in chemical reaction systems. *Adv. Complex Syst.*, **2011**, *14*, 77–96.

[17] Behre, J.; de Figueiredo, L. F.; Schuster, S.; Kaleta, C. Detecting structural invariants in biological reaction networks. *Methods Mol. Biol.*, **2012**, *804*, 377–407.

[18] Benkö, G.; Centler, F.; Dittrich, P.; Flamm, C.; Stadler, B. M. R.; Stadler, P. F. A topological approach to chemical organizations. *Alife*, **2009**, *15*, 71–88.

[19] Deb, K. *Multi-Objective Optimization using Evolutionary Algorithms*. Wiley, Chichester, NY, 2001.

[20] Stadler, P. F.; Flamm, C. Barrier trees on poset-valued landscapes. *Genet. Program. Evol. M.* **2003**, *4*, 7-20.

[21] Flamm, C.; Stadler, B. M. R.; Stadler, P. F. Saddles and barrier in landscapes of generalized search operators, in: C. R. Stephens; M. Toussaint; D. Whitley; P. F. Stadler (eds), *Foundations of Genetic Algortithms IX*, volume 4436 of *Lecture Notes Comp. Sci.*, pages 194–212, Berlin, Heidelberg, 2007. Springer. 9th International Workshop, FOGA 2007, Mexico City, Mexico, January 8-11, **2007**.

[22] Gitchoff, P.; Wagner, G. P. Recombination induced hypergraphs: a new approach to mutation-recombination isomorphism. *Complexity*, **1996**, *2*, 37–43.

[23] Shpak, M.; Wagner, J. P. Asymmetry of configuration space induced by unequal crossover: implications for a mathematical theory of evolutionary innovation. *Artif. Life*, **2000**, *6*, 25–43.

[24] Stadler, B. M. R.; Stadler, P. F.; Shpak, M.; Wagner, G. P. Recombination spaces, metrics, and pretopologies. *Z. Phys. Chem.* **2002**, *216*, 217– 234.

[25] Changat, M.; Klavžar, S.; Mulder, H. M. The all-path transit function of a graph. *Czech. Math. J.*, **2001**, *51*, 439–448.

[26] Stadler, P. F.; Wagner, G. P. The algebraic theory of recombinationc spaces. *Evol. Comp.* **1998**, *5*, 241–275.

[27] Stadler, P. F.; Seitz, R.; Wagner, G. P. Evolvability of complex characters: Population dependent Fourier decomposition of fitness landscapes over recombination spaces. *Bull. Math. Biol.* **2000**, *62*, 399–428.

[28] Adeleke, S. A; Neumann, P. M. *Relations Related to Betweenness: Their Structure and Automorphisms*. American. Mathematical Society, New York, 1998.

[29] Wallace, D. Separation spaces. *Ann. Math.*, **1941**, *43*, 687–697.

[30] Hammer, P. C. Extended topology: Connected sets and Wallace separations. *Portug. Math.* **1963**, *22*, 77–93.

[31] Klotz, T.; Kobe, S. "Valley Structures" in the phase space of a finite 3D Ising spin glass with $\pm i$ interactions. *J. Phys. A-Math. Gen.*, **1994**, *27*, L95–L100.

[32] Garstecki, P.; Hoang, T. X.; Cieplak, M. Energy landscapes, supergraphs, and "folding funnels" in spin systems. *Phys. Rev. E*, **1999**, *60*, 3219–3226.

[33] Doye, J. P.; Miller, M. A.; Welsh, D. J. Evolution of the potential energy surface with size for Lennard-Jones clusters. *J. Chem. Phys.* **1999**, *111*, 8417–8429.

[34] Flamm, C.; Fontana, W.; Hofacker, I.; Schuster, P. RNA folding kinetics at elementary step resolution. *RNA*, **2000**, *6*, 325–338.

[35] Flamm, C.; Hofacker, I. L.; Stadler, P. F.; Wolfinger, M. T. Barrier trees of degenerate landscapes. *Z. Phys. Chem.*, **2002**, *216*, 155–173.

[36] Hallam, J.; Prügel-Bennett, A. Large barrier trees for studying search. *IEEE Trans. Evol. Comput.*, **2005**, *9*, 385–397.
[37] Tøstesen, E. Partly melted DNA conformations obtained with a probability peak finding method. *Phys. Rev. E*, **2005**, *71*, 061922.
[38] Stadler, B. M. R.; Stadler, P. F. Combinatorial vector fields and the valley structure of fitness landscapes. *J. Math. Biol.*, **2010**, *61*, 877–898.
[39] Stadler, P. F.; Stadler B. M. R. Genotype phenotype maps. *Biological Theory*, **2006**, 3: 268–279.
[40] Schuster, P.; Fontana, W.; Stadler, P. F.; Hofacker, I. L. From sequences to shapes and back: A case study in RNA secondary structures. *Proc. Roy. Soc. Lond. B*, **1994**, *255*, 279–284.
[41] Fontana, W.; Schuster, P. Continuity in Evolution: On the Nature of Transitions. *Science*, **1998**, *280*, 1451–1455.
[42] Fontana, W.; Schuster, P. Shaping Space: The Possible and the Attainable in RNA Genotype-Phenotype Mapping. *J. Theor. Biol.*, **1998**, *194*, 491–515.
[43] Cupal, J.; Hofacker, I. L.; Stadler, P. F. Dynamic programming algorithm for the density of states of RNA secondary structures, in: R. Hofstädt; T. Lengauer; M. Löffler; D. Schomburg (eds), *Computer Science and Biology 96 (Prooceedings of the German Conference on Bioinformatics)*, pages 184–186, Leipzig, Germany, **1996**. Universität Leipzig.
[44] Stadler, B. M. R.; Stadler. P. F. Generalized topological spaces in evolutionary theory and combinatorial chemistry. *J. Chem. Inf. Comput. Sci.* **2002**, *42*, 577–585.
[45] Cupal, J.; Kopp, S.; Stadler, P. F. RNA shape space topology. *Alife*, **2000,** *6*, 3–23.
[46] Stadler, B. M. R.; Stadler, P. F.; Wagner, G. P.; Fontana, W. The topology of the possible: Formal spaces underlying patterns of evolutionary change. *J. Theor. Biol.* **2001**, *213*, 241–274.
[47] Lewontin, R. C. Adaptation. *Sci. Am.*, **1978**, *239*, 156–169.
[48] Wagner, G.; Stadler, P. F. Quasi-independence, homology and the unity of type: A topological theory of characters. *J. Theor. Biol.*, **2003**, *220*, 505–527.
[49] Hellmuth, M.; Imrich, W.; Klöckl, W.; Stadler, P. F. Approximate graph products. *Eur. J. Comb.* **2009**, *30*, 1119–1133.
[50] Hellmuth, M.; Imrich, W.; Klöckl, W.; Stadler, P. F. Local algorithms for the prime factorization of strong product graphs. *Math. Comp. Sci.* **2009**, *2*, 653–682.
[51] Stadler, B. M. R.; Stadler, P. F. The topology of evolutionary biology, in: G. Ciobanu; G. Rozenberg (eds), *Modeling in Molecular Biology*, Natural Computing Series, **2004**, 267-286.
[52] Dress, A. W. M.; Moulton, V.; Steel, M.; Wu, T. Species, clusters and the "tree of life": A graph-theoretic perspective. *J. Theor. Biol.* **2010**, *265*, 535–542.
[53] Steel, M.; Hein, J. Reconstructing pedigrees: A combinatorial perspective. *J. Theor. Biol.*, **2006**, *240*, 360–367.
[54] Riester, M.; Stadler, P. F.; Klemm, K. FRANz: Fast reconstruction of wild pedigrees. *Bioinformatics*, **2009**, *25*, 2134–2139.
[55] Kunin, V.; Ogldovsky, L.; Darzentas, N.; Ouzounis, A. The net of life: reconstructing the microbial phylogenetic network. *Genome Res.*, **2005**, *15*, 954–959.
[56] Day, M. M. Convergence, closure, and neighborhoods. *Duke Math. J.* **1944**, *11*, 181–199.

[57] Gniłka, S. On continuity in extended topologies. *Ann. Soc. Math. Pol., Ser. I, Commentat. Math.*, **1997**, *37*, 99–108.

[58] Hammer, P. C. General topology, symmetry, and convexity. *Trans. Acad. Sci., Arts, Letters*, **1955**, *44*, 221–255.

[59] Hammer, P. C. Extended topology: Set-valued set functions. *Nieuw Arch. Wisk. III*, **1962**, *10*, 55–77.

[60] Brissaud, M. M. Les espaces pretopologiques. *C. R. Acad. Sc. Paris Ser. A*, **1975**, *280*, 705–708.

[61] Gniłka, S. On extended topologies. I: Closure operators. *Ann. Soc. Math. Pol., Ser. I, Commentat. Math*, **1994**, *34*, 81–94.

[62] Kuratowski, C. Sur la notion de limite topologique d'ensembles. *Ann. Soc. Polon. Math.*, **1949**, *21*, 219–225.

[63] Hammer, P. C. Extended topology: Continuity I. *Portug. Math.* **1964**, *25*, 77–93.

[64] Soltan, W. P. *An Introduction in Axiomatic Theory of Convexity*. Shtiintsa, Kishinev, **1984**. [Russian].

[65] Malitza, M. Topology, binary relations, and internal operations. *Rev. Roumaine Math. Pures Appl.* **1975**, *4*, 515–519.

[66] Calude, C.; Căzănescu, V. E. On topologies generated by Mosil resemblance relations. *Discrete Math.*, **1979**, *25*, 109–115.

[67] Alexandroff, P. Diskrete Räume. *Math. Sb. (N.S.)*, **1937**, *2*, 501–518.

[68] Imrich, W.; Klavžar, S. *Product Graphs: Structure and Recognition*; Wiley, New York, 2000.

[69] Stadler, B. M. R.; Stadler, P. F. Higher separation axioms in generalized closure spaces. *Commentationes Math. Warszawa, Ser. I*, **2003**, *43*, 257–273.

[70] Harris, J. M. Continuity and separation for point-wise symmetric isotonic closure functions. Technical report, **2005**. arXiv:math.GN/0507230.

[71] Habil, E. D.; Elzenati, K. A. Connectedness in isotonic spaces. *Turk. J. Math*, **2006,** *30*,247–262.

[72] Hell, P.; Nešetřil, J. *Graphs and homomorphisms*. Oxford University Press, Oxford, UK, 2004.

[73] Hammack, R.; Imrich, W.; Klavžar, S. *Handbook of Product graphs*. CRC Press, Boca Raton, 2nd edition, 2011.

[74] Davey, B. A.; Priestley, H. A. *Introduction to Lattice and Order*. Cambridge Univ. Press, Cambridge UK, 1990.

[75] Changat, M.; Klavžar, S.; Mulder, H. M.; Vijayakuma, A. *Convexity in discrete structures*. Ramanujan Mathematical Society Lecture Notes Series, Vol. 5. International Press of Boston, **2010**.

[76] Imrich, W.; Stadler, P. F. A prime factor theorem for a generalized direct product. *Discussiones Math. Graph Th.*, **2006**, *26*, 135–140.

[77] de Vries, J. *Elements of topological dynamics*. Kluwer, Dordrecht, NL, 2010.

[78] Jonsson, J. *Simplicial Complexes of Graphs*. Springer, Berlin, 2007.

[79] Forman, R. Combinatorial vector fields and dynamical systems. *Math. Z.*, **1998**, *228*, 629–681.

[80] Fialkowski, M.; Bishop, K. J. M.; Chubukov, V. A.; Campbell, C. J.; nd Grzybowski, B. A. Architecture and evolution of organic chemistry. *Angew. Chem. Int. Ed.*, **2005**, 44, 7263–7269.

[81] Bishop, K. J. M.; Klajn, R.; Grzybowski, B. A. The core and most useful molecules in organic chemistry. *Angew. Chem. Int. Ed.*, **2006**, *45*, 5348–5354.

[82] Grzybowski, B. A.; Bishop, K. J. M.; Kowalczyk, B.; Wilmer, C. E. The 'wired' universe of organic chemistry. *Nat. Chem.*, **2009**, *1*, 31– 36.

[83] Gothard, C. M.; Soh, S.; Gothard, N. A.; Kowalczyk, B.; Wei, Y.; Baytekin, B.; Grzybowski, B. A. Rewiring chemistry: Algorithmic discovery and experimental validation of one-pot reactions in the network of organic chemistry. *Angew. Chem. Int. Ed.*, **2012**, *51*, 7922–7927.

[84] Andersen, J, L.; Flamm, C.; Merkle, D.; Stadler, P. F. Inferring chemical reaction patterns using graph grammar rule composition. *J. Syst. Chem.*, **2013**, *4*, 4.

Subject Index

Subhash C. Basak, Guillermo Restrepo & José L. Villaveces (Eds)
Copyright © 2015 Bentham Science Publishers Ltd. Published by Elsevier Inc. All rights reserved.
10.1016/B978-1-68108-053-6.50018-1

K

L

M

N

S

T

FSC
www.fsc.org
MIX
Papier | Fördert
gute Waldnutzung
FSC® C083411

Kolibri 360 GmbH
Ferdinand-Jühlke-Straße 7
99095 Erfurt, Deutschland
produktsicherheit@kolibri360.de